Contributors to This Volume

G. BIAGINI

L. P. EVERHART, JR.

RICHARD G. HAM

T. C. HSU

ROBERT T. JOHNSON

HELMUT KINZEL

R. LASCHI

MILTON A. LESSLER

HARALD LORENZEN

N. M. MARALDI

JÜRGEN PFAU

POTU N. RAO

HORST SENGER

P. SIMONI

EDUARD J. STADELMANN

JACK D. THRASHER

G. S. VENKATARAMAN

KLAUS WERTHMÜLLER

I. B. ZBARSKY

Methods in
Cell Physiology

Edited by

DAVID M. PRESCOTT

DEPARTMENT OF MOLECULAR, CELLULAR AND
DEVELOPMENTAL BIOLOGY
UNIVERSITY OF COLORADO
BOULDER, COLORADO

VOLUME V

1972

ACADEMIC PRESS • New York and London

ACADEMIC PRESS, INC.
111 Fifth Avenue, New York, New York 10003

United Kingdom Edition published by
ACADEMIC PRESS, INC. (LONDON) LTD.
24/28 Oval Road, London NW1

LIBRARY OF CONGRESS CATALOG CARD NUMBER: 64-14220

PRINTED IN THE UNITED STATES OF AMERICA

CONTENTS

8. Comparison of a New Method with Usual Methods for Preparing Monolayers in Electron Microscopy Autoradiography
N. M. Maraldi, G. Biagini, P. Simoni, and R. Laschi

9. Continuous Automatic Cultivation of Homocontinuous and Synchronized Microalgae
Horst Senger, Jürgen Pfau, and Klaus Werthmüller

10. Vital Staining of Plant Cells
Eduard J. Stadelmann and Helmut Kinzel

11. Synchrony in Blue-Green Algae
Harald Lorenzen and G. S. Venkataraman

LIST OF CONTRIBUTORS

Numbers in parentheses indicate the pages on which the authors' contributions begin.

G. BIAGINI, Centro di Microscopia Elettronica, Universita di Bologna, Italy, (289)

L. P. EVERHART, JR., Department of Molecular, Cellular and Developmental Biology, University of Colorado, Boulder, Colorado (219)

RICHARD G. HAM, Department of Molecular, Cellular and Developmental Biology, University of Colorado, Boulder, Colorado (37)

T. C. HSU, The Section of Cell Biology, Department of Biology, The University of Texas M. D. Anderson Hospital and Tumor Institute, Houston, Texas (1)

ROBERT T. JOHNSON, Department of Zoology, University of Cambridge, Cambrige, England (75)

HELMUT KINZEL, Pflanzenphysiologisches Institut der Universität Wien, Austria (325)

R. LASCHI, Centro di Microscopia Elettronica, Universita di Bologna, Italy (289)

MILTON A. LESSLER, Department of Physiology, Ohio State University, College of Medicine, Columbus, Ohio (199)

HARALD LORENZEN, Pflanzenphysiologisches Institut der Universität Göttingen, Germany (373)

N. M. MARALDI, Centro di Microscopia Elettronica, Universita di Bologna, Italy (289)

JÜRGEN PFAU, Botanisches Institut der Universität Marburg, Marburg, Germany (301)

POTU N. RAO,[1] The Eleanor Roosevelt Institute for Cancer Research, and Department of Biophysics and Genetics, University of Colorado Medical Center, Denver, Colorado (75)

HORST SENGER, Botanisches Institut der Universität Marburg, Marburg, Germany (301)

P. SIMONI, Centro di Microscopia Elettronica, Universita di Bologna, Italy (289)

EDUARD J. STADELMANN, Department of Horticultural Science, University of Minnesota, St. Paul, Minnesota (325)

JACK D. THRASHER, Department of Anatomy, UCLA School of Medicine, Los Angeles, California (127)

G. S. VENKATARAMAN, Division of Microbiology, Indian Agricultural Research Institute, Delhi, India (373)

KLAUS WERTHMÜLLER, Botanisches Institut der Universität Marburg, Marburg, Germany (301)

I. B. ZBARSKY, Institute of Developmental Biology, Academy of Sciences of the USSR, Moscow, USSR (167)

---

[1] *Present address:* Department of Developmental Therapeutics, The University of Texas M. D. Anderson Hospital and Tumor Institute at Houston, Houston, Texas.

PREFACE

Volume V of this treatise continues to present techniques and methods in cell research that have not been published or have been published in sources that are not readily available. Much of the information on experimental techniques in modern cell biology is scattered in a fragmentary fashion throughout the research literature. In addition, the general practice of condensing to the most abbreviated form materials and methods sections of journal articles has led to descriptions that are frequently inadequate guides to techniques. The aim of this volume is to bring together into one compilation complete and detailed treatment of a number of widely useful techniques which have not been published in full detail elsewhere in the literature.

In the absence of firsthand personal instruction, researchers are often reluctant to adopt new techniques. This hesitancy probably stems chiefly from the fact that descriptions in the literature do not contain sufficient detail concerning methodology; in addition, the information given may not be sufficient to estimate the difficulties or practicality of the technique or to judge whether the method can actually provide a suitable solution to the problem under consideration. The presentations in this volume are designed to overcome these drawbacks. They are comprehensive to the extent that they may serve not only as a practical introduction to experimental procedures but also to provide, to some extent, an evaluation of the limitations, potentialities, and current applications of the methods. Only those theoretical considerations needed for proper use of the method are included.

Finally, special emphasis has been placed on inclusion of much reference material in order to guide readers to early and current pertinent literature.

DAVID M. PRESCOTT

CONTENTS OF PREVIOUS VOLUMES

Volume I

Volume II

Volume III

Volume IV

Chapter 1

Procedures for Mammalian Chromosome Preparations

T. C. HSU

*The Section of Cell Biology, Department of Biology, The University of Texas
M. D. Anderson Hospital and Tumor Institute, Houston, Texas*

I.   Introduction

Since chromosome analyses of mammalian cells have been used in recent years for studies on a variety of biological problems, it is considered appropriate to present the various techniques in one chapter for easy reference. However, for every technique different laboratories have their own modifications and idiosyncrasies. It is, therefore, not feasible to cover all the variations. This chapter simply describes some of the procedures currently used in this laboratory. This does not mean that our systems are the best—it only proves that we have our own idiosyncrasies.

II.   Animal Materials

A.   Procurement of Animals

If the experimental materials involve only the standard laboratory animals (e.g., laboratory mice, Norway rats, and Syrian and Chinese hamsters) their procurement is no problem since they are commercially available. However, if the animals to be used are not standard laboratory strains, one must use other means to obtain them. There are a number of ways to do this, all of which can be pursued simultaneously.

1.   BUYING

Many animals can be purchased from dealers, and there are numerous dealers who are anxious to sell any kind of animal to you from lions and chipmunks to skunks. The problem is that prices are high, and, in most cases, there are no accurate locality data for each specimen. Since one must sometimes deal with dealers, a catalog should be handy for easy consultation. "Zoos and Aquariums in America" by William Hoff, American Association of Zoological Parks and Aquariums, Oblebay Park, Wheeling, West Virginia, will prove useful.

2.   BEGGING

This is naturally the best way of getting what one wants at very low cost. It is financially impractical if not impossible to buy an elephant just to get a piece of tissue for chromosome studies. Thus, zoos are useful for such purpose providing that the investigator can enlist the cooperation of zoo directors and veterinarians. Other potential sources such as game farms, primate centers, special laboratories, are all worth exploring.

Many mammalogists collect their own research material in the field, and they invariably trap animals in which they are not interested. Most of them are kind enough to help other investigators by collecting additional material.

3.   TRADING

If one is working on the chromosomes of one mammalian group, he should not restrict his work to that taxon alone. He should get material whenever he can even if he is not interested in the animals. He can use this material for trading—someone else may find delight in it. Simply save the slides and the carcasses or complete the karyotypes. In either case, trading is profitable for both sides.

4.   STEALING

This is the most difficult of all methods because one must find places where stealing can be accomplished. Also one needs good accomplices. When someone goes out on a field trip, your accomplice would keep an eye out for the animals you want and ship them to you or take biopsies (without the consent of the owner, of course, otherwise it would not be called stealing). However, do not rely on this method too much—your accomplice might keep the spoils for himself.

5.   FIELD TRIPS

Perhaps the most direct and most reliable way of acquiring the desired animal material is to get it by yourself. Besides, field trips are delightful interludes from smog and noise-filled city life. Of course, for the non-mammalogists, a few trips with experienced field workers will be most useful and educational. For chromosome work, the following list of equipment and supplies might be useful.

  a. General Supplies
  Money
  Camping equipment, gifts (in some countries), some personal
    belongings
  b. Supplies for Animal Collecting
  Live traps (Sherman type for rodents, Havahart for larger animals),
    baits (oatmeal, peanut butter, bananas, sardines, etc.), mist nets
    (for bats), firearms, spades, holding cages, cloth bags, leather
    gloves, and flashlights and batteries (head band flashlights are
    most desirable)
  Metofane (a good anesthetic, buy through a veterinarian)
  Kit for making museum study specimens (rulers, notebook paper,

pen, labels, scissors, forceps, cotton, wire, needles, thread, corn-
meal, pliers)
Mothballs
c. *Supplies for Cytological Preparations*
Microscope (there is a portable type that is very good)
Centrifuge (in areas without electricity, a hand-cranked type)
Dissecting board and instruments
Centrifuge tubes, Kahn tubes, various vials, staining jars, other
glassware
Slides, cover slips, slide boxes
Alcohol, acetic acid, distilled water, disinfectants, Colcemid or
Velban
Growth medium (in sterile vials)
Syringes and needles, Pasteur pipettes, nipples
Adhesive tapes (various kinds), cotton and gauze, razor blades, sur-
gical scalpels, and holders
Stains, mounting medium
Dry ice chest with dry ice
Notebooks, labels, marking pens, pencils

B.   Tissues for Primary Culture

Practically any tissue can be used to initiate primary growth *in vitro*
if the purpose is to start a culture instead of a particular culture. However, even to start a culture, some tissues are better than others because
they yield faster growth and are less troublesome to handle. In most
cases, a pair of forceps and a pair of scissors will suffice. Disinfect
these by wiping with 75% alcohol. The following tissue sources are
recommended.

1.   BLOOD

For some mammalian species, especially man, short-term lymphocyte
cultures are excellent for chromosome studies and standard procedures
can be followed. For others one must experiment with various conditions
before good results can be obtained.

2.   LUNG

We found that lungs are excellent tissues to start cultures. Take narrow
strips along the periphery of the lung lobes so that large bronchi or
trachea can be excluded. We seldom found microbial contamination.
However, for insurance and safety, always try to take two or more

separate biopsies from each animal and keep them in separate containers. When an animal is found dead, lung tissue is usually contaminated.

3.   SPLEEN

This tissue is not as good as lung tissue. Like the lung, it is useless for animals found dead.

4.   EAR

Ear cultures grow much slower than lung cultures. They are useful because the animals do not have to be sacrificed. There is no necessity to shave the hair off the ear unless it is a large animal. Simply rub the ear vigorously with alcohol and cut off a fragment. Ear fragments are good even for animals found dead.

5.   MUSCLE AND CARTILAGE

For animals that have died after a few hours, these tissues, though slow to initiate growth, have less chance of becoming contaminated than some faster growing tissues.

6.   GONADS

Testicular biopsy can be used to initiate cell growth *in vitro* without killing the animal. Also, it is the only material for meiotic figures.

7.   SKIN

For large and expensive animals, skin biopsy is the only sensible material to initiate cell cultures. The tail is probably the easiest (and safest) place to take tissue samples. Proceed as follows:

a. Secure the animal. It is not advisable to anesthetize an animal about which you know nothing regarding its response to the drug. Use a squeeze cage or a tranquilizer to immobilize the animal.

b. Shave. Using a clipper and then razor blades, shave an area of skin as closely as possible.

c. Disinfect skin with Phisohex (Winthrop Lab., New York). Rub vigorously. *Never* use iodine tincture or compounds containing mercury.

d. Rinse with 70% alcohol to remove Phisohex.

e. Repeat application of Phisohex and rinse.

f. Use a sharp, sterile needle to pierce skin horizontally and lift the skin up. Use a sterile blade to slice the skin immediately below the needle. The size of the skin biopsy thus depends upon the base where the blade starts to cut. It is always safe to take two or more biopsy specimens to avoid occasional microbial contamination, since no assurance can be made that the skin so treated is absolutely free from contaminants.

All biopsy specimens should be placed immediately in a sterile vial containing growth medium.

Usually, biopsy or autopsy specimens can survive for long periods of time if (1) the container is tightly capped, (2) the tissue fragments are not too large, and (3) the medium does not become exhausted. The containers can be shipped through regular postal service; for long distance, ship by air.

C. Preservation of Animal Specimens

Every mammalogist knows the importance of preserved specimens and how to prepare them. Cytologists and cytogeneticists without training in mammalogy or in taxonomy may consider a deer mouse a deer mouse and would dispose of the carcass after taking the tissues. Biochemists are often guilty of the same practice. These researchers often do not realize that there are many species and subspecies of deer mice and that taxonomists are not always in agreement regarding the identities of some taxa. How can a cytogeneticist or biochemist claim that he actually studied species P_3 instead of P_4 when he does not have the animal preserved as a voucher specimen?

An example to illustrate how cytology can help in a taxonomic study can be found in the work of A. L. Gardner. Gardner found two kinds of karyotypes in the large American opossum *Didelphis marsupialis* ($2n = 22$), one with 22 acrocentrics and the other with 12 of the 22 chromosomes submetacentrics. After establishing the karyological differences, he carefully examined animals of known karyotype and discovered a number of morphological differences. He was then able to apply the morphological information to specimens already in museums and correctly reidentify them. Now the large North American opossums are known to be two species, *Didelphis marsupialis* and *Didelphis virginiana,* respectively.

If a cytogeneticist wishes to make permanent study specimens, he should learn the procedure from an expert museum worker. For convenience, however, he does not have to prepare specimens if the animals are turned over to a mammalogist who will prepare them and deposit them in a museum collection. The following points are simply guidelines for the cytologist who does not desire to retain or prepare study specimens of the animals.

1. Always record complete locality information as well as ecological, reproductive, and other useful data for each specimen.

2. Always freeze the animals immediately after the tissues are taken

if the specimen is not to be prepared at the same time. Place the animals in plastic bags to prevent dehydration in the freezer. Do not thaw and refreeze animals.

3. Never remove or crush the skull of an animal. Many of the key taxonomic characters are found in dental and cranial structures.

4. Never remove any more of the animals' tissue than is necessary. If taking ear biopsies, leave one ear intact for measurements.

5. Always try to protect the skin and hair from blood and other substances that might alter the texture or color pattern.

6. Always be certain that any specimen, regardless of condition, is ultimately deposited in a suitable natural history museum.

D.   Breeding Rodents

Breeding standard laboratory rodents such as mice and rats requires no discussion here since volumes of information are available. However, sometimes one finds that it is necessary to breed some particular rodent species which cannot adapt to the laboratory routine used for *Mus musculus*. The difficulty increases when the animals of an institution are managed by veterinarians who think mice are mice whether they eat grasshoppers or not and that one system should be good enough for everybody.

Most burrowing rodents like to hide and make their own den. Actually, the laboratory mice and rats prefer to do the same, but they can tolerate being exposed, whereas other rodents, e.g., the Chinese hamsters, *Microtus*, etc. would become very unhappy. When they are unhappy, they fight instead of making love so that injuries instead of babies result. Even if babies are occasionally born, the mother frequently eats them fearing that humans would take them away from her.

The system described below probably applies to most cricetids and murids. Use large-sized plastic cages ($12 \times 16 \times 6''$) with enough shavings ($1\frac{1}{2}$ inches deep). Place one pair to each cage and let nature take its course. Change cages once a week. Frequent changes make the animals uncomfortable. With Chinese hamster, one can use the community mating system, viz., place two or more pairs to each cage. Generally speaking, pair mating is preferred.

Handle all animals with hands whenever feasible. Do not use instruments (such as forceps) to transfer animals. The latter procedure would make the animals mean and belligerent, not only to humans, but to themselves. If one plays with young animals with one's hands a few times, they soon get the idea that human hands do not harm them, and one can transfer them even without gloves. They breed better when they are treated "unscientifically." If one is afraid, gloves should be used.

Give laboratory rat chow and supplement the chow with seeds (e.g., sunflower seeds, bird seeds). Occasionally, give some fresh vegetables. Use regular light—turn on lights in the morning and off at night. Change the water bottle at the same time the cages are changed except when the bottles leak. It is a good policy to tighten the stopper and the spout when using a fresh bottle.

When females are pregnant, give them a little cotton to make a nest for the young. Unless absolutely necessary, skip cage changes for 1 week when there are infants. Weaning animals can be separated and start a new breeding cage. We found that brother–sister matings are not successful if they have never been separated. However, matings will be more successful if they are separated for a period of time. Generally the males are not removed from the cages even when there are babies.

The above-mentioned system can be applied to the Chinese hamster, *Peromyscus, Microtus,* etc. For *Microtus* it is desirable to supply the animals with fresh vegetables; we give them lettuce and apples.

We were fortunate to have some success in breeding a species of pocket mice, *Perognathus baileyi.* We kept one pair originally caught in the field and obtained two generations in the laboratory. Plastic cages are not suitable since the animals may chew a hole in the cage and escape. The cages we use for these animals are medium-sized aquariums, the bottom of which is filled with sand. We place two small wood boxes in each aquarium, one at each end. Cut a hole at one side of the box as entrance and exit. The animals will plug it with cotton and seeds. Two houses (boxes) are necessary because with one house constant fighting occurs.

For desert animals such as *Perognathus,* no water supply is needed. Occasionally, they should receive some slices of apples, carrots, lettuce, etc. Change the cage every 3–6 months. Do not let anyone talk you into changing cages more frequently. This is detrimental.

III. Tissue and Cell Cultures

A. Procurement of Cell Lines

In many fields of cellular research, it is desirable to use some of the "standard" cell lines. There are a number of mammalian cell lines certified and preserved in the American Type Culture Collection (ATCC), and they can be purchased at a nominal cost. Obtain a catalog and read through the descriptions carefully. You may find just what you need. The address of ATCC is The American Type Culture Collection, 12301 Parklawn Drive, Rockville, Maryland 20852.

B.   Sterilization and Disinfection

Needless to say, for tissue and cell cultures, an autoclave is an absolute necessity. If funds are no problem, lamina-flow rooms or hoods are most desirable. Otherwise, one must rely on less effective systems.

1.   ROOM DISINFECTION

A fogging machine (Center Chemical Co., Hospital Division, P.O. Box 1182, Houston, Texas or local suppliers) is effective in sterilizing culture areas. The disinfectant Cento Chlorocide (diluted 1:250 with water) can be purchased from the same supplier. Depending upon the size of the room, the machine will fill the room with mist of the disinfectant in a few minutes. The mist is allowed to settle before the room is used. It is also advisable to fog a room after working with questionable materials such as virus, bacteria, and wild animals.

2.   SANITIZER

Sanitizer (deodorant and disinfectant available from the Chemical Supply Co., P.O. Box 7123, Houston, Texas) is used to disinfect the work bench. The solution is diluted approximately 1:8 for routine use. The entire culture area is wiped with this solution. The table top used for culturing should be covered with a towel dampened with it. These damp towels also trap dust particles. One towel is used per day. If the towel dries, it can be rewetted. Bottle tops may be placed face down on the towel, and bottle lips may be wiped with the same towel.

3.   FILTERS

Filters are necessary for sterilizing culture media and other solutions that cannot withstand heat. It is advisable that filters of different sizes are available. The users must be familiar with the advantages and disadvantages of different types of filters.

The asbestos filter pads, if used, should be detoxified with dilute HCl and rinsed with distilled water before use.

4. Other Disinfectants

Ethyl alcohol is still one of the most effective and most reasonable of disinfectants. We use only alcohol to disinfect all our instruments (forceps, scissors, etc.). For animal skin biopsy we use Phisohex and alcohol. This is particularly convenient on field trips where supplies should be kept to a minimum. During long expeditions to jungles, deserts, and other areas where ethyl alcohol cannot be purchased, local hard spirits will do.

C. Solutions

1. Hanks' Solution

Substance	Amount (gm/liter)
NaCl	8.00
KCl	0.40
$CaCl_2$	0.14 (Dissolve separately in small quantity of distilled water and add slowly)
$MgSO_4 \cdot 7 H_2O$	0.20
$Na_2HPO_4 \cdot 2H_2O$	0.06
KH_2PO_4	0.06
Glucose	1.00
$NaHCO_3$	0.35
Neomycin (200 mg/ml, Mycifradin, Upjohn)	0.5 ml
Phenol red (1% solution, Difco)	1.0 ml

Adjust volume to 1000 ml with double-distilled water. Filter through Seitz filter, if a sterile solution is necessary, and store at 4°C. If frozen, the salts tend to precipitate.

2. Rinsing Solution

The rinsing solution is Hanks' solution without $CaCl_2$ and $MgSO_4$. It is used in the subculturing procedure. After the old medium has been decanted, 10–15 ml (for one T-60 flask) of the rinsing solution is added to each flask. The flask is gently shaken with this solution to remove the excess old medium which contains proteins and peptides. If free proteins are removed with the rinsing solution, trypsin or pronase will act more efficiently in dislodging the monolayer cells from the glass surface. We also use this solution to replace saline.

3. TRYPSIN SOLUTION (0.02%)

The trypsin used in our laboratory is lyophilized 3X crystallized (Worthington Biochemical Corp., Freehold, New Jersey). The type we purchase is nonsterile because it is much cheaper than the sterile compounds. Use impure trypsin at your own risk. Many batches of impure trypsin contain toxic or undesirable contaminants.

Dissolve trypsin in rinsing solution, adjusting the pH to 8.5. Filter through a Scitz filter and dispense in suitable quantities. Store in the freezer.

4. PRONASE (0.1%) (B GRADE, CALIFORNIA BIOCHEMICAL CORP.)

Substance	Amount (gm/liter)
NaCl	9.0
Pronase	1.0
Neomycin (200 mg/ml, Mycifradin, Upjohn)	0.5 ml
Phenol red (1% Difco)	1.0 ml

Adjust volume to 1000 ml with double-distilled water and adjust pH to approximately 8.5. Filter solution through a Seitz filter and store in the freezer.

The above concentration is too strong for routine subculture, but works well for removing the entire monolayer for cytological, biochemical, and biophysical studies when the cultures are to be terminated. For such purposes, filtration is not necessary.

5. RNASE STOCK PREPARATION (2 MG/ML BOVINE PANCREATIC, TYPE 1-A, SIGMA)

Dissolve the crystallized RNase in 0.15 M NaCl and adjust to pH 5.0. To destroy contaminating DNase, heat in a boiling water bath for 20 minutes. If watched closely, it may be boiled directly.

Dispense in small quantities in Kahn tubes and store in the freezer.

6. COLCEMID (CIBA PHARMACEUTICAL CO., SUMMIT, NEW JERSEY)

Prepare stock solution (2 μg/ml): Colcemid, 200 μg, 95% ETOH, 1.0 ml. Dissolve Colcemid in ethanol and then add 99 ml of rinsing solution. Sterilize through either a Seitz or Millipore filter. Store in the freezer.

When freshly prepared, a final concentration of 0.05–0.06 μg/ml of medium or body weight works well. However, we found that the potency of Colcemid solution in arresting mitoses reduces after storage, even in the refrigerator. For this reason, the next solution, Velban, may be used to replace Colcemid completely for both animals and cell cultures.

7. VELBAN

Velban (vinblastine sulfate, Eli Lily Co., Indianapolis, Minnesota) is a compound similar to Colcemid in that it arrests cells in metaphase. It may be used for collecting metaphase cells from bone marrow of small animals. The solution is injected subcutaneously. It appears stable at tropical temperatures for a number of days.

In our laboratory, Velban is diluted to 0.25 mg/ml of distilled water and the injection dose of this stock solution is 0.01 ml/gm body weight. This concentration is much lower than the manufacturer's suggested dosage. Higher concentrations may result in overcontracted chromosomes.

For tissue culture, use 0.01 ml/ml medium.

8. TM

This is the solution devised by Maio and Schildkraut for isolating metaphase chromosomes. It is hypotonic with respect to Na^+, but the high concentration of divalent cations prevents chromosome disintegration. It is also used to maintain membrane integrity. We use this solution frequently on a number of occasions: Tris, $0.02\,M$; $CaCl_2$, $0.001\,M$; $MgCl_2$, $0.001\,M$; $ZnCl_2$, $0.001\,M$. Adjust pH to 7.0.

9. MYCOSTATIN (ELI LILY CO.)

This antibiotic is very useful for laboratories which routinely cultivate primary tissues, especially those collected in the field. Prepare 100,000 units/ml in rinsing solution as the stock solution (actually suspension) and store in the refrigerator. The compound is rather stable even at room temperature. It is excellent for fungal contamination carried in biopsies. At a final concentration of 1000 units/ml of medium and incubation for 24 hours, it is effective against heavy fungal growth with no detrimental effect on cell survival. It is not advisable to use Mycostatin after the suspension is prepared for longer than 1 month.

10. SSC

The following formula is for 10X concentration: NaCl, $1.5\,M$; trisodium citrate, $0.15\,M$. Dissolve in distilled water and adjust pH to 7.0.

For preparing lower concentrations of SSC, simply dilute the 10X solution with distilled water.

11. HYPOTONIC SOLUTION

Hypotonic solutions can be prepared in a variety of ways. Generally, a 0.8–1.0% sodium citrate solution is very good for direct bone marrow preparations as well as for many cell cultures. However, when cells are removed from culture flasks by trypsin or pronase solutions, the cells sometimes clump if citrate solution is used. A diluted growth medium (growth medium diluted two or three times with distilled water) will eliminate clumping.

Sometimes the cell membrane becomes highly susceptible to breakage after hypotonic solution treatment. We find that adding a portion of TM (see p. 12) to the hypotonic solution helps to strengthen the membrane.

D. CO_2

For laboratories where CO_2 incubators are available no discussion is necessary. However, a simple devise can be used to replace CO_2 incubators unless large numbers of petri dishes are routinely used in cloning experiments for quantitative data collection.

In large cities where tanked gasses are available, one can order tanks containing a mixture of 10% CO_2 and 90% air. The size of the tank is determined by the frequency of cultures set up in a laboratory. In our laboratory, we use the E-size tanks, which are light enough to be carried around by a special cart, but large enough to last for 1 or 2 months.

The tank should be equipped with a flow-adjusting yoke and the outlet attached to a rubber of Teflon tubing. Each time a sterile Pasteur pipette (with cotton at its neck) should be used to allow gas to flow *gently* into each culture flask to replace the original air. There is no danger of overadjusting, but the operation requires only a few seconds for each flask.

E. Growth Media

1. SUPPLEMENTS

The availability of commercially prepared media makes the business of medium preparation a great deal simpler than a decade or so ago. We would only stress that minimal medium formulas should be replaced with richer media such as Ham's F10, McCoy's 5a, or others equally good.

In our laboratory the powdered supplements of a modified 5a formula are prepared and packaged in 5-liter quantities by a commercial company. The powdered medium is frozen until ready to use. It is dissolved in glass-distilled water and filtered. We then add 20% sterile fetal bovine serum to make the final volume to 6.25 liters.

For laboratories consuming small amounts of growth medium, it may be worth the expense to purchase ready-made liquid medium to save labor and trouble. For those laboratories that do not have access to frozen medium, we append here our medium formula.

a. Solution A. Amino Acids

	Amount per		
	1 liter (mg)	5 liters (gm)	50 liters (gm)
L-Tryptophan	3	0.015	0.15
L-Phenylalanine	16	0.080	0.80
L-Tyrosine	18	0.090	0.90
L-Arginine · HCl	126	0.630	6.30
L-Histidine · HCl · H$_2$O	21	0.105	1.05
L-Lysine · HCl	36	0.180	1.80
L-Cysteine · HCl	41	0.205	2.05
L-Methionine	15	0.075	0.75
L-Isoleucine	39	0.195	1.95
L-Leucine	39	0.195	1.95
L-Valine	18	0.090	0.90
L-Threonine	18	0.090	0.90
L-Asparagine · H$_2$O	64	0.320	3.20
Glycine	8	0.040	0.40
L-Serine	26	0.130	1.30
L-Alanine	13	0.065	0.65
L-Proline	17	0.085	0.85
L-Hydroxyproline	20	0.100	1.00
L-Aspartic acid	20	0.100	1.00
L-Glutamic acid	20	0.100	1.00

Dissolve the 50-liter quantity in 2500 ml double-distilled water and bottle 5-liter concentrates in separate containers. Freeze at 20°C.

b. Solution B. Glutamine

	Amount per		
	1 liter (mg)	5 liters (gm)	50 liters (gm)
L-Glutamine	438	2.190	21.90

Freshly prepared each time.

c. Solution C. Vitamins

	Amount per		
	1 liter (mg)	5 liters (gm)	50 liters (gm)
Thiamine · HCl	0.6	0.003	0.030
Riboflavin	0.2	0.001	0.010
Pyridoxine · HCl	0.5	0.0025	0.025
Pyridoxal · HCl	0.5	0.0025	0.025
Nicotinic acid (niacin)	0.5	0.0025	0.025
Nicotinamide (niacinamide)	0.5	0.0025	0.025
Ca–pantothenate	0.4	0.002	0.020
Biotin	0.2	0.001	0.010
Folic acid	0.2	0.001	0.010
Choline chloride	5.0	0.025	0.250
Inositol	36.0	0.180	1.800
p-Aminobenzoic acid	1.0	0.005	0.050
Ascorbic acid	0.5	0.0025	0.025
Glutathione	0.5	0.0025	0.025

Dissolve the 50-liter quantity in 1000 ml double-distilled H_2O and bottle in separate containers. Freeze at $-20°C$.

d. Solution D. Vitamin B_{12}

	Amount per		
	1 liter	5 liters	50 liters
Vitamin B_{12}	0.01 mg	0.05 mg	0.50 mg
If Squibb 1000 gm/ml solution:	0.01 ml	0.05 ml	0.5 ml

e. Solution E. Minerals and Glucose

	Amount per		
	1 liter (gm)	5 liters (gm)	50 liters (gm)
NaCl	6.46	32.3	323
KCl	0.4	2.0	20
$MgSO_4 · 7 H_2O$	0.20	1.0	10
$NaH_2PO_4 · H_2O$	0.58	2.9	29
KH_2PO_4	0.08	0.4	4
$NaHCO_3$	2.20	11.0	110
Glucose	3.0	15.0	150

Dissolve the above minerals each time medium is prepared. Each compound should be dissolved separately and then added to the solution.

f. *Solution F*. $CaCl_2$

	Amount per		
	1 liter (gm)	5 liters (gm)	50 liters (gm)
$CaCl_2$	0.1	0.50	5.0

g. *Solution G*. Lactoalbumin Hydrolyzate

	Amount per		
	1 liter (gm)	5 liters (gm)	50 liters (gm)
Lactoalbumin hydrolyzate	0.5	2.5	25

h. *Solution H*. Phenol Red (1% Solution Difco)

	Amount per		
	1 liter (ml)	5 liters (ml)	50 liters (ml)
Phenol red	1	5	50

i. *Solution I*. Antibiotics (Neomycin 5 gm/25 ml) (Mycifradin)

	Amount per		
	1 liter (mg)	5 liters (gm)	50 liters (gm)
Neomycin	0.5 ml	2.5 ml	25 ml

Dissolve each 5.0 gm Mycifradin (Upjohn) with 25 ml sterile H_2O.

2. SERUM

In our laboratory we use fetal calf serum. Routinely we use 20% serum in the final medium. Our advice is that each batch of serum be tested for toxicity and contamination. Some supply companies collect and prepare the serum more conscientiously than others. Generally, when one company is found to be good, persist with its product. However, test as many samples from various companies as feasible in case one company changes hands and the product quality deteriorates.

Most companies supply sterile fetal calf serum. No filtration is necessary.

3. PROCEDURE FOR MEDIUM PREPARATION

The following directions are specifically for a 5-liter supplement solution (without serum) prepared in one's own laboratory:

1. Use 2000 ml of double-distilled water in a 6-liter flask.

2. Add one bottle (250 ml) of the amino acid stock solution. Rinse bottle with double-distilled water and add the rinsing solution to the mixture. Stir the mixture with a magnetic stirrer.

3. Add glutamine to the solution and stir until dissolved.

4. Add one bottle (100 ml) of the vitamin stock solution. Rinse bottle and continue to stir the solution.

5. Add vitamin B_{12} separately.

6. The minerals must be dissolved and added separately each time the media is made. Be sure to rinse the beaker well after each mineral has been dissolved. A stock solution should not be made for the minerals.

7. After all the minerals have been added, dissolve $CaCl_2$ in at least 100 ml of distilled water before adding to the mixing flask.

8. Dissolve lactoalbumin hydrolyzate and add to the mixing flask.

9. Add neomycin and phenol red.

10. Bring the final volume to 5000 ml with distilled water.

11. Pass CO_2 gas through a coarse sintered glass until the color of the solution becomes yellowish orange (pH 7.2).

12. Filter with a Seitz filter.

13. Dispense asceptically into sterile bottles, and leave enough room for the serum. Add the serum and seal the bottle caps with tape.

14. Keep the container in the refrigerator.

For frozen powder, simply dissolve the powder in the appropriate amount of double-distilled water and readjust pH. Add serum.

F.   Primary Cultures

1. Cut the tissues into fragments as fine as is practical. A surgical scalpel (size 21) is most suitable since its blade is large and curved so that cutting is somewhat easier than it is with straight blades (such as size 11). Use a piece of sterile glass block for cutting.

2. Transfer the tissue fragments to a culture vessel. The T-series or plastic flasks are ideal because they have good optical quality for observation of cellular growth. However, screw-capped medicine bottles can be used as substitutes if T-flasks are not available. Add a small amount of growth medium to each flask. Gas the flasks with 10% CO_2 and stopper tightly. Whenever feasible, set up a duplicate culture.

3. Add a small amount of growth medium to each flask the next day. Thereafter, feed the cultures every other day or at least three times per week.

4. For lung cultures cell emigration and growth should be noticeable 48 hours following the initial culture. For ear or skin cultures, cell growth will require a few more days. When cellular growth is obvious, especially when some fragments attach to the floor of the flask, it is advisable to shake the flasks everyday to dislodge the tissue fragments so that they will settle on different places and start new colonies. This method minimizes contact inhibition and disperses cellular growth. For skin and ear cultures, tissue fragments seldom attach to the floor of the culture flask, and cells are generally well scattered without forming compact colonies. It is not necessary to hold the tissue fragments in place by cellophane sheets. In fact it is desirable for the fragments to float, as long as they are not removed when feeding.

5. In case cell colonies are extremely dense but the flask contains only a few colonies, it is advisable to disperse the colonies each time prior to feeding. Decant the medium and place a drop or two of the trypsin solution directly over each colony. Leave the flask in room temperature for a few minutes before adding fresh growth medium. Shake the flask vigorously and reincubate. The small amount of trypsin left in the flask should not affect cellular growth. This process may be repeated several times.

6. Depending upon the initial amount of tissue fragments, a T-60 flask should have sufficient cellular growth for cytological preparations or for subculturing in 1 to 2 weeks. The best harvest time for cytological preparations is approximately 16–18 hours after feeding. For subculturing, feeding time is not important.

7. For tissues with yeast or fungus contamination or when such contamination develops in the initial phase of primary culture, Mycostatin is extremely useful for decontamination. It is effective even when fungal growth is heavy. Use Mycostatin (p. 12) and incubate for 1 day. Feed the cultures the next day to remove Mycostatin. There will be precipitates in the flask but they are harmless. For bacterial contamination it is best to consider the culture a loss.

G. Subculturing

Although various laboratories have used Pronase solution or EDTA solution to dislodge monolayer cells for subculturing, we suggest the use of trypsin solution, at least for the initial subculture series. When cultures are plentiful, one may try other agents.

The cells of most primary cultures attach to the culture flasks rather firmly so that they are difficult to remove by trypsin solution. Rinse the flask with rinsing solution and add a small amount of trypsin to barely cover the floor of the flask. Incubate for not more than 10 minutes. Add growth medium into the flask without removing the trypsin solution, and shake the flask vigorously to suspend those cells dislodged by trypsin. Transfer the cell suspension, including tissue fragments, to a fresh flask as subculture, and refeed the original flask. Gas both the new and the old cultures with CO_2. Inspect the cultures frequently to determine time for harvest or for the next subculture.

After it is determined that the cells can withstand trypsin well, subculturing may be done by larger amount of trypsin solution to remove all the cells in a flask. The cells are collected as a pellet after centrifugation. The trypsin solution is then decanted, and a small amount of growth medium is added to the centrifuge tube to suspend the cells. The contents can be divided into appropriate aliquots for further propagation.

H. Long-Term Cultures

Each cell line has its own characteristics. It is therefore not possible to make general rules. With experience, each cell line will respond to the routine procedure with anticipated results.

It is fallacious to "split" one culture into two every 2 days when the generation time is less than 20 hours, because invariably the cell populations become overcrowded and finally die. On the other hand, it is too cumbersome to count the cells for routine inoculations. A good rule of thumb is to inspect the size of the pellet and estimate the number of cells. Gas each subculture with 10% CO_2.

Probably the most fastidious cell line we have is strain Don C. The cells must be subcultured daily from one to three or four cultures. If subculturing is delayed for as short a time as 6 hours, the cultures will recover very poorly or even terminate. However, if the above routine is followed, the cell strain will yield a tremendous amount of growth in a very short time. The parent line of Don C, Don, is not as fastidious, but a subculture schedule of three times per week is certainly insufficient.

Many laboratories complain that the Don line either changes its chromosome composition or die after a few subcultures. Quite possibly they do not follow a good routine.

I. Cloning

Cloning is important when one uses clone counts as an assay method for obtaining quantitative data and when one tries to select single-cell derivatives for genetically homogeneous cell populations. This is especially important in growing tumors because all tumor tissues contain normal cells as contaminants and the normal cells may outgrow the tumor cells in culture. Early cloning to obtain pure tumor cell lines ensures the perpetuation of the desired cellular material.

The popular cloning method requires growing cells in petri dishes. To grow cells in petri dishes the cultures must be kept in an incubator which provides 100% relative humidity and 5% CO_2 in the atmosphere. Thus a CO_2 incubator is ideal for such purposes. However, for laboratories which do not employ cloning experiments as routine procedure, CO_2 incubators are troublesome luxuries. For small number of petri dish cultures, a good substitute for CO_2 incubator is the desiccator. Desiccators can keep moisture as well as CO_2, and they are easy to clean and to dry when not in use. There will be no mildew problem.

Naturally there are numerous methods for obtaining clones. We have tried most of them but found the following two procedures more reliable:

1. Place broken slivers of cover slips in plastic petri dishes (60 mm). Make cell suspension in conditioned medium and introduce the suspension into the dishes. The number of cells per dish should be adjusted. Incubate the dishes. After the cells settle, inspect the dishes with an inverted microscope. Select the glass chips which hold one cell on each, and place one chip into a small container (such as T-3 or T-9 flask). Add a small amount of the conditioned medium, gas with CO_2, and reincubate. Colonies, if grown, can be propagated as clonal strains.

2. BBL MicroTest II tissue culture plate (Falcon Plastics) is used for isolating clonal derivatives. The plate contains 96 optically clear, flat-bottom cells, each 6 mm in diameter. The sterile microplates are supplied in individual packages with separate sterile lids, also individually packaged. The wells are seeded with a cell suspension properly diluted in conditioned medium to contain an average of one cell per drop. One drop per well is dispensed with a Pasteur pipette; an additional two drops of growth medium is then added to each well.

Approximately 24 hours after seeding, the plates are examined with an inverted microscope. Those wells containing only one cell each are identified by marking the cover directly over the well with a China-marking pencil. The medium should be changed on day 4 after seeding, and the plate reexamined on day 8. The wells previously marked and now containing a colony of cells can be trypsinized for propagation of clone strains.

J. Cell Preservation

Ideally, each cell line should be frozen in a proper medium at a proper rate of temperature drop and preserved in liquid nitrogen. This is the way to ensure a high yield of live cells when the frozen ampoules are thawed. However, most laboratories freeze cells for rainy days. Whether the recovery rate is 50 or 95% of cells is not too important. Thus the crude system used in our laboratory, one preservative medium for all, no controlled freezing rate, dry ice chest for storage, etc., suffices. In many cases the recovery may be poor, but rarely did we encounter no recovery.

1. FREEZING

a. The monolayer of cells is removed from the flask in the same manner as for subculturing. The cells are collected and centrifuged. Content of approximately one full T-60 flask is used for each freezing ampoule.

b. The trypsin solution is decanted and 1 ml of the preservative medium (5% glycerol in McCoy's 5a growth medium) is introduced.

c. The cells are suspended by a pipette and transferred into a 2-ml ampoule (Wheaton Glass Co., Millville, New Jersey. Cat. No. 12486) which is then sealed with an oxygen torch.

d. Sealed ampoules should remain at room temperature for approximately 2 hours in order to allow equilibrium of glycerol between the cells and the medium.

e. Six ampoules can be placed in a powder box, which is in turn placed in a compartmentalized metal or plastic rack inside the dry-ice chest.

2. THAWING

a. Remove the ampoule from the freezer and immediately place it in a beaker containing warm water (temperature not more than 37°C). Shake the ampoule rapidly until the medium is thawed.

b. Open the ampoule aseptically and add 1 ml of fresh medium with a Pasteur pipette.

c. Mix the preservative medium with fresh medium and transfer the suspension into a T-flask.

d. Incubate for 1 hour.

e. Add 7–10 ml of growth medium; gas with 10% CO_2 + 90% air, and incubate at 37°C.

f. After 12–18 hours, change with fresh medium to remove the glycerol-containing medium.

g. Most cell cultures require 3–7 days after thawing to form a monolayer heavy enough for subculture.

IV. Cytological Preparations

A. Stains

1. ACETIC ORCEIN

Some laboratories are very fastidious about the orcein they use. They claim that different batches from the same company may be different in staining quality and capacity. This is true when one uses bright-field microscopy because usually orcein does not stain the chromosomes intensely enough. If one uses phase-contrast microscopy, where only dilute orcein solution is needed, then any batch from any supplier will suffice.

For bright-field microscopy, a 2% acetic orcein is recommended. Actually this is an oversaturated solution (boil orcein in 50% acetic acid, filter when cool), and precipitates occur after storage. Thus in applying the stain, filtering the necessary amount before use would eliminate undesirable particles in the final preparations. For phase microscopy, a 1% solution is used. It is not necessary to filter before use.

2. GIEMSA

Ordinarily, commercially prepared stock Giemsa solution (e.g., Giemsa blood stain by W. H. Curtin Co.) works well. This is prepared as shown in the following tabulation.

Distilled water	50 ml
Citric acid (0.1 M)	1.5 ml
Adjust pH with 0.2 M Na_2HPO_4 to 6.8–7.2	
Methanol	1.5 ml
Giemsa stock solution	5 ml

If one desires to prepare his own stock Giemsa solution, he could use the following formula:

> Giemsa powder   1 gm
> Glycerol   45 ml
> Heat in water bath (60°C) for 2–3 hours
> Add methanol   33 ml

Let stand overnight before use.

3.   CARBOFUCHSIN

Prepare as follows.

> Solution A
>   Basic fuchsin   3 gm
>   70% ethanol   100 ml
> Solution B
>   Solution A   10 ml
>   5% phenol solution in water   90 ml
> Staining solution
>   Solution B   45 ml
>   Glacial acetic acid   6 ml
>   37% Formaldehyde   6 ml

Let stand overnight before use. The color of the staining solution should be purple with a greenish surface film. Stain for 15 minutes. Rinse several times in absolute ethanol and air dry.

B.   Bone Marrow Preparations

Even for laboratories with cell culture facilities, bone marrow is still an excellent material for cytological preparations, especially for small animals that can be sacrificed. There are several reasons for its popularity.

1. It is direct. The cells are taken directly from the body, so that an intervening period of cell culture is eliminated. This also alleviates possible changes in the chromosome constitution *in vitro*.

2. It is economical. No tissue culture facilities and no tissue culture expenses are required. This is especially true for field trips.

3. It is quick. One can process many specimens a day and have the slides immediately ready for observation.

Naturally, it also has some shortcomings:

1. The quality of the preparations is not always good.

2. The bone marrow samples may contain no mitotic divisions.

3. It is impractical for large animals since they would require a large amount of Colcemid or Velban.

In our laboratory we routinely use bone marrow but for safety we also take lung or ear biopsies. Velban treatment does not interfere with the growth of these tissues in culture. When the bone marrow preparations prove unsatisfactory, the biopsies are then used to set up primary cultures.

The procedure for bone marrow preparations of small mammals follows:

1. Dissolve contents of Velban (10-gm vial, Eli Lilly) in 40 ml of water. This solution can be kept in the refrigerator without losing strength. On field trips when no refrigeration is available, keep it in a cool place.

2. Inject 0.01 ml of Velban solution for every 2 gm of body weight. Incubate *in vivo* for not more than 2 hours, preferably 1 hour. This dose may still be too strong for some animals, but it is good for a start.

3. Kill the animal by cervical dislocation, heart pressure, or anesthesia. Make a longitudinal incision in the abdomen and cut out the femurs of both hind legs. Remove muscles and trim off portions of epiphyses to expose the marrow. The animal specimens should be saved (see p. 6). In the case of bats, use the humerus.

4. Prepare a hypotonic solution (see p. 13). Load a syringe (5 ml ideal) with this solution and attach a small needle. For animals the size of a rat, use a No. 22 needle; the size of a mouse, a No. 25 needle; for the smallest rodents and shrews, a No. 27 needle. Push the hypotonic solution through the femur so that the marrow plug will be flushed out. Collect marrow from both femurs (and sometimes include tibiae). Fix the pellet with 50% acetic acid for squash (p. 26) or with Carnoy (3 parts methanol and 1 part glacial acetic acid) for air-dried preparations (p. 25).

5. It is desirable to prepare slides with freshly fixed cells, but on field trips it is sometimes necessary to make a few critical slides and preserve the duplicate specimens in order to save time.

C.   Harvesting Cell Cultures for Chromosome Preparations

Cultures ready for chromosome preparations should be fed 18–24 hours prior to harvest to ensure active growth and ample mitotic divisions. Depending upon the amount of mitosis present in the culture, Colcemid or Velban treatment should vary from 2 to 4 hours. Use a final concentration of 0.06 μg Colcemid. Decant medium, rinse cultures with the rinsing solution, and add approximately 10 ml of pronase solution. At room temperature, the cells should be free from monolayer cultures in a maximum of 5 minutes. It is advisable to inspect the cultures in pronase with a microscope to determine whether all the cells have detached. Prolonged pronase treatment should be avoided.

Centrifuge the cell suspension. Use a clinical centrifuge with one speed. The cells should form a visible pellet. If the size of the pellet is small, pool the duplicate cultures together in the next step.

Prepare a hypotonic solution by mixing growth medium 1 part and distilled water 3 parts (see p. 13). Suspend the cells in a small amount of this solution initially to facilitate easy suspension of cells and then add solution to approximately 7 ml. Do not use violent force to suspend the cells. Use a Pasteur pipette attached to a small nipple, blowing in air to suspend the cells.

Leave the cells in the hypotonic solution for approximately 5 minutes before centrifugation.

Fixation of the cells depends upon the method for chromosome preparation. For air-dried technique, use 3 parts of methyl alcohol and 1 part of glacial acetic acid mixture. This mixture should be freshly prepared. For squash preparations, use 50% acetic or propionic acid. Do not disturb the cell pellet for at least 20 minutes. The cells, after being suspended in the hypotonic solution, are literally bloated. Their membranes are easily broken by agitation. After fixation, however, they can be resuspended.

D. Air-Dried Preparations

1. After fixation for at least 20 minutes, suspend the cells in the fixative.

2. Centrifuge.

3. Decant the original fixative and add fresh fixative.

4. Resuspend the cells. Be sure not to draw the cell suspension up into the pipette, for the cells are now very sticky. Blow in air instead.

5. Repeat this procedure twice.

6. Suspend the cells in fresh fixative once more. This time use 2–3 ml of fixative.

7. Place a small drop of the cell suspension on a clean slide. Blow or fan it dry. Then place another drop, repeat the process. The number of drops to be placed on a slide depends upon the thickness of the cell suspension. It is a good idea to inspect the first slide after the addition of each drop. Igniting the fixative to dry the slides is a popular way of making this type of preparation. When flame-dried preparations are made, put enough drops of cell suspension on a slide instead of drying drop by drop. The flame-dried preparations are useful for the hetero-chromatin staining procedure (p. 33).

8. The air-dried or flame-dried slides may be stained with Giemsa, carbofuchsin, or acetic orcein.

E. Squash Preparations

The old-fashioned squash preparations are still useful in many ways. The reason most investigators prefer the air-dried technique, however, is that they found a large number of broken cells in their squash preparations. Actually, if this defect can be corrected, squash preparations are superior for the following reasons.

1. They are quick. The squash technique avoids all the washing and centrifugation processes. After fixation, one can make slides within 1 or 2 minutes.

2. They give excellent photographs for duplicate photography without cumbersome procedures.

3. Where phase microscopy is used, the intactness of the cellular outline can be determined in squash preparations. Thus one avoids counting incomplete metaphases. In squash preparations, when intactness of the cellular outline is determined by low magnification lens, one seldom encounters an incomplete count.

4. For solid tissues, squash is the most sensible method.

5. When the quality of the preparation is poor, some manipulations can be made to improve the quality to such an extent that at least the data will not be lost.

It is true that without proper instructions and some practice, the air-dried or flame-dried preparations are easier to make than are the squash preparations. We hope that the following discussion would point out some of the common mistakes in making squashes, so that the readers may correct their procedure if they want to try this type of preparation.

The trouble commonly encountered with squash preparations is that many cells rupture. This is caused by several defects in the process of preparation:

1. Suspending the cells in the hypotonic solution too violently.
2. Suspending the cells in the fixative too soon.
3. Squashing too hard at the beginning.
4. Tapping the cover slip.
5. Pressing with an up-and-down motion only.

For phase-contrast microscopy, use 1% acetic orcein. Over-stained chromosomes will give a golden glow and lose microscopic details. When unsatisfactory staining quality occurs, one may either add stronger staining solution to intensify the stain or add 50% acetic acid to dilute the stain. For bright-field microscopy, however, use the strong 2% stain.

After the cell pellet is fixed for at least 20 minutes, decant and drain the fixative. The amount of stain to be added varies according to the size of the cell pellet. It is suggested that one drop of acetic orcein be used for a tiny pellet and two or three drops for a larger pellet. If the cell suspension is too thick, one can always add more stain later. It is a good practice to make a tentative slide and view it microscopically to determine whether the suspension and the stain are appropriate. Each cell should have enough room to expand fully, but the cell density should not be so sparse that the cytologist would have to scan large areas to find a mitotic figure. Nevertheless, it is good practice to make the suspension somewhat thinner because the quality of the slides is usually better when the cell density is low.

Use a Pasteur pipette to suspend the cells. Place a small drop on a clean slide, and place a cover slip gently over the cell suspension. We find 22 × 30 mm cover slips most ideal, but other sizes should serve equally well. If air pockets are seen under the cover slip, do not worry; most air pockets will be driven out in the process of squashing. Even if they remain, there are plenty of areas in which to find good cells.

The next step is the key to squash preparations. Some persons think that squashing is an art and is not possible for everyone. This is wrong. Anyone can make a good squash preparation if he corrects certain mistakes that cause the cells to break.

The proper procedure is to place a piece of absorbent paper (e.g., Whatman No. 1 filter paper) over the slide. Hold the thumb at the middle of the cover slip and roll the long axis of the thumb very gently toward one side of the cover slip, squeezing out a little stain each time. Remember that when pressure is applied to the cover slip, the stain beneath the cover slip will be forced to the sides. If it is not absorbed, it will rush back to the center as the pressure is released, thereby creating a violent current which breaks the cells. If the pressure is gentle and the liquid squeezed to the edge of the cover slip is absorbed, there is little chance for runback. The cells will become flatter, but not broken. Putting a minute drop of cell suspension on each slide is also good practice.

Repeat the process (always from the center of the cover slip toward the sides), and each time squeeze out a little excess stain. Increase the pressure until no stain can be pressed out under very strong pressure. This is the time to inspect the slide to determine (1) whether the sample contains any mitotic figure, (2) whether the suspension is too crowded, and (3) whether the quality of the preparation is good. Use a phase-contrast microscope with an objective between 16 and 20X for scanning. If no mitotic figure is present, discard the entire sample. If the cells are too crowded, add one or two more drops of stain to dilute the suspension. If the quality of the metaphase spreads is good, make a few more slides.

If the chromosomes of most metaphases seem to be crowded (this is especially true in bone marrow preparations), one may borrow an old botanist's technique for squashing, viz., to heat the slides. Pass the slide over an alcohol or a gas flame with a dodging motion. The slide should be so heated that the liquid is just below its boiling point. Now quickly cover the slide with a filter paper and press with a strong pressure, rolling the thumb as before. This heating before the final press will greatly improve the quality of the metaphases in a large number of cells.

Seal the edge of the cover slip with Krönig cement. When kept at refrigerator temperatures, well-sealed slides will last at least 1 year. At room temperature, they deteriorate within a few months.

F.   Male Meiosis

We found that regular squash preparations of testicular biopsy are quite satisfactory. It is not necessary to treat the material with a hypotonic solution prior to fixation. Mince a small portion of the testicular tissue in an isotonic medium, and fix it with 50% acetic acid after centrifugation. The size of the pellet should not exceed 0.1 ml or the fixation will not be ideal. Unless the testis comes from very young animals, all stages of spermatogenesis and spermiogenesis can be found.

G.   Making Temporary Slides Permanent

1. Place temporary slides on a cake of flat dry ice and leave them for at least 20 minutes.

2. Use a single-edge razor blade or a surgical scalpel to quickly scrape off the sealing cement. Do one slide at a time.

3. Place the corner of the blade at one corner of the cover slip. With one quick flip, the cover slip should be detached from the slide, but the cells should remain.

4. Rinse the slides in two changes of commercial alcohol and dry them in air.

5. The slides can be restained with any stain.

H.   Karyotyping

There is really not much to be said about karyotyping, but some general suggestions can be made. At present everyone uses photographic karyotyping, so instruction in the art of camera lucida drawing is no longer needed. The important factor is the photograph. Needless to say, the better the negative, the better the print.

For human karyotype, the technique is more or less standardized. For wild animals, we often have to make a number of changes about matching difficult chromosome pairs or the arrangement of the chromosomes. Glueing the chromosomes with rubber cement or any permanent paste or taping over the chromosomes with transparent tape leaves something to be desired.

We have been using the double adhesive tape (Scotch Brand No. 400, skip split liner, double-coated tissue 1 inch, Minnesota Mining & Mfg. Co., St. Paul, Minnesota) for several years and find it most satisfactory. This tape has adhesive on both sides, but one side is protected by a backing. The backing has a slit in the middle, so that it can be ripped off easily. Do not buy the kind without the split.

Spray the mounting board with a clear plastic (Crystal Clear Spray Coating No. 1303, Krylon, Inc., Norristown, Pennsylvania) which would prevent the fibers of the paper board being picked up by the tape when one moves the chromosomes around. Stick the roughly cut chromosomes on the upper side of the tape. To facilitate arrangement, first stick only the tip of the cut chromosome so that the chromosomes can be moved easily with a pair of forceps. When the chromosomes are roughly arranged, trim the chromosomes and place them over the tape. It is unnecessary to trim each chromosome according to its contour. Trim them in rectangulars and arrange them the way one desires. Simply stick these finished chromosomes on a plasticized board; the chromosomes can be moved at will with a pair of forceps.

Even though this cannot be considered conventional, most cytologists place the short arm on top and long arm on bottom. Some persons, however, place the chromosomes the other way around. There is no rule against such a practice. But it may be my personal prejudice that the chromosomes so arranged appear unstable as if all of them will fall like dominos. This seems especially true with acrocentrics.

I.   Autoradiography

The procedure described below uses DNA synthesis as an example since the routine method is the same whatever the experiment. In our laboratory, we use the old-fashioned AR-10 stripping film. This is because we know this film well and hesitate to change. However, we shall present hints here for both strippers and dippers.

1.   EQUIPMENT AND SUPPLIES

Darkroom, illuminated by red safelight (Kodak filter Wratten series 2)
Film cutting board (Glass Engineering Co., 3601 White Oak, Houston, Texas)
Two trays (5 × 7 inch) for immersion of film plates, dark colored or

stainless steel. When white plastic is used the film plate cannot be seen

One tray (8 × 10 inch)

Clips to attach slides

Camel's hair brush

Drying rack for slides

Drying box with an electric blower (without heat)

Staining dishes for developer, fixative, and stain

Staining racks

Slide boxes (black plastic each holding 25 slides)

Kodak AR-10 fine grain autoradiographic stripping film

Kodak developer D-19B

Acid fixer

Dehydrating agent (e.g., Drierite)

Gauze squares or tea bag (embedding bag)

Mounting medium (e.g., Permount)

Brush to apply mounting medium

Razor blades

2. Continuous Labeling with Thymidine-^{3}H

This method is excellent for studying late replicating patterns. Use duplicate cultures if the length of the G_2 phase has not been determined. Add H^3-TdR to cultures at a final concentration of 1 μCi/ml. Beginning hour 2 and at every hour thereafter, Colcemid is introduced to one culture, which is harvested an hour later. Thus cultures are fixed at hours 3, 4, 5, and 6, after labeling, each with 1-hour Colcemid treatment. Squash or flame-dried preparations are made for each sample. Some slides should be sealed and stored in the refrigerator. Some should have the cover slips removed immediately for pilot autoradiographs.

3. Pulse Labeling

Introduce thymidine-^{3}H (2–5 μCi/ml) to the cultures for the desired period of exposure. Wash and reincubate cultures at the termination of labeling period with prewarmed fresh or conditioned medium containing 2–5 μg/ml nonradioactive thymidine and reincubate.

4. Pilot Slides

It is always a safe procedure to use spare preparations as pilot slides to test appropriate exposure time and to determine the best sample for detailed analysis. Slides with frosted edges are preferred, because one can label them with a pencil. The slides are to be immersed in various solutions, which would fade ink or dissolve wax if these are used for writing.

5. DUPLICATE PHOTOGRAPHY

There are two systems for duplicate photography, viz., photographing cells before autoradiography and photographing cells after autoradiography. The advantage of the former is that the cell and cellular components display their best photographic quality. The disadvantage is that much time and photographic film are wasted because a number of the selected cells eventually show undesirable autoradiographic features such as no label or too heavy a label. The second method avoids these problems because the investigator selects the desired autoradiographic patterns before critically examining the underlying cellular material. After removal of the silver grains, the cellular components will then be rephotographed. This method is more effective than the first one, but the quality of the photographs suffers somewhat. In either case, coordinates of the subjects photographed must be recorded in the notebook.

6. PROCEDURE FOR STRIPPING FILMS

Kodak stripping film AR-10 is cut into 40×40 mm squares with the aid of a plastic cutting board which is actually a frame with pegs on its rims to serve as a guide. Use a new scalpel to cut the film, one blade per plate. Each plate yields 12 film squares. (In extremely dry areas, the cutting is best performed in a tray containing 70% alcohol.)

The plate is immersed in a tray containing 70% alcohol. After 3 minutes it is transferred to a second tray containing absolute alcohol. At this stage, the cut film squares are still attached to the glass plate. With a pair of small forceps a single square is lifted and removed from the plate. The film must loosen easily without distortion or curling up. It is then dropped, with the emulsion side down, into a third tray containing a solution prepared as follows:

Sucrose (white granulated sugar)	200 gm
KBr	0.1 gm
Water	1 liter

For use, dilute the above stock solution with 9 volumes of water and mix thoroughly. Use a temperature of 16–21°C (61–70°F). This method minimizes the background grains caused by static electricity.

With one hand the experimenter dips a slide into the sucrose–bromide solution and with the other hand maneuvers the film square into position, touching one end of the film near the frosted end of the slide. When the slide is lifted out of the water, the film will cover the preparation and its overlapping ends will fold to the underside of the slide. The slide is then put on a glass plate and attached to it with a clip. For about 3 minutes this plate is placed back into the tray with distilled water; this is conveniently done in batches of five slides. After they are taken out of the water, the slides are removed from the glass plate, one by one, and the film square is straightened out by using a camel's hair brush and fingers. The slides are placed into a rack for drying. For this purpose, we use a slide box ($22 \times 16 \times 3$ cm) with large openings sawed in the bottom and lid so that only a frame structure remains.

The drying rack is inserted into a lightproof drying box which is provided with a small electric blower without heat to circulate the air through the box at a slow speed. The built-in light trap system also prevents dust from settling on the wet slides. The drying must be done at room temperature and usually takes about 3 hours. As soon as the slides are dry, they are transferred to a small slide box (black plastic or metal) which is taped around the lid to prevent light leakage. In this box, space must be provided for about 1 teaspoon of a dehydrating agent, such as silica gel desiccant, wrapped in some gauze or a tea bag. The box is kept in a refrigerator until the slides are ready to be developed. Pilot slides are placed in separate boxes so that most of the slides remain undisturbed if the exposure time must be extended.

Before the slides are developed, the reverse side of each slide should be painted with a mounting medium (e.g., Permount, Euparal) in order to prevent any shifting of the film squares during development and subsequent staining. The mounting medium can be diluted with xylol (1:5), and the slides are dried within a matter of minutes in the darkroom. However, when thicker mounting medium is used, the slides may be dried in the same drying box previously described. For developing and staining, the slides are conveniently placed in a rack holding some 10 or 20 slides to facilitate moving them simultaneously from one fluid to another. Kodak developer D-19B is adjusted to 20°C. The developing time of 5 minutes recommended by the manufacturer for conventional use is too long for chromosome autoradiography. We recommend 2 or 3 minutes. This yields considerably smaller grains and better resolution. After fixation in acid fixer for 2 minutes, the slides are then washed in a 20°C running water bath for 5 minutes. For drying the slides, the rack and the drying box are again used. Drying in 37°C incubator may be substituted.

The dry slides are stained with Giemsa (see p. 22) for 7 minutes. They are then dipped twice in distilled water, dried at room temperature, and mounted in Permount. Use a razor blade to scrape off the ends of the film on the back of the slides.

7. LIQUID EMULSION

The liquid emulsion is simpler to use than the stripping film because the operation and drying require less time. There are several types of liquid emulsion available in the market, and the current popular one is the Ilford Nuclear Research Emulsion K5. This emulsion produces finer grains than the NTB series.

The Ilford emulsion is a semisolid and should be dissolved in the darkroom with warm (approximately 40°C) water. The amount of water used depends upon the thickness of the final emulsion desired. As a start, try 1 part emulsion:3 parts water. The emulsion must be completely dissolved before use. In case a high background occurs, try adding 1% glycerol to the final solution.

Dip the preparations into a jar containing the emulsion (a 50-ml staining jar is ideal). Pull out and let the liquid drain back into the jar. The emulsion on the back of the slides may be wiped off at this time. Dry the preparations at room temperature in the darkroom, and pack the slides for exposure (see instruction for stripping film above).

8. DEGRAINING

This method works well with both stripping film and liquid emulsion film.

a. Place the slide in a Coplin jar containing 10% potassium ferricyanide (stock solution of 20% diluted 1:1 with distilled water).

b. Leave for 1 minute. Drain, do not rinse.

c. Transfer the slide to a Coplin jar containing 25% sodium thiosulfate (stock solution of 50% diluted 1:1 with distilled water).

d. Leave for 1 minute and drain.

e. Rinse twice with distilled water for about 30 to 60 seconds each.

f. Drain and air-dry thoroughly.

The stock solutions of ferricyanide and hypo seem to be stable enough to be kept ready for use. Also, the treatments may remove some stain; therefore, the slide should be restained. The time during which the slides are exposed to the above solutions should be kept at a minimum. If, however, all the grains are not removed, the above procedure may be repeated at least once without damaging the chromosomes. Sometimes it also helps to wet the slides with distilled water prior to ferricyanide treatment.

J. Constitutive Heterochromatin

1. Subbed slides. If flame-dried preparations are used, "subbing" slides is not necessary. If squash preparations are used, subbing prevents the

loss of cells during alkaline treatment. This is done by dipping alcohol-cleaned slides into a solution containing 0.1% gelatin and 0.01% chrome alum (chromium potassium sulfate). Dry thoroughly and store in a dry place.

2. Make chromosome preparations with either the flame-dried or squash method without staining. Flip cover slips off the squashed slides, and wash off the fixative with ethyl alcohol.

3. Treat cells on slides with 0.2 N HCl at room temperature for 30 minutes. For heterochromatin staining only, this step may be omitted.

4. Rinse several times in distilled water and air-dry.

5. Treat the slides with pancreatic RNase (100 μg/ml in 2XSSC) at 37°C in a moist chamber for 60 minutes. Boil RNase before use (see p. 11). For the formula of SSC, see p. 12. This step should be performed in moist chambers to conserve RNase. RNase solution is placed on the slide and a cover slip is used to cover the solution. For heterochromatin staining only, this step may be omitted.

6. Rinse several times in 2XSSC, 70% ethanol, 95% ethanol, and air-dry.

7. Treat the slides with 0.07 N NaOH for 2 minutes. This step is critical. In case the chromosomes of the final preparations appear bloated, use SSC instead of water to prepare this solution.

8. Rinse in several changes of 70% ethanol and several changes of 95% ethanol. It is important to remove NaOH as rapidly as possible to prevent excessive DNA denaturation. If the NaOH solution is prepared with SSC, rinse in SSC.

9. Incubate the slides in 2XSSC or 6XSSC at 65°C overnight. Again, use moist chambers.

10. Rinse several times in 70% ethanol and 95% ethanol.

11. Stain with Giemsa, rinse, dry, and mount. It is important to test various aliquots of the Giemsa stain because some batches work better than others.

To illustrate constitutive heterochromatin, Fig. 1 presents the pattern of heterochromatin in a male eastern harvest mouse, *Reithrodontomys fulvescens*. Note heterochromatin in the centromeric areas of all acrocentrics, the short arm of the X, and the Y.

K. *In Situ* Hybridization

In *in situ* hybrids one tries to determine the nuclear and chromosomal locations of a particular species of DNA or RNA. Thus the DNA of the nuclei and chromosomes in the cytological preparations can be considered the immobile receptor material and the tritium-labeled molecules, introduced for hybridization, the mobile component.

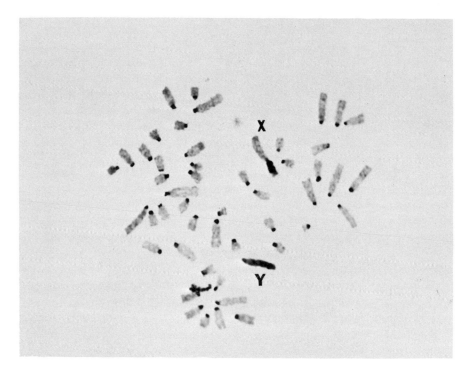

FIG. 1. A metaphase of a male eastern harvest mouse, *Reithrodontomys fulvescens*, showing the distribution of constitutive heterochromatin.

It is not within the scope of this chapter to describe in detail the methodology for obtaining the mobile component. Generally speaking, the mobile component can be prepared by two major schemes:

1. ^{3}H-Labeled DNA or RNA fractions directly isolated from tissues or cells. This method is inferior because it depends upon the synthetic activity of the cells, and thus requires a cell culture system. The demand for a large amount of nucleic acid molecules necessarily requires an incredible number of cell cultures. Furthermore the radioactivity of the nucleic acids so isolated is never very high for effective autoradiography.

2. Synthesis of ^{3}H-labeled complementary RNA (cRNA) using DNA fractions as template. This method, though indirect, is effective because one may use any DNA fraction as template for cRNA synthesis. It requires the synthesis of RNA in a cellfree system using *E. coli* RNA polymerase and ^{3}H-labeled nucleotides. Only ATP, UTP, and CTP are tritium-labeled, however. Nonradioactive GTP is added in appropriate amount.

The ^{3}H-cRNA is then used for *in situ* hybridization experiments.

The hybridization procedure is essentially the same as the procedure for constitutive heterochromatin staining (p. 33) except that cRNA is cluded in step 9. After overnight incubation at 65°C, the preparations are rinsed in 2XSSC, and RNase is applied to destroy the loose RNA molecules. Autoradiographs are prepared from the final preparation. When cRNA with a radioactivity in the dpm range of 10^7 is used, the exposure time requires approximately 1 to 2 weeks.

ACKNOWLEDGMENT

I would like to thank Dr. Frances E. Arrighi and Miss Linda Shirley for their assistance.

Chapter 2

Cloning of Mammalian Cells[1]

RICHARD G. HAM

*Department of Molecular, Cellular and Developmental Biology,
University of Colorado, Boulder, Colorado*

I. Introduction

Three conditions must be satisfied in order to establish a new population of cultured cells from a single isolated cell.

(1) It must be possible to isolate single cells from the parental culture or tissue without loss of viability.

[1] Studies on clonal growth of cells taken directly from adult rabbits have been supported by National Science Foundation Grants GB-4392, GB-7007, and GB-18502 and by American Cancer Society Grant E-505.

37

(2) Culture conditions must be such that single isolated cells will multiply *in vitro.*

(3) The progeny of the selected cell must be kept physically separated from all other cells.

This chapter is divided into two major parts: one, a theoretical discussion of the principles and problems involved in satisfying each of the three conditions necessary for the establishment of clonal populations; and two, a step-by-step description of the specific experimental procedures that are recommended. Additional detailed information on recommended equipment, media, and supplies for cloning is provided in a series of appendixes at the end of the chapter.

No attempt is made in this chapter to review in detail the literature of cloning or to discuss all of the possible procedures that could be used. Techniques that are described are mostly those that have been used in the laboratory of the author. For additional information on other approaches to cloning, the reader is referred to standard texts on cell culture (e.g., Paul, 1970), to the recent review of clonal techniques for embryonic chicken cells by Cahn *et al.* (1967), and to the heading "Clone Cells" in current issues of "Index Medicus."

II. Theoretical Considerations

A. Definitions

A *clone* is a population of cells consisting entirely of the progeny of a single cell. The term *cloning* has been used somewhat indiscriminately in the literature of mammalian cell culture to refer to any experimental procedure in which single cells multiply to form identifiable colonies. In this chapter, use of the term will be restricted to the establishment of a new culture composed exclusively of the mitotic descendants of a single cell. All other procedures involving formation of colonies from populations of isolated single cells will be described as *clonal growth.*

B. Special Requirements for Growth of Isolated Cells

1. MEDIUM

In their native habitat, mammalian cells are crowded tightly together and are exposed to a relatively small volume of extracellular fluid per cell. Under these conditions the cells work together to provide for themselves a precisely regulated environment that remains essentially constant over a wide range of external conditions.

In monolayer and other high density cultures the volume of medium per cell is also relatively small (generally less than 10^{-5} ml/cell). The cells in such cultures retain an appreciable ability to control their environment, an ability commonly called "conditioning" of the culture medium.

In typical clonal cultures, however, the volume of medium per cell is much larger (generally greater than 10^{-2} ml/cell). Thus, little if any conditioning is possible and the cells are almost totally dependent upon the culture environment that is provided for them.

Historically, cloning of mammalian cells was first accomplished by conditioning the culture medium either through use of a capillary tube to restrict the volume of medium surrounding a single cell (Sanford et al., 1948) or through use of a "feeder layer" of cells that had been prevented from multiplication by massive doses of irradiation (Puck and Marcus, 1955). Subsequent improvements of culture media and techniques have eliminated the need for these extreme measures but it is still necessary to consider for clonal growth a number of special requirements that can be ignored for larger numbers of cells.

One major consideration is freedom from toxicity. In the case of toxic substances that are taken up by or bound to cells (e.g., heavy metal ions, antibodies), each cell in a clonal inoculum is potentially exposed to three or four orders of magnitude more of the toxic material than each cell in a monolayer inoculum. Thus, in the preparation of media for clonal growth, extreme caution must be taken to use pure chemicals and to avoid all potentially toxic materials (Fisher and Puck, 1956).

Care must also be exercised to avoid toxic excesses of essential nutrients. Studies seeking to determine optimum concentrations of nutrients in medium F10 and in medium F12 have shown repeatedly that too much of a nutrient is as bad as too little (Ham, 1963a, 1965). In the most extreme case studied, deficiency, optimum growth, and toxic excess of linoleic acid all occurred within a tenfold concentration range (Ham, 1963b).

Another major consideration is the frequent requirement for exogenous supplies of substances that are actually being made by the cells in relatively large amounts. The reason behind such requirements appears to be that the cells within the mammalian body do not rely on permeability barriers for certain metabolic intermediates. Instead, the substances accumulate in the rather limited extracellular volume, and the intracellular concentration is maintained by equilibration with the extracellular "pool." A similar phenomenon can also occur in crowded cultures, but under clonal growth conditions any intermediates that escape from the cells become so greatly diluted that they are no longer available for essential biosynthetic reactions. Thus it becomes necessary to provide exogenous supplies of substances such as pyruvate, carbon dioxide, and "nonessential" amino acids to obtain satisfactory growth of most cell types in low density culture (Neuman and McCoy, 1958; Herzenberg and Roosa, 1960; Eagle and Piez, 1962; Ham 1963a, 1965).

The selection of a specific medium for clonal growth must ultimately be based on what works best for the cell population in question. In cases where clonal growth has been previously described in the literature, it is generally desirable to start with the medium that has already been shown to work. In other cases, the starting point should be a medium designed specifically for clonal growth, such as F12M (Appendix A; Ham, 1965), F10 (Ham, 1963a), or Eagle's minimum essential medium with optional supplements for clonal growth (Lockart and Eagle, 1959; Eagle, 1959; Eagle and Piez, 1962). Medium F12 (Ham, 1965) has proved very effective for some types of cells, but is not recommended for general use because of the toxicity of its high concentration of zinc to certain cell populations. Medium F12M (Ham, 1965; Appendix A) is identical to F12 except for a reduced concentration of zinc.

2. Serum and Other Supplements

Although clonal growth of some mammalian cells has been achieved in defined or semidefined media, it is generally desirable, when possible, to use a supplement of whole serum or its macromolecular components. The presence of serum tends to compensate for minor variations or toxicities in the medium and results in more uniform and reproducible growth. In addition, growth rate and plating efficiency are usually much better in serum-containing media.

Fetal bovine serum at a concentration of 10% of the total medium is generally a good starting point for determining the optimum serum supplementation for clonal growth of a particular type of cell. In some cases, mixtures of serum are found to be superior. For example, a mixture of fetal bovine serum and human placental cord serum is favorable for clonal growth of human diploid cells (Ham, 1963a; Kao and Puck, 1970). Other additives, such as chicken embryo extract or its macromolecular fraction are also sometimes beneficial. However, care must be taken to avoid toxic excesses of such substances (Coon and Cahn, 1966; Cahn, 1968; Ham *et al.*, 1970).

3. pH Control

The problem of pH control is critical for clonal growth, largely because of the necessity for supplying carbon dioxide exogenously. Although carbon dioxide is commonly thought of as a waste product, it is also an essential biosynthetic intermediate that must be supplied from external sources when cells are maintained under conditions where metabolically produced carbon dioxide cannot accumulate. The usual procedure is to maintain a fixed concentration of carbon dioxide in the gas phase over the cultures and to add enough sodium bicarbonate to the culture medium to prevent acidification. This arrangement works very well as long as the carbon dioxide tension is maintained. However, when the cultures are removed from the incubator or the incubator door is opened, the pH rises rapidly due to loss of dissolved carbon dioxide. This problem can be partially overcome by installing devices to control the carbon dioxide content of the incubator or to add extra carbon dioxide after each opening of the door (cf., Appendix D; Ham and Puck, 1962a). Recent investigations in this laboratory indicate that additional benefits can be derived by including HEPES (N-2-hydroxyethylpiperazine-N'-2-ethane sulfonic acid) buffer (Good, et al., 1966) in the medium at a concentration twice that of the bicarbonate (Ham and Sattler, 1971). The HEPES buffer greatly reduces the tendency of the medium to become alkaline when it is removed from the carbon dioxide atmosphere for a short time.

Very recent reports by Ceccarini and Eagle (1971a,b) add critical importance to precise control of pH in all cultures. Eagle (1971) recommends combining several organic buffers for accurate pH control with minimum toxicity.

4. HUMIDIFICATION

The range of osmotic pressures that can be tolerated by cultured cells is very small. Most clonal growth experiments require relatively long times and are done in petri dishes with loose fitting lids that will not prevent evaporation. Thus, it is essential to keep the humidity inside the incubator very close to saturation at all times to avoid loss of water from the medium through evaporation.

Simple humidification devices such as open pans of water are usually not adequate since the water tends to be cooled by evaporation whenever there is a major need for humidification, such as after the incubator door has been opened. When this happens the water vapor in the gas mixture inside the incubator equilibrates with the liquid water at a reduced temperature and the humidity becomes less than saturated with respect to the incubation temperature. Because of its large mass, the water in the pan tends to heat very slowly, keeping the humidity low for prolonged periods. This problem becomes particularly severe in a dry climate where a near total loss of moisture can occur each time the incubator door is opened. One of the most effective solutions is to place a thermostated immersion heater in the water pan to maintain it at the desired incubation temperature at all times (cf. Appendix D).

Another major difficulty in maintaining satisfactory humidification is condensation. It is impossible to maintain adequate moisture in an incubator that contains any surface that is cooler than the incubation temperature. Thus, thorough insulation of all parts of the incubator including the door is very important to successful clonal growth.

5. Culture Surface

The culture surface is another important aspect of the culture environment for clonal growth. Although clonal growth of some types of cells can be achieved in suspension cultures in semisolid media, the usual clonal growth experiments are done with the cells firmly attached to a solid substrate. The most commonly used culture surfaces are modified polystyrene and glass. Disposable polystyrene "tissue culture" petri dishes and flasks manufactured by Falcon Plastic Company have been used for many years with excellent results. (The process for production of the modified polystyrene surface remains a carefully guarded trade secret.)

Although there have been numerous reports of cell growth on a variety of other surfaces, glass remains the most widely used alternative to modified polystyrene. The key to consistent clonal growth on a glass surface is careful and thorough washing. Specific procedures for the preparation of glass petri dishes and microscope slides for cell growth are given in Appendix C.

A thin coating of collagen is sometimes added to glass or plastic surfaces to enhance growth or the expression of differentiated properties (Ehrmann and Gey, 1956; Hauschka and Konigsberg, 1966).

Clonal growth of established cell lines, virally transformed cells, and normal chondrocytes, granulocytes, and mast cells can sometimes be achieved by use of semisolid media containing low concentrations of agar, agarose, fibrin, methyl cellulose, or agar mixed with collagen (Puck et al., 1956; Pluznik and Sachs, 1965; Robinson et al., 1967; Schindler, 1964; Macpherson and Montagnier, 1964; Walters et al., 1970; Horowitz and Dorfman, 1970; Sanders and Smith, 1970). For such cells, the selection of clonal populations can frequently be achieved by the application of procedures similar to those used for bacterial cells (cf. Section III,E).

C. Preparation of Viable Single Cells

Cells that are to be used for clonal growth experiments must be handled more gently than those for monolayer or other high density cultures. Procedures that are satisfactory for subculturing large cell populations frequently fail to yield cells that are "viable" under clonal growth conditions.

The damage caused to cell membranes by various procedures for dispersing cells was studied by Magee *et al.* (1958), who reported that procedures developed for clonal growth (dilute trypsin in saline without divalent ions; Puck *et al.*, 1956) caused far less leakage of cytoplasmic contents than other procedures commonly in use at that time for larger cell populations. Because of the relatively large volume of medium per cell that is used in clonal growth experiments (cf. Section II,B), any materials that leak out of cells under clonal growth conditions are extensively diluted and are unlikely to be available for reutilization by the cells (the situation is analogous to the loss by diffusion of pyruvate, carbon dioxide, and other essential metabolic intermediates discussed above).

Suspensions of single cells for clonal growth can be prepared from monolayer or suspension cultures of established cell lines, from primary monolayer cultures, from monolayer outgrowths of primary explants, from organ cultures, or directly from freshly excised tissue. Most of the common procedures use enzymes or chelating agents to weaken the bonds between cells, followed by a mild mechanical agitation to ensure complete separation of the cells.

Like many other aspects of clonal culture, the choice of a cell dispersing agent must ultimately be based upon what is most effective for the type of cell in question. For the preparation of cell suspensions from monolayers (established cell lines, primary monolayer cultures, and outgrowths from explants) dilute trypsin is widely used. The popularity of trypsin appears to rest both on its effectiveness and on the fact that it is fully neutralized by serum proteins. There have been claims that Pronase is actually more effective for dispersing cells from monolayer cultures, but it is not inhibited by serum and the cells must be washed before good attachment and growth can be obtained (for additional discussion, cf. Foley and Aftonomos, 1970).

The preparation of viable cell suspensions from tissue fragments (or organ cultures) is still very much of an art with few firmly established rules. In this laboratory, the best results have generally been obtained with collagenase or with Pronase depending on the type of tissue (Ham and Murray, 1967; Ham and Sattler, 1968).

A complete discussion of the conflicting claims about the relative effectiveness of agents such as trypsin, pancreatin, Viokase, elastase, collagenase, papain, Pronase, aspergillin, EDTA, and sodium tetraphenylboron for dispersing cells is beyond the scope of the current chapter. For additional information the reader is referred to Rinaldini (1958), Rappaport and Howze (1966), Cahn *et al.* (1967), and Paul (1970).

A crude measurement of "viability" can be made by determining the ability of freshly prepared cell suspensions to exclude macromolecular dyes such as trypan blue, erythrosin B, eosin Y, or nigrosin (cf. Paul, 1970 for recommended procedures). The ultimate test of "viability" in clonal culture is whether or not each cell is able to multiply and form a colony. Unfortunately, this test is dependent upon many other variables besides the treatment received by the cells during preparation of the inoculum.

D.   Selection of Clonal Populations

Techniques for the establishment of clonal populations fall into two broad categories: (1) those that isolate a single cell in a separate container and allow it to grow into a new population and (2) those that allow colony formation to occur before selection and physical isolation of the new clone. Advocates of single cell isolation cite well-documented cases of satellite colony formation (Abbott and Holtzer, 1968) and tend to discount all studies of clonal growth done without physical isolation of the cells. On the other hand, advocates of the colony selection technique maintain that it can be quite reliable if done properly and that the advantages of convenience and the ability to select colonies with desired properties more than compensate for the slight risk involved.

The principal concern in experiments involving growth of more than one clone in the same petri dish is that of contamination. Two major sources of contamination must be considered: (1) formation of a colony from a clump of two or more cells and (2) incorporation of a floating cell or a nearby attached cell into a developing colony. The problem of multiple cell origin can be minimized by careful microscopic examination of the cell inoculum to be certain that it is free from aggregated cells and by frequent stirring of the cell suspension as the cultures are being inoculated. It is also possible to remove most clumps by straining the cells through Nitex monofilament screen cloth (Cahn *et al.*, 1967). The chances of a colony being derived from two closely adjacent cells can be minimized by using a small inoculum to ensure good separation of colonies and by rejecting any colony that is unusually large or gives the appearance of being formed by the fusion of two overlapping circles.

The severity of the problem of loose, floating cells varies greatly with the type of cell and the culture techniques. Under conditions where the petri dishes are handled and the medium changed frequently, the detachment of cells and formation of satellite colonies (or the contamination of other colonies) is a common occurrence. However, with most cell types little difficulty is encountered under conditions where the dishes are not handled and the medium is not changed during the growth period. Under these conditions the concept of "plating efficiency" appears to be valid and the selection of colonies seems to be a reasonably safe operation.

Semisolid media can sometimes be used to eliminate the problem of floating cells. However, such media do not eliminate the possibility of a colony arising from a clump of cells rather than from a single cell.

Some laboratories attempt to get around the dangers inherent in colony selection statistically by selecting a colony, using its cells for a second clonal inoculum, and then selecting a daughter colony. In most cases this sequential selection appears to be quite satisfactory. However, if absolute certainty is required in clone selection, a procedure involving the isolation of a single cell into a separate container should be utilized.

Cell isolation techniques are somewhat more complex and time consuming than colony isolation methods although several relatively easy techniques have been developed (cf. Section III,D). The most serious drawbacks, however, are that they do not allow for easy selection of colonies with desirable properties and that there is no assurance that the cell selected will grow into a colony (which can be a serious problem for cultures with low plating efficiencies).

The selection of a clonal population does not ensure genetic uniformity, particularly in the case of established cell lines. Most established lines have undergone major chromosomal upheavals and are hyperdiploid. In many cases their chromosome numbers are variable, with a statistical distribution around stem line numbers (Chu and Giles, 1958). Thus, there is no assurance that a particular cell in a "clonal" line will have the same number of copies of a particular gene as other cells in the same population.

Techniques are beginning to be developed that permit mammalian cell clones with particular genetic properties to be selected. One such technique involves incorporation of 5-bromodeoxyuridine followed by irradiation with short wavelength visible light. This procedure kills all cells except those that lack certain biosynthetic activities and are therefore unable to incorporate BUdR when placed in minimal media (Puck and Kao, 1967). Similarly, several procedures are now available for selecting "hybrid" cells that have incorporated genetic information from two parental lines of cells (Littlefield, 1966; Kao et al., 1969; Kao and Puck, 1970).

E.   Current State of the Art

The preceding discussion has stressed what is known about cloning mammalian cells. In order to balance the presentation it is necessary to call attention also to what is *not* yet known.

One of the major difficulties with mammalian cell culture is the fact that it is often regarded primarily as a tool to be used in the study of other cellular problems. There is very little tendency to tamper with a technique that "works," and, thus, there is generally very little investigation of why it works or how it could be improved.

Most widely used cell culture media were originally developed for dense populations of established cell lines. In recent years, some attention has been given to the requirements for clonal growth of established lines (cf. Section II,B) and relatively reliable methods and media have been developed. However, even for established lines, good clonal growth generally remains dependent on serum or other undefined additives.

The nutritional needs for clonal growth of unaltered cells taken directly from the mammalian (or avian) body remain largely unstudied and unknown. Recent interest in the study of differentiation *in vitro* has led to direct clonal growth of a number of types of differentiated cells (Konigsberg, 1963; Cahn and Cahn, 1966; Coon, 1966; Pluznik and Sachs, 1965; Robinson *et al.*, 1967; Ham and Murray, 1967; Ham and Sattler, 1968; Coon, 1969). However, this growth has generally been accomplished through minor modifications of media and techniques originally developed for established cell lines (which had undergone extensive nutritional adaptation and selection before their nutritional requirements were studied). At the present time no media have been described that are formulated specifically to satisfy the nutritional needs for clonal growth of primary cultures.

Thus for many types of cells it is still necessary to perfect new culture techniques or media before clonal experiments can be undertaken.

III.   Procedures for Establishing Clonal Populations

A.   General Introduction to Experimental Procedures

This section consists of detailed descriptions of specific experimental procedures for achieving clonal growth and isolating clonal populations of mammalian cells. Procedures for establishing clonal growth of several cell types in petri dishes are presented in Section B, followed by procedures for subculturing a single colony in Section C. Special techniques involving isolation of single cells and use of semisolid media are presented in Sections D and E.

The word "sterile" has generally been omitted from these procedures in order to avoid endless repetition. However, it must be remembered that the media used for mammalian cell culture are very rich and will grow almost anything more rapidly than mammalian cells. Thus all media, solutions, bottles, dishes, and pipettes must be thoroughly sterilized and sterile techniques must be employed in all operations involving the growth of mammalian cells in culture.

Formulas for preparation of most of the media and enzyme solutions used are given in the Appendixes. Where this is not the case, specific literature references are cited.

B. Colony Formation on Solid Surfaces

1. CLONAL GROWTH OF ESTABLISHED CELL LINES

The following procedures will result in colony formation with a good plating efficiency for cells of many established cell lines including HeLa, Mouse L, and Chinese hamster ovary (adapted from Puck *et al.*, 1956, and Ham and Puck, 1962b).

1. Two or three days before the clonal growth experiment is to be done, subculture the parental monolayer culture in 75 cm^2 plastic tissue culture flasks (Falcon No. 3024) each containing 20 ml of culture medium. (Note that all procedures described below are given in terms of this particular flask. A variety of other culture vessels including petri dishes, milk dilution bottles, and roller tubes or bottles can be used equally well, provided that appropriate modifications are made in the amounts of reagents used.) The medium used should support optimum growth of the monolayer culture and the size of the inoculum should be adjusted such that the cells will be in a state of active multiplication and the culture surface will be nearly covered with cells at the time of the clonal experiment.

2. One or two hours before the experiment, prepare a mixture consisting of 90% medium F12M and 10% fetal bovine serum and place 5 ml in each of several 60 × 15 mm plastic tissue culture petri dishes (Falcon Plastic No. 3002). Place the dishes of medium in the tissue culture incubator to equilibrate with the 5% carbon dioxide atmosphere. (If the incubator has removable shelves, the dishes can be kept on a shelf throughout the inoculation procedures described below.)

3. Examine the monolayer cultures with an inverted microscope and

select one with a uniform population of cells that is free of loose or granular cells and also free of areas of clumped or multilayered cells. In addition be certain that the medium over the cells is not cloudy or excessively acidic. (The phenol red indicator in the medium is orange at the desired pH and becomes yellow when excessively acidic.)

4. Remove the used culture medium from the flask with a 25-ml pipette.

5. With a 10-ml pipette add 5 ml of 0.05% trypsin solution (prewarmed to 37°C) to the culture flask. Keeping the pipette all the way in the culture flask, gently rock the flask so that the trypsin flows over the cells; then remove as much of the trypsin as possible with the pipette. Using a fresh pipette, repeat the entire rinse a second time. (The purpose of the rinsings is to remove the last traces of serum from the culture flask. The serum contains a trypsin inhibitor that would prevent effective release of the cells from the culture surface. Some cell lines may require only one rinse.)

6. Add 5 ml of trypsin solution to the culture flask and allow it to remain there. Screw the cap on the culture flask tightly (so that the trypsin solution will not be excessively acidified by the CO_2 atmosphere of the cell culture incubator), and incubate the flask at 37.5°C with the trypsin in contact with the cell monolayer for 5 to 15 minutes. The exact time of incubation varies with the cell type and with the condition of the cells. The trypsin digestion should be ended as soon as a majority of the cells have been freed from the culture surface. Excessive exposure to trypsin causes the cells to become very sticky and form long strings that are difficult to separate into a uniform cell suspension.

7. While waiting for the cells to be released, place 9 ml of neutralized (pH 7.3) culture medium (or culture medium prepared without sodium bicarbonate) in a sterile test tube with a slip-on metal or plastic cap. Also have additional tubes and neutralized medium ready for diluting the cell suspension. If monolayer subcultures are to be made, place 20 ml of complete culture medium in each flask to be used for a subculture.

8. As soon as the majority of the cells can be shaken loose from the culture surface draw them up and down a 5-ml pipette a few times to break up the clumps and obtain a smooth single cell suspension.

9. Quickly add 1.0 ml of the cell suspension to the 9.0 ml of neutralized medium that was previously placed in a capped test tube. Mix the diluted cell suspension by drawing up and down with a 10-ml pipette.

10. Add an appropriate amount of the concentrated cell suspension from the trypsinized flask to the flasks previously prepared for subculturing and incubate them.

11. Resuspend the concentrated cell suspension in the original culture

flask by drawing it up and down a pipette. Using a 0.1-ml pipette held as nearly horizontal as possible, quickly place one drop of the cell suspension from the culture flask in each side of a hemocytometer (American Optical No. 1492). Count all cells in one large square (9 large squares per side). One large square represents 10^{-4} ml of cell suspension. This count multiplied by 10^4 yields the number of cells per milliliter in the suspension. If the count is very high (over 200), it is permissible to count the diluted suspension from the dilution tube. At least two squares, one in each half of the counting chamber, should be counted. If the cell count is low, sufficient squares should be counted to obtain a total count of at least 200 cells in order that the count may have reasonable statistical validity.

12. Using neutralized medium, and sterile, capped test tubes, dilute the cell suspension from the first dilution tube to a final concentration of 1000 cells/ml. It is generally desirable to make sequential stepwise dilutions of the cell suspension in order to avoid working with very small volumes of the concentrated suspension or very large volumes of the final suspension.

13. Add 0.1 ml of the final cell suspension (100 cells) to each petri dish previously prepared, using a separate 0.1-ml pipette for each dish. Gently swirl each dish immediately after the cells are added to be certain that they are uniformly distributed throughout the medium.

14. After all dishes are inoculated swirl them again gently and place them in the cell culture incubator. (If the dishes are being handled on an incubator shelf the entire shelf can be swirled gently. Care must be taken not to spill medium on the shelf, since spilled medium will support mold growth inside the incubator.)

15. Incubate at a temperature of 37.5°C in an atmosphere of 5% carbon dioxide in air saturated with water vapor. Do not disturb the dishes during the incubation period. If it is desirable to monitor growth, prepare an additional dish and keep it on a separate shelf in the incubator so that the clonal cultures will not be disturbed when the shelf is slid out to examine the sample culture.

16. After 7 to 10 days the colonies will be large enough to allow a single colony to be selected and subcultured (see Section C, below). For determination of plating efficiency, colonies can be fixed with 10% formalin, and stained with 0.1% crystal violet in water for approximately 10 minutes.

2. Clonal Growth of Chondrocytes from Adult Rabbits

The following procedures have been used extensively in this laboratory for clonal growth of chondrocytes prepared directly from the ear cartilage of adult rabbits. With minor modification they can also be used for other cartilagenous tissues in the rabbit and for other species (adapted from Ham and Sattler, 1968).

1. Wash the ear of a freshly killed rabbit with 70% ethyl alcohol.

2. Cut out a portion approximately 2.5 × 2.5 cm from near the center of the ear where the cartilage is relatively flat and uniform in thickness. (If it is not convenient to do the experiment at the time the rabbit is killed, the entire ear can be removed and stored in a refrigerator for up to 24 hours without detrimental effects. Cartilage is a nonvascular tissue and its cells are quite stable for relatively long periods without any obvious supply of nutrients or oxygen.)

3. Remove the skin with sterilized dissecting instruments and place the tissue in a glass petri dish containing neutralized medium F12M plus 10% fetal bovine serum. (The dissecting instruments can be sterilized conveniently at the time of the experiment by dipping them in 95% ethanol and burning off the alcohol.)

4. Carefully dissect away the thin layer of connective tissue and remove any adhering blood vessels, leaving a piece of clean shiny white cartilage.

5. Transfer a portion of the cartilage approximately 1 cm square to a petri dish containing 5 ml of dissociating medium FBC (collagenase plus fetal bovine serum; cf. Appendix B).

6. Cut the isolated cartilage into small pieces (1–2 mm cubes)* with a scalpel.

7. Incubate in dissociating medium FBC for 15 minutes at 37.5°C.

8. Draw the fragments vigorously up and down a wide-mouth 10-ml pipette.

9. Draw off the dissociating medium with a pipette, allowing the tissue fragments to remain in the petri dish.

10. Add 5 ml of fresh dissociating medium FBC and continue the incubation at 37.5°C until a sufficient number of cells have been liberated. (This will normally require several hours; the exact time is dependent both on the age of the rabbit and on the concentration of collagenase.)

11. While the cartilage is digesting, prepare the growth medium (90% medium F12M and 10% fetal bovine serum) and place 5 ml in each of several 60 × 15 mm tissue culture petri dishes (Falcon No. 3002). Place the dishes of medium in the tissue culture incubator to equilibrate. [For optimum expression of differentiated properties by the chondrocyte colonies it may be desirable to add chicken embryo extract macromolecular fraction to the medium at a protein concentration of 50 μg/ml. This fraction can easily be prepared by passing commercial embryo extract through a Sephadex G-25 column (Ham *et al.*, 1970).]

12. When sufficient numbers of loose cells appear in the dissociation

mixture, add an equal volume of complete serum-containing growth medium, then draw the tissue fragments up and down in a wide-mouth 10-ml pipette, and transfer the resulting cell suspension and remaining fragments to a sterile 15-ml conical centrifuge tube.

13. Centrifuge briefly to remove the tissue fragments. (In this laboratory the centrifugation is done in a model CL International clinical centrifuge with No. 221 head. The instrument is set at speed setting 3 for 10 seconds, but does not reach equilibrium speed during this period.)

14. Transfer the remaining suspension to another sterile 15-ml conical centrifuge tube and centrifuge gently for 3 minutes to pack the cells at the bottom of the tube. (In this laboratory, setting 3 of the model CL clinical centrifuge, estimated to be about 1200 rpm, is used.)

15. Discard the supernatant solution and resuspend the cell pellet in 1 ml of complete growth medium.

16. Count a sample of the suspension in the hemocytometer (see previous procedure). A suspension containing between 10^5 and 10^6 cells/ml is normally obtained. The suspension contains cells that vary greatly in size, but is essentially free of red blood cells.

17. Dilute the cell suspension with complete growth medium to a concentration of 1000 cells/ml and add an inoculum of 100 cells (0.1 ml) to each of the petri dishes of medium previously prepared.

18. Incubate at 37.5°C in an atmosphere of 5% carbon dioxide in air saturated with water vapor.

19. After about 10 days the colonies will be large enough to permit a single colony to be subcultured (see Section C). The best expression of cartilage-like differentiation occurs around 14 days. At that time differentiated colonies can be detected with the naked eye by their three-dimensional, nearly hemispherical appearance. The presence of cartilage-like matrix material in living colonies can be detected at an earlier stage with an inverted phase contrast microscope. The matrix appears as a bright refractive halo around the cells in the differentiated colonies.

20. The following procedures can be used to stain the differentiated colonies: Discard the culture medium and fix the colonies with 10% Formalin for 1 hour. Rinse with tap water. Treat for 3 minutes with a solution of 3% acetic acid. Stain for 10 minutes in a solution containing 0.5 gm alcian green (Aldrich Chemical Company) in 100 ml of 3% acetic acid. Rinse with distilled water. Treat for 3 minutes with 3% acetic acid. Wash in tap water for 3 minutes. Counterstain for one-half to 1 minute in a solution containing 0.25 gm metanil yellow (C.I. No. 13065) in 100 ml distilled water to which 0.1 ml glacial acetic acid has been added. Rinse with tap water and allow to dry. Areas with cartilage-like matrix are stained a brilliant green against a bright yellow background of nondifferentiated cells (Ham and Sattler, 1968).

3. CLONAL GROWTH OF CHOROID CELLS FROM ADULT RABBITS

The following procedure yields clonal growth with high plating efficiency for cells taken directly from the choroid coat of the eye in adult rabbits. With minor modifications it can also be applied to many other tissues in the rabbit (adapted from Ham and Murray, 1967).

1. Prepare the culture medium (90% medium F12M, 5% fetal bovine serum, 5% rabbit serum) and place 5 ml in each of several 60 × 15 mm plastic tissue culture petri dishes. Place the dishes in the tissue culture incubator to equilibrate.

2. Remove an intact eyeball from a freshly killed rabbit.

3. Rinse the eyeball with sterile saline and trim off excess tissue.

4. Cut the eyeball into hemispheres, such that one-half contains essentially all of the retina and choroid while the iris and cornea remain in the other half, which is discarded.

5. Working in a petri dish under sterile T saline (Appendix B), gently peel the neural retina away from the choroid and discard it.

6. Using blunt forceps, gently push the choroid off the sclera.

7. Transfer the choroid and the adhering retinal epithelium to fresh sterile saline in another petri dish and agitate with forceps to rinse off extra red blood cells.

8. Place the choroid in a 60-mm petri dish containing 5 ml of 0.1 mg/ml Pronase in T saline (Appendix B) and cut it into small pieces with sharp dissecting scissors. (Choroids from both eyes of the same rabbit are generally used to increase the cell yield.)

9. Incubate at 37°C for 10–20 minutes.

10. Gently pipette the Pronase solution and fragments to free adhering cells from the chunks.

11. Draw the suspension into a vertically held pipette and allow the larger fragments to settle to the tip. Discharge them with a minimum volume of solution into a petri dish. Transfer the remaining cell suspension to a sterile 15-ml conical centrifuge tube containing 2 ml of complete growth medium (prepared in step 1).

12. Add an additional 5 ml of Pronase solution to the remaining fragments and digest for an additional 1 minute.

13. Pipette the suspension gently and add it, including the fragments, to the centrifuge tube containing the previous suspension.

14. Centrifuge the combined suspensions gently to remove the tissue fragments. [In this laboratory an International clinical centrifuge, model CL with No. 221 head is run 20 seconds at speed setting 3 (approx. 1200 rpm) followed by 30 seconds at speed setting 1 (approx. 400 rpm).]

15. Transfer the supernatant to another sterile 15-ml conical centrifuge tube and centrifuge at setting 3 for 3 minutes.

16. Discard the supernatant and resuspend the cells in 1.0 ml of complete growth medium.

17. Count a sample of the cell suspension in the hemocytometer (see step 11, p. 48). Red blood cells, which can be distinguished by their smaller size, greater refractivity, and characteristic morphology, are not counted. (Difficulties in distinguishing red blood cells can be eliminated by mixing a sample of 0.1 ml of cell suspension with 0.1 ml of 0.1 N HCl to lyse the red blood cells before counting. Most other types of cells remain intact in the counting chamber under these conditions. It is important to remember that a twofold dilution occurs when this method is used.)

18. Dilute the cell suspension to a concentration of 1000 cells/ml with complete growth medium. Add 100 cells (0.1 ml) to each of the previously prepared dishes of medium.

19. Incubate at 37.5°C in an atmosphere of 5% carbon dioxide in air saturated with water vapor.

20. After about 10 days the colonies will be large enough to permit a single colony to be subcultured (see Section C). For determination of plating efficiency the colonies can be fixed with 10% formalin and stained with 0.1% crystal violet in water.

4. DISCUSSION

The procedures for obtaining clonal growth that have been described utilize three of the most commonly used cell dispersing agents, trypsin, Pronase, and collagenase. Experience in this laboratory has shown trypsin to be the most satisfactory for established cell lines, while Pronase and collagenase have been found more effective for the preparation of cell suspensions directly from fresh tissue.

Minor modifications of the Pronase and collagenase techniques have made possible clonal growth of primary cultures from a variety of tissues in the adult rabbit (Ham et al., 1971). The two tissues cited as examples are among those that yield the best plating efficiencies. For tissues that yield a lower plating efficiency, it is necessary to increase the number of cells inoculated into each petri dish in order to obtain a sufficient number of colonies.

Some difficulties may be encountered when the techniques for clonal growth of primary cultures described above are applied to species other than the rabbit. Limited experience in this and other laboratories suggests considerable species specificity in nutritional requirements for clonal growth. Bovine articular cartilage (from tail vertebrae) yields excellent growth of differentiated colonies with no modification of the procedure described for rabbit cartilage. A variety of cell types from the mouse, on the other hand, fail to yield satisfactory clonal growth under these conditions. Cells from infant rats have been reported to yield good clonal growth with techniques similar to those described (Coon, 1968, 1969, and personal communication). Closely related procedures for clonal growth of a variety of cells taken directly from chicken embryos have been described in detail by Cahn et al. (1967).

C.   Subculturing Single Colonies

1.   USE OF CLONING CYLINDERS

1. Place a thin layer of Dow-Corning silicone stopcock grease in a glass petri dish and sterilize it by autoclaving.

2. Place stainless steel cloning cylinders (about 6 mm in diameter and 12 mm high with 1 mm walls) in a glass petri dish and sterilize by autoclaving or heating in a dry air oven.

3. Examine a petri dish containing living colonies with an inverted microscope. Phase optics are helpful but not essential, particularly for low magnifications. Contrast with bright field optics can be enhanced by nearly closing the illuminator diaphragm and by increasing the distance between the condenser and the cells. Some resolution is lost in this procedure but it is usually not critical at the low magnification involved. Partially closing the condenser diaphragm will also enhance contrast, but is a less desirable approach in that it increases the depth of focus to a point where dirt or scratches on the outside of the petri dish become a major distraction.

4. Select an average-sized colony that is well separated from neighboring colonies. (Extra large colonies should be avoided because of the danger that they might have originated from clumps of cells rather than single cells.)

5. Mark the position of the colony by making a circle with a sharpened wax pencil on the bottom of the petri dish. Verify with the microscope that the colony is well centered within the circle.

6. Remove the growth medium with a pipette and rinse the colonies twice with sterile saline. Leave only enough saline in the petri dish to keep the cells moist.

7. Using forceps sterilized by dipping in 95% ethanol and flaming, touch the end of a stainless steel cloning cylinder to the sterilized silicone stopcock grease and then set the cylinder over the selected colony, pressing it down to obtain a good seal with the stopcock grease.

8. Using the inverted microscope verify that the cylinder is centered over the colony.

9. Using a small Pasteur pipette or a hypodermic syringe, deliver a few drops of 0.05% trypsin into the open top of the cylinder.

10. Incubate for 5 to 10 minutes at 37.5°C.

11. Agitate the contents of the cylinder to release the cells from the petri dish. This may be accomplished conveniently by forcing the trypsin–cell mixture several times through a 20-gauge needle on a 1.0-ml tuberculin syringe, being careful not to dislodge the cylinder.

12. Withdraw the entire contents of the cylinder into a syringe and transfer it to a petri dish or culture flask containing a rich medium that will support clonal growth of the cells in question.

13. Incubate the subculture until good colony formation occurs and then trypsinize and subdivide as needed, maintaining the cloned cells either at clonal or monolayer densities.

2. SCRAPING

The following procedure is crude, but often highly effective for obtaining clones quickly and easily.

1. Remove the culture medium and rinse the petri dish with sterile saline.

2. Add 5 ml of sterile saline to the petri dish.

3. Hold a sterile Pasteur pipette near a well isolated colony and draw the colony into the Pasteur pipette as it is gently pushed loose from the culture surface with a sterilized rubber policeman.

4. Place the colony in a test tube containing a small amount of 0.05% trypsin and incubate at 37.5°C for 5 to 10 minutes. (In some cases it is possible to eliminate the trypsin treatment and disperse the cells entirely by mechanical agitation. However, this may result in decreased viability for some types of cells.)

5. Agitate gently to separate the cells. Add the resulting cell suspension to growth medium in a petri dish or culture flask.

This procedure is most effective for large colonies that can be seen easily with the naked eye or with a low power dissecting microscope.

3. PRIMARY CULTURES

The phenomenon of limited lifetime *in vitro* (Hayflick, 1965) sometimes restricts the usefulness of clones derived from primary cultures. In such cases the cloning procedure uses up enough of the available cell generations so that extensive experimentation cannot be carried out with clonal populations that are obtained from primary cultures. The best procedure generally is to harvest a colony and use the resulting cell suspension immediately for the desired clonal experiment without attempting to grow up a large clonal population.

D.   Isolation of Single Cells

When it is necessary to be absolutely certain that a "clonal" population has originated from a single cell it is desirable to place the single cell in a separate container so there is no possibility of contamination from other cells during growth of the clonal population. Three of the most convenient approaches to this problem are described in detail immediately below and brief references to additional methods are given in the discussion that follows.

1.   USE OF A CLONING CYLINDER

In this approach, all steps described above for obtaining clonal growth in a plastic tissue culture petri dish are followed. After the cells have attached, but before the first division (generally between 6 and 18 hours after inoculation), well isolated single cells are located with the inverted microscope and their locations marked on the bottom of the petri dish with a sharpened wax pencil. A cloning cylinder (steps 2 and 7, Section C, above) is then put in place around each selected cell without removal of the culture medium (the medium does not interfere with the silicone grease seal and growth will proceed quite normally inside the cylinder). After microscopic verification that only one cell was included inside the cylinder, colony formation is allowed to occur normally. After adequate cell multiplication, the medium is removed with a sterile hypodermic needle and syringe and the colony is rinsed and trypsinized as described in Section C above. Similar procedures have been used by Puck *et al.* (1956) and Ham and Puck (1962b) for isolation of single cells.

2.   USE OF A DROP OF MEDIUM CONTAINING A SINGLE CELL

In this approach, a dilute cell suspension is prepared in complete growth medium. A small drop of medium statistically expected to contain one cell is placed in each of several dry plastic tissue culture petri dishes. The location of the drop is marked on the outside of the petri dish bottom with a wax pencil. An inverted microscope is used to determine which dishes actually received a drop containing a single cell. Those dishes are then placed in a well humidified incubator until the cell has attached. Additional medium is then added and the cell is allowed to multiply.

A method similar to this was used by Abbott and Holtzer (1968) to demonstrate conclusively that multiple colonies can arise from a single cell under certain culture conditions.

3. USE OF FRAGMENTS OF BROKEN COVER SLIPS

In this approach, single cells are allowed to attach to small glass fragments which are then transferred to new containers of medium, resulting in an inoculum of one cell per container. This procedure is based on a technique originally introduced by Schenck and Moskowitz (1958) and subsequently modified by Freeman et al. (1964) and by Mondal and Heidelberger (1970).

1. Clean cover slips for clonal growth (Appendix C).

2. Grind gently with a clean mortar and pestle.

3. Select the fragments that pass through a 10 mesh per inch screen but are retained on a 20 mesh per inch screen. (In this and preceeding steps after the initial cleaning, the culture surfaces should not be touched. Alternately it is possible to clean the fragments after grinding by sequential washing in tap water, 70% ethanol, and ether—cf. Freeman et al., 1964).

4. Place the selected fragments in a glass petri dish and sterilize in a dry air oven.

5. Transfer enough fragments to a plastic tissue culture petri dish (Falcon No. 3001 or No. 3002) so the bottom is half covered.

6. Add culture medium to the dish. Verify that all fragments have sunk to the bottom and that they do not overlap.

7. Inoculate the culture with cells at a clonal density (about 30 cells in a No. 3001 dish or 100 cells in a No. 3002 dish).

8. After the cells are attached but before the first division, examine the dishes with an inverted microscope and mark the locations of fragments containing one attached cell.

9. Using fine curved tip forceps sterilized by dipping in 95% ethanol and flaming, transfer each fragment containing a single attached cell to a fresh petri dish containing medium. (Be sure that the fragment is placed in the dish cell side up.)

10. Using the inverted microscope, verify that the cell has remained attached to the fragment and that only one cell was transferred with the fragment.

11. Incubate until a large colony is formed, then use dilute trypsin to disperse the cells and inoculate additional cultures as desired.

4. OTHER METHODS

The first reported cloning of a mammalian cell was accomplished by placing a single cell in a small volume of medium in a capillary tube that had been previously coated with a thin layer of plasma clot (Sanford et al., 1948). Numerous investigators have isolated single cells in small droplets of medium under liquid paraffin (Lwoff *et al.*, 1955; Wildy and Stoker, 1958; Paul, 1970). The individual wells in plastic "Microtest" plates have also been used as culture chambers for isolated cells (Cooper, 1970; Robb, 1970). Replica plating techniques have also been developed using similar vessels (Goldsby and Zipser, 1969). These methods, together with those described in detail above, offer a wide range of possible techniques for the establishment of clonal populations with little or no chance of contamination by unwanted cells.

E.   Semisolid Media

Several techniques involving the use of semisolid media have been developed for cloning cells that are capable of growth in suspension. Many variations exist, but the following procedure modified from that of Puck *et al.* (1956) is representative.

1. Dissolve 10 gm of agar in 1 liter of 0.145 M NaCl (85.0 gm/liter). Sterilize by autoclaving. (Note that some batches of agar are more toxic than others. If necessary the powdered agar can be partially purified by extraction with cold water and/or acetone prior to use.)

2. Add 1.0 ml of the melted agar to each 60-mm petri dish and allow it to solidify. Cool the remaining agar to 44°C and maintain it at that temperature in a water bath.

3. Prepare a suspension of cells for clonal growth (cf. Section III,B, above) and dilute to a concentration of approximately 25 cells/ml in complete growth medium warmed to about 35°C.

4. Quickly mix 20 ml of 1% agar (at a temperature of 44°C) with each 80 ml of the cell suspension and add 4 ml of the resulting cell suspension in 0.2% agar to each petri dish previously prepared with an agar base layer.

5. Incubate 7 to 10 days in the tissue culture incubator. Under these conditions colony growth will be three dimensional and each clone will have a spherical appearance. Individual clones can be harvested and transferred to new containers either with a small pipette or with a bacterial loop.

A variation on this technique mixes equal volumes of 1% agar (prepared in distilled water, autoclaved and cooled to 44°C) with double-strength culture medium (also maintained at 44°C). Five milliliters of this mixture is added to each dish to provide a thick base of 0.5% agar containing nutrients. A cell suspension in 0.33% agar is then prepared by mixing two parts of 0.5% agar with one part of medium containing cells. Each petri dish receives 1.5 ml of the cell suspension on top of the nutrient agar base (Macpherson and Montagnier, 1964; Paul, 1970).

Modifications of the agar suspension technique have been utilized by several investigators for clonal growth of cells from bone marrow, spleen, and cartilage (Robinson *et al.*, 1967; Pluznik and Sachs, 1965; Horowitz and Dorfman, 1970).

F.   Modification of Current Techniques

In order to minimize confusion, the procedures described above have been made specific for particular types of cells. Considerable trial and error adjustment will frequently be necessary to obtain optimum results with other types of cells. In some cases suggestions for modification have been included parenthetically or added after the procedure. However, in most cases, a major need for innovation remains, particularly as the techniques are applied to primary cultures.

Investigators who are not thoroughly familiar with the intricacies of mammalian cell culture are cautioned to master the procedures as described before attempting major variations. Since cell culture is still partly an art, small changes in procedure will often cause large changes in results that cannot be predicted easily. Once a procedure has been made to work, a series of controlled experiments can (and frequently should) be carried out to determine optimum conditions for clonal growth and colony selection for the cells being used. There is an acute need for precise definition of the detailed requirements for clonal growth of many additional kinds of mammalian cells.

ACKNOWLEDGMENTS

I am indebted to Louann W. Murray and Gerald L. Sattler for their collaboration in developing the clonal techniques for primary cultures and for their assistance in the preparation of this manuscript. To Dr. Theodore T. Puck, who pioneered many of the techniques described in this chapter, special thanks are due, both for the training I received in his laboratory and for his suggestions and criticism of the manuscript.

Appendix A.   Preparation of Medium F12M

The method of preparation currently used for Medium F12M has been modified only slightly from the originally published method (Ham, 1965). A series of stock solutions are prepared as follows (concentrations in the final medium are also listed for convenience):

Stock No. 1 (100X)

Component	Molecular weight	Stock solution		Final medium	
		gm/liter	moles/liter	mg/liter	moles/liter
Arginine·HCl	210.7	21.07	1×10^{-1}	210.7	1×10^{-3}
Choline	139.6	1.396	1×10^{-2}	13.96	1×10^{-4}
Histidine·HCl·H$_2$O	209.6	2.096	1×10^{-2}	20.96	1×10^{-4}
Isoleucine *allo*-free	131.2	0.3936	3×10^{-3}	3.936	3×10^{-5}
Leucine	131.2	1.312	1×10^{-2}	13.12	1×10^{-4}
Lysine·HCl	182.7	3.654	2×10^{-2}	36.54	2×10^{-4}
Methionine	149.2	0.4476	3×10^{-3}	4.476	3×10^{-5}
Phenylalanine	165.2	0.4956	3×10^{-3}	4.956	3×10^{-5}
Serine	105.1	1.051	1×10^{-2}	10.51	1×10^{-4}
Threonine	119.1	1.191	1×10^{-2}	11.91	1×10^{-4}
Tryptophan	204.2	0.2042	1×10^{-3}	2.042	1×10^{-5}
Tyrosine	181.2	0.5436	3×10^{-3}	5.436	3×10^{-5}
Valine	117.1	1.171	1×10^{-2}	11.71	1×10^{-4}

Gentle heating with stirring is helpful in dissolving the components of Stock No. 1. Freshly thawed Stock No. 1 frequently contains a precipitate, which will dissolve on gentle warming. Stock No. 1 is stored at 4°C for up to 2 months or frozen at −20°C for longer periods.

Stock No. 2 (100X)

Component	Molecular weight	Stock solution		Final medium	
		gm/liter	moles/liter	mg/liter	moles/liter
Biotin	244.3	0.00073	3×10^{-6}	0.00733	3×10^{-8}
Calcium pantothenate	258.3[a]	0.0258	1×10^{-4a}	0.258	1×10^{-6a}
Niacinamide	122.1	0.003663	3×10^{-5}	0.03663	3×10^{-7}
Potassium chloride	74.55	22.365	3×10^{-1}	223.65	3×10^{-3}
Pyridoxine·HCl	205.7	0.006171	3×10^{-5}	0.06171	3×10^{-7}
Thiamine·HCl	337.3	0.03373	1×10^{-4}	0.3373	1×10^{-6}

[a] Molar concentrations for calcium pantothenate are expressed in terms of pantothenate. The "molecular weight" listed is for a half molecule containing one pantothenate residue.

Stock No. 2 is stored at 4°C up to 2 months, or frozen at −20°C for longer periods.

Stock No. 3 (100X)

Component	Molecular weight	Stock solution		Final medium	
		gm/liter	moles/liter	mg/liter	moles/liter
Folic acid	441.4	0.1324	3×10^{-4}	1.324	3×10^{-6}
$Na_2HPO_4 \cdot 7 H_2O$	268.1	26.81	1×10^{-1}	268.1	1×10^{-3}

Stock No. 3 is stored at 4°C up to 2 months, or frozen at —20°C for longer periods. The disodium phosphate provides alkaline conditions necessary to dissolve the folic acid. In experimental media where the folic acid is prepared separately, great care must be taken to be certain that the folic acid has dissolved. Under neutral or slightly acidic conditions folic acid tends to form a fine particulate suspension which is not easily seen in the medium, but which is retained on the sterilizing filter. Once the folic acid has been dissolved completely, however, it will normally remain in solution in the final medium even under slightly acidic conditions.

Stock No. 4 (100X)

Component	Molecular weight	Stock solution		Final medium	
		gm/liter	moles/liter	mg/liter	moles/liter
$FeSO_4 \cdot 7 H_2O$	278.02	0.0834	3×10^{-4}	0.834	3×10^{-6}
$MgCl_2 \cdot 6 H_2O$	203.33	12.20	6×10^{-2}	122.0	6×10^{-4}
$CaCl_2 \cdot 2 H_2O$	147.03	4.411	3×10^{-2}	44.11	3×10^{-4}

A small amount of concentrated hydrochloric acid is added to stock No. 4 (one drop per 100 ml or 0.5 ml/liter) to prevent the gradual precipitation of ferric hydroxides from the solution. Stock No. 4 is stored in tightly stoppered bottles at room temperature. Stock No. 4 is normally not added to the final medium until immediately before use.

Stock No. 5 (1000X)

Component	Molecular weight	Stock solution		Final medium	
		gm/liter	moles/liter	mg/liter	moles/liter
Phenol Red	376.36	1.242	3.3×10^{-3}	1.242	3.3×10^{-6}

Stock No. 5 is stored at room temperature in tightly stoppered bottles.

Stock Nos. 6a, 6b, and 6c (100X each)

The original stock No. 6, which contained four relatively reactive components is no longer prepared as a single stock solution. Currently glutamine, riboflavin, and sodium pyruvate are prepared as separate 100X stocks and added individually when the final medium is made. Glucose is weighed out and added directly to the final medium.

Stock	Component	Molecular weight	Stock solution		Final medium	
			gm/liter	moles/liter	mg/liter	moles/liter
6a	Glutamine	146.2	14.62	1×10^{-1}	146.2	1×10^{-3}
6b	Sodium pyruvate	110.1	11.01	1×10^{-1}	110.1	1×10^{-3}
6c	Riboflavin	376.4	0.003764	1×10^{-5}	0.03764	1×10^{-7}

All three of these stocks should be stored frozen at $-20°C$ if kept for an appreciable time. In addition, stock No. 6c (riboflavin) must be protected from light.

Stock No. 7 (100X)

Component	Molecular weight	Stock solution		Final medium	
		gm/liter	moles/liter	mg/liter	moles/liter
Cysteine·HCl·H₂O	175.6	3.512	2×10^{-2}	35.12	2×10^{-4}

Stock No. 7 must be discarded if it contains a precipitate. Deterioration of the solution can be minimized by keeping it tightly stoppered to reduce air oxidation. Stock No. 7 should be kept frozen at $-20°C$ for storage of more than a few days.

Stock No. 8 (100X)

Component	Molecular weight	Stock solution		Final medium	
		gm/liter	moles/liter	mg/liter	moles/liter
Asparagine	150.1	1.501	1×10^{-2}	15.01	1×10^{-4}
Proline	115.1	3.453	3×10^{-2}	34.53	3×10^{-4}
Putrescine	161.1	0.01611	1×10^{-4}	0.1611	1×10^{-6}
Vitamin B_{12}	1356.9	0.1357	1×10^{-4}	1.357	1×10^{-6}

Stock No. 8 is stored at $4°C$ for up to 2 months, or frozen at $-20°C$ for longer periods.

Stock No. 9 (100X)

Component	Molecular weight	Stock solution		Final medium	
		gm/liter	moles/liter	mg/liter	moles/liter
Alanine	89.1	0.891	1×10^{-2}	8.91	1×10^{-4}
Aspartic acid	133.1	1.331	1×10^{-2}	13.31	1×10^{-4}
Glutamic acid	147.1	1.471	1×10^{-2}	14.71	1×10^{-4}
Glycine	75.1	0.751	1×10^{-2}	7.51	1×10^{-4}

Stock No. 9 is prepared as follows: The aspartic and glutamic acids are added to slightly less than the final volume of water. One ml/liter of phenol red indicator solution (stock 5) is added, and 1.0 N sodium hydroxide is added with stirring just rapidly enough to keep the solution neutral (orange) as the aspartic and glutamic acids dissolve. When no solids remain and a stable orange colored solution is achieved the alanine and glycine are dissolved in the solution and the rest of the water is added to yield the final volume. Stock No. 9 is stored at 4°C for up to 2 months, or frozen at −20°C for longer periods.

Stock No. 10 (100X)

Component	Molecular weight	Stock solution		Final medium	
		gm/liter	moles/liter	mg/liter	moles/liter
Hypoxanthine	136.1	0.4083	3×10^{-3}	4.083	3×10^{-5}
myo-Inositol	180.2	1.802	1×10^{-2}	18.02	1×10^{-4}
Lipoic acid	206.3	0.02063	1×10^{-4}	0.2063	1×10^{-6}
Thymidine	242.2	0.07266	3×10^{-4}	0.7266	3×10^{-6}
$CuSO_4 \cdot H_2O$	249.71	0.000249	1×10^{-6}	0.00249	1×10^{-8}

Stock No. 10 is prepared as follows: The hypoxanthine is dissolved in a small volume of boiling water (5–10% of the final volume of the stock solution). The lipoic acid is dissolved in a few drops of 1.0 N NaOH and then diluted and added to the stock solution. These two components dissolve readily when these procedures are followed, but are nearly impossible to dissolve if added directly to the final solution. The other three components dissolve readily. Stock No. 10 is stored at 4°C for up to 2 months, or frozen at −20°C for longer periods.

Stock No. 11-M (1000X)

Component	Molecular weight	Stock solution		Final medium	
		gm/liter	moles/liter	mg/liter	moles/liter
$ZnSO_4 \cdot 7\, H_2O$	287.55	0.144	5×10^{-4}	0.144	5×10^{-7}

Stock No. 11-M is stored at room temperature in tightly stoppered bottles. Stock No. 11-M is normally not added to the final medium until immediately before use. The zinc sulfate has been prepared as a separate stock so that either F12M or F12 can be prepared from the stock solutions. (For F12 six times as much of the stock is used.) If this flexibility is not needed the zinc sulfate can be included in stock 4 at $5 \times 10^{-5}\, M$ (0.0144 gm/liter).

Stock No. 12 (1000X)

Component	Molecular weight	Stock solution		Final medium	
		gm/liter	moles/liter	mg/liter	moles/liter
Linoleic acid	280.4	0.08412	3×10^{-4}	0.08412	3×10^{-7}

Stock No. 12 is prepared in absolute ethanol and is stored at $-20°C$, in tightly stoppered bottles to minimize oxidation. Stock No. 12 should be prepared fresh at least once a month. Stock No. 12 is normally not added to the final medium until immediately before use. Stock No. 12 is frequently omitted entirely from medium which is to be used with a serum supplement.

Components Added Directly to the Final Medium

The following components are weighed individually and added directly to the final medium without the preparation of stock solutions, cf. "Preparation of Final Medium" below for details.

Component	Molecular weight	Concentration in final medium	
		mg/liter	moles/liter
Glucose	180.16	1801.6	1.0×10^{-2}
Sodium chloride	58.45	7599	1.3×10^{-1}
Sodium bicarbonate	84.02	1176	1.4×10^{-2}

Preparation of Final Medium

In this laboratory medium F12M is generally prepared in 20-liter lots by adding the necessary stocks and chemicals to triple-distilled water. Stocks 4, 11, and 12 are normally not added to the medium until just before it is to be used. Procedures are given below for the preparation of 1 liter and 20 liters of medium.

	For 1 liter	For 20 liters
Triple-distilled water	(800 ml)	(16,000 ml)
add with stirring:		
Stock No. 1	10 ml	200 ml
Stock No. 2	10 ml	200 ml
Stock No. 3	10 ml	200 ml
Stock No. 5	1 ml	20 ml
Stock No. 6a	10 ml	200 ml
Stock No. 6b	10 ml	200 ml
Stock No. 6c	10 ml	200 ml
Stock No. 7	10 ml	200 ml
Stock No. 8	10 ml	200 ml
Stock No. 9	10 ml	200 ml
Stock No. 10	10 ml	200 ml
Glucose	1.802 gm	36.032 gm
Sodium chloride	7.599 gm	151.98 gm
Adjust to pH 7.3–7.4 (orange) with 1.0 N		
NaOH, then add: sodium bicarbonate	1.176 gm	23.520 gm
Add triple-distilled water to final volume	988 ml	19760 ml

(Note: In practice the volumes can be adjusted to 1.00 liter and 20.00 liters, respectively, without detrimental effects.)

The medium (minus stocks 4, 11, and 12) is then filtered through a type GS Millipore filter (0.22 μ pore size) which has been prerinsed with distilled water to remove detergents. For 1 liter a 47-mm filter in a type XX10-047-30 Pyrex filter holder is used. For 20 liters, a 142-mm filter in a type YY30-142-00 stainless steel filter holder is used, together with a pressure vessel. The medium is stored frozen at −20°C, in volumes convenient for a single experiment (normally 200–500 ml).

When the medium is ready to be used it is thawed by placing the bottle in a 37°C shaking water bath (a regular water bath is satisfactory but slower). Sterile stocks 4, 11, and 12 are then added. Stocks 4 and 11 are sterilized by Millipore filtration or by autoclaving. Stock 12 can also be sterilized by Millipore filtration (the membrane wrinkles somewhat but does not rupture or dissolve), although more commonly it is assumed to be self-sterilizing. The amounts of stocks added are as follows:

	per 100 ml	per 500 ml
To prepare medium F12M		
Stock No. 4	1.0 ml	5.0 ml
Stock No. 11M	0.1 ml	0.5 ml
Stock No. 12	0.1 ml	0.5 ml
To prepare medium F12		
Stock No. 4	1.0 ml	5.0 ml
Stock No. 11M	0.6 ml	3.0 ml
Stock No. 12	0.1 ml	0.5 ml

Modification of Procedures

Many of the reasons for the currently used procedures are historical (e.g., nutrients are grouped in particular stocks that reflect the nutrient requirements of different cell lines, or because they form convenient groups for the study of interactions of related substances). However, practical considerations are also involved in several cases (e.g., grouping disodium phosphate with folic acid to provide alkaline conditions for dissolving the folic acid; including the potassium chloride in stock No. 2 so that it would not crack bottles as readily when it was frozen; keeping the divalent ions away from the phosphate or other substances which they might precipitate from concentrated solutions). It is quite possible to use other procedures to prepare medium F12M, but great care must be taken to be certain that the end result is the same (i.e., that everything dissolves and that nothing is precipitated or inactivated).

Appendix B.   Preparation of Enzyme Solutions

Preparation of T Saline

In this laboratory enzyme solutions for cell dispersal are prepared in "T saline" (trypsinizing saline), which contains no divalent cations, and has the same concentrations of sodium, potassium, chloride, phosphate, and glucose as medium F12. Its composition is as follows:

Component	Moles/liter	gm/liter	gm/20 liter
NaCl	0.130	7.60	152.0
KCl	0.0030	0.224	4.48
Na$_2$HPO$_4$·7 H$_2$O	0.0010	0.268	5.36
Glucose	0.010	1.80	36.0
Phenol red[a]	0.0000033	0.0012	0.024

[a] Phenol red can conveniently be added as 0.1 ml of stock 5 (from the preparation of medium F12M) per liter of T saline.

A final pH adjustment is normally not made until after the enzymes are added. However, in cases where the saline is to be used alone it should be adjusted to pH 7.3–7.4 (deep orange) with 1.0 N HCl.

Trypsin Solution

A solution of 0.05% trypsin is prepared by dissolving 0.50 gm/liter of trypsin (1:300, Nutritional Biochemical Corp.) in T saline. The enzyme dissolves slowly and best results are obtained with mechanical stirring for 2–3 hours at room temperature or overnight in the cold room. The solution is then neutralized to pH 7.3–7.4 (deep orange) with 1.0 N NaOH and sterilized by passage through a 0.22 μ Millipore filter (type GS). Difficulties due to clogging the filter can be minimized by first passing the solution through 0.45 μ filters under nonsterile conditions, changing the filter membrane each time the flow rate begins to decline. The sterilized trypsin solution is stable in the refrigerator for several days. For long-term storage it should be frozen at or below −20°C.

Pronase Solution

A solution of 0.01% Pronase is prepared by dissolving 0.10 gm/liter of Pronase (Calbiochem) in T saline. The Pronase is dissolved, neutralized, and sterilized as described above for trypsin. The solutions should be stored frozen at or below −20°C.

Collagenase Solution (FBC)

A solution containing 0.25% collagenase and 10% fetal bovine serum (FBC) is prepared as follows: 0.5 gm of collagenase (Worthington, CLS) is dissolved in 180 ml of T saline. The enzyme dissolves slowly and is stirred, neutralized, and sterilized by filtration as described for trypsin.

After sterilization, 20 ml of sterile fetal bovine serum is added aseptically and mixed thoroughly with the enzyme solution. The fetal bovine serum–collagenase mixture (designated FBC) is then divided into 10-ml aliquots (enough for one experiment) and stored frozen in small bottles at or below —20°C until ready to be used. More rapid tissue dissociation can be obtained by doubling the concentration of collagenase. However, the benefits obtained must be weighed against increased cost and increased difficulty in sterilizing the collagenase solution.

Appendix C.   Cleaning of Glassware

Petri Dishes

The following procedure, developed in the laboratory of T. T. Puck, has been found to yield petri dishes that are consistently free of toxicity and satisfactory for single-cell plating (adapted from Ham and Puck, 1962b).

1. Place dishes into warm 1% solution of 7X (a laboratory glassware cleaner formulated for low toxicity to tissue cultures—Linbro Chemical Co.) and brush each for 15 to 30 seconds, using a hard-bristle China brush.

2. Rinse five times with tap water.

3. Autoclave for 30 minutes submerged in a 1% solution of disodium EDTA (Sequesterene Na$_2$—Geigy Industrial Chemicals).

4. Rinse five times with tap water.

5. Autoclave for 30 minutes submerged in a 1% solution of 7X.

6. Brush until clean (be certain that old colonies have been removed completely).

7. Rinse five times with tap water.

8. Wipe thoroughly with Miracloth (Miracloth Corp.).

9. Rinse overnight in running tap water.

10. Rinse three times with single-distilled water.

11. Rinse once with double-distilled water.

12. Drain dry.

13. Assemble tops and bottoms, place in metal cans (e.g., one-pound coffee cans), and sterilize in the dry air oven for 5 hours at 175°C.

Note that petri tops, which do not come into contact with the cells or culture medium, are merely washed thoroughly with detergent and rinsed with distilled water.

Cover Slips

As received from the manufacturer, cover slips are generally covered by a thin film of grease that renders them unsatisfactory for cell attachment and growth. The following procedure will overcome this difficulty:

1. Fill a glass petri dish with acidified alcohol (1.0 ml, 1.0 N HCl/ 100 ml 95% ethanol).

2. Drop cover slips individually into the acidified alcohol and allow them to soak for at least 30 minutes. (Be certain that the cover slips are not sticking together and that they are exposed to the acidified alcohol on both sides.)

3. Using cover slip forceps, transfer the cover slips one at a time to a glass petri dish filled with 95% ethanol. After all cover slips are in the alcohol, gently agitate the dish.

4. Transfer each cover slip to a glass petri dish filled with distilled water and agitate as before.

5. Lift each cover slip out of the water with the forceps and stand it on edge on an absorbant paper or cloth towel to dry (the cover slip can be leaned against the sides of petri dishes or any clean vertical surface).

6. Polish the cover slips with microscope lens paper, place them in a dry glass petri dish, and sterilize in the dry air oven for 5 hours at 175°C. (The cover slips must be thoroughly dry or they will stick together after they have been heated. Autoclaving is unsatisfactory since it tends to leave the cover slips wet and stuck together.)

7. Using sterile forceps, transfer individual cover slips to glass or plastic petri dishes as needed. The remainder can be stored for relatively long periods in the glass petri dish that was used for their sterilization.

Pipettes

The following procedures will consistently yield pipettes that drain well and are free of toxic materials.

1. Immediately after the pipettes are used, soak them in a 1% solution of Alconox (a laboratory cleaner made by Alconox Inc.).

2. Remove the cotton plugs from the pipettes and wash them for 30 minutes, tip up, in a recirculating pipette washer (Virtis) that repeatedly fills them with a boiling solution of 1% Alconox and then empties them. (Normally the pipettes are loaded into a mesh rack that can be transferred from first wash to second wash and on to the rinses without unloading.)

3. Wash for 30 minutes in a similar manner with boiling 1% 7X.

4. Rinse for 1 hour in a pipette rinser that repeatedly fills them with running tap water and then drains them through an automatic siphon.

5. Rinse three times with distilled water and once with triple-distilled water.

6. Dry the pipettes in air, tip down on clean cloth towels.

7. Insert cotton plugs (if desired) and sterilize the pipettes in metal cans in the dry air oven at 175°C for 5 hours.

Care must be taken to keep silicone grease strictly away from the pipette washing procedures. A very small amount of silicone grease can coat an entire batch of pipettes and cause uneven draining and drop formation that is extremely difficult to correct. Only corrosion-resistant hard glass pipettes can be cleaned by the method recommended above. Others will soon become badly etched.

Satisfactory results have also been obtained using "Impact" detergent in a laboratory glassware washer (BetterBuilt) equipped with a pipette basket and a distilled water rinse cycle.

Appendix D.   Incubation

The incubator used for cloning experiments should accomplish the following: (1) maintain a uniform temperature of $37.5 \pm 0.5°C$; (2) maintain near-saturated (greater than 95%) humidity; and (3) maintain an atmosphere of 5% carbon dioxide. All three of these environmental conditions must be kept constant and each must be capable of rapid recovery without overshooting after the incubator door has been opened.

Control of Temperature

The best temperature control can usually be achieved with a water-jacketed incubator. The water jacket provides a large heat reserve for rapid recovery and also heats the incubator chamber uniformly from all sides. Sensitivity of control can usually be improved greatly by placing a sensitive thermostat (such as Aminco No. 4-235F with No. 4-5300 super-sensitive relay) in the water jacket, wired in series with the regular control thermostat of the incubator. The regular thermostat is then set about one degree high so that it functions only to cut off heat in the event that the sensitive thermostat fails.

Incubators with air heaters can also be used successfully if they are properly designed. Essential features include forced circulation to maintain temperature uniformity and close coupling between the heater and the control thermostat to prevent overheating after the incubator has been cooled. The intensity and direction of air circulation must also be controlled carefully to prevent contamination of petri dish cultures by airborne spores.

Control of Humidity

Humidification is generally accomplished by evaporation of water contained in a pan at the bottom of the incubator. In models with continuous gassing, the incoming gas is usually bubbled through the water pan, while in models with controlled injection of carbon dioxide, the forced-air circulation is directed over the water pan.

In either case, the water in the pan tends to be cooled by evaporation, particularly when the door has been opened and the humidity inside the incubator is low. This problem is best overcome by placing a thermostatically controlled immersion heater in the water pan to keep the water at the desired incubation temperature at all times. Caution must be exercised in choice of materials, however, since the warm, moist, slightly acidic atmosphere in the incubator is highly corrosive.

A heater in the water pan also helps reheat the incubator quickly after it has been cooled. Care must be exercised not to set the heater too high, however. If the pan becomes appreciably hotter than the rest of the incubator, water will evaporate from the pan and condense on all other parts of the incubator. This increases the danger of contamination of the cultures due to the formation of liquid "bridges" between the petri dish covers and bottoms.

Control of Carbon Dioxide Tension

The most common approach to the control of carbon dioxide tension is continuous-flow gassing with a mixture of 5% carbon dioxide in air. Most commercial incubators are equipped (or can be equipped as an optional extra) with devices to regulate the flow of air and carbon dioxide into the incubation chamber. One of the problems with continuous-flow gassing is the relatively slow rate of turnover and the consequent long recovery time after the incubator has been opened. This time can be reduced greatly by injecting enough pure carbon dioxide to return the chamber to a 5% concentration after each opening of the door. Quick recovery controls are available commercially and should be utilized on any clonal growth incubator that is opened frequently.

Continuous monitoring of carbon dioxide tension and controlled injection of pure carbon dioxide is desirable but has never become very popular because of the complexity and lack of reliability of the equipment required. One approach that is quite successful is based on measuring of the pH of a bicarbonate solution continuously equilibrated with samples of gas withdrawn from the incubator chamber (Ham and Puck, 1962a).

A modernized version of this control equipment can be constructed as follows:

1. Use an expanded scale pH meter with switching circuits (e.g., Corning model 10C) to control carbon dioxide flow into the incubator. Wire the high pH switch to a solenoid valve in the carbon dioxide line (preferably through a relay) such that carbon dioxide is injected into the incubator whenever the meter registers a pH higher than 7.4. (For best results the carbon dioxide should be injected into the intake port of the blower used for air circulation within the incubator.)

2. Construct a pH sampling cell from a small flat jar with a silicone rubber stopper as shown in Fig. 1. Fit the pH electrode(s) through a hole(s) in the stopper (either combination or separate electrodes can be used). Fill the jar one-half full with a solution of sodium bicarbonate and phenol red at the same concentrations as in medium F12M ($1.4 \times 10^{-2} M$ NaHCO$_3$, $3.3 \times 10^{-6} M$ phenol red). Insert an inlet tube (¼ inch polypropylene tubing) through a hole in the stopper and nearly to the bottom of the cell (into the liquid). Insert an outlet tube (also ¼ inch polypropylene tubing) through a hole in the stopper into the air space above the liquid and connect it via a condensation trap to a vacuum pump or reliable vacuum line. Adjust the amount of vacuum with a screw-type pinch clamp so that air from the inlet tube bubbles gently through the bicarbonate solution. (Do not restrict the inlet tube as this will reduce the partial pressure of carbon dioxide and alter the pH reading.)

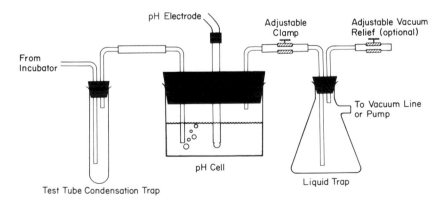

FIG. 1. pH cell and condensation traps for use in regulating the CO$_2$ concentration in a cell culture incubator.

3. Connect the inlet tube via a small condensation trap (conveniently made from a test tube and a small two-hole stopper) to the sampling port of the incubator (or directly to the inside of the incubator chamber through a gassing port). The gas sample drawn out of the incubator equilibrates continually with the bicarbonate solution such that the meter registers a pH similar to that of medium F12 inside the incubator (the medium and the solution in the pH cell contain the same concentration of bicarbonate). Whenever the carbon dioxide tension falls within the incubator the indicator solution becomes more alkaline and the switching circuit adds more carbon dioxide to the incubator.

Since the gas sample coming from the incubator is saturated with humidity at a temperature higher than that of the pH cell it is important to include the condensation trap in the sample line. If the trap is omitted or not emptied regularly, the condensate will dilute the bicarbonate solution in the pH cell. When this happens the pH of the solution in the cell becomes more acid and insufficient carbon dioxide is added to the incubator.

With regular attention to the condensation trap (daily, 7 days a week) and occasional recalibration of the pH meter and changing of the solution in the pH cell (once or twice a month) such a system will provide reliable control of carbon dioxide tension over a long period. Use of a larger condensation trap will reduce the frequency of needed attention, but it also increases the response time of the control system and is generally not desirable.

Additional details concerning incubator design and operation have been described elsewhere (Ham and Puck, 1962a,b).

REFERENCES

Abbott, J., and Holtzer, H. (1968). *Proc. Nat. Acad. Sci. U. S.* **59**, 1144.
Cahn, R. D. (1968). *In* "The Stability of the Differentiated State" (H. Ursprung, ed.), p. 58. Springer Verlag, New York.
Cahn, R. D., and Cahn, M. B. (1966). *Proc. Nat. Acad. Sci. U. S.* **55**, 106.
Cahn, R. D., Coon, H. G., and Cahn, M. B. (1967). *In* "Methods in Developmental Biology" (F. H. Wilt and N. Wessels, eds.), p. 493. Thomas Y. Crowell, New York.
Ceccarini, C., and Eagle, H. (1971a). *Proc. Nat. Acad. Sci.* **68**, 229.
Ceccarini, C., and Eagle, H. (1971b). *Nature New Biol.* **233**, 271.
Chu, E. H. Y., and Giles, N. M. (1958). *J. Nat. Cancer Inst.* **20**, 383.
Coon, H. G. (1966). *Proc. Nat. Acad. Sci. U. S.* **55**, 66.
Coon, H. G. (1968). *J. Cell. Biol.* **39**, 29a.
Coon, H. G. (1969). *Carnegie Inst. Wash. Yearb.* **67**, 419.
Coon, H. G., and Cahn, R. D. (1966). *Science* **153**, 1116.
Cooper, J. E. K. (1970). *Texas. Rep. Biol. Med.* **28**, 29.
Eagle, H. (1959). *Science* **130**, 432.
Eagle, H. (1971). *Science* **174**, 500.
Eagle, H., and Piez, K. (1962). *J. Exp. Med.* **116**, 29.

Ehrmann, R. L., and Gey, G. O. (1956). *J. Nat. Cancer. Inst.* **16**, 1375.
Fisher, H. W., and Puck, T. T. (1956). *Proc. Nat. Acad. Sci. U. S.* **42**, 900.
Foley, J. J., and Aftonomos, B. (1970). *J. Cell. Physiol.* **75**, 159.
Freeman, A. E., Ward, T. G., and Wolford, R. G. (1964). *Proc. Soc. Exp. Biol. Med.* **116**, 339.
Goldsby, R. A., and Zipser, E. (1969). *Exp. Cell Res.* **54**, 271.
Good, N. E., Winget, G. D., Winter, W., Connolly, T. N., Izawa, S., and Singh, R. M. M. (1966). *Biochemistry* **5**, 467.
Ham, R. G. (1963a). *Exp. Cell Res.* **29**, 515.
Ham, R. G. (1963b). *Science* **140**, 802.
Ham, R. G. (1965). *Proc. Nat. Acad. Sci. U. S.* **53**, 288.
Ham, R. G., and Murray, L. W. (1967). *J. Cell. Physiol.* **70**, 275.
Ham, R. G., and Puck, T. T. (1962a). *Proc. Soc. Exp. Biol. Med.* **111**, 67.
Ham, R. G., and Puck, T. T. (1962b). *In* "Methods in Enzymology" (S. P. Colowick and N. O. Kaplan, eds.), Vol. V, p. 90. Academic Press, New York.
Ham, R. G., and Sattler, G. L. (1968). *J. Cell. Physiol.* **72**, 109.
Ham, R. G., and Sattler, G. L. (1971). In preparation.
Ham, R. G., Murray, L. W., and Sattler, G. L. (1970). *J. Cell. Physiol.* **75**, 353.
Ham, R. G., Murray, L. W., and Sattler, G. L. (1971). In preparation.
Hauschka, S. D., and Konigsberg, I. R. (1966). *Proc. Nat. Acad. Sci. U. S.* **55**, 119.
Hayflick, L. (1965). *Exp. Cell Res.* **37**, 614.
Herzenberg, L. A., and Roosa, R. A. (1960). *Exp. Cell Res.* **21**, 430.
Horowitz, A., and Dorfman, A. (1970). *J. Cell. Biol.* **45**, 434.
Kao, F. T., and Puck, T. T. (1970). *Nature* (*London*) **228**, 329.
Kao, F. T., Chasin, L., and Puck, T. T. (1969). *Proc. Nat. Acad. Sci. U. S.* **64**, 1284.
Konigsberg, I. R. (1963). *Science* **140**, 1273.
Littlefield, J. W. (1966). *Exp. Cell Res.* **41**, 190.
Lockart, R. Z., Jr., and Eagle, H. (1959). *Science* **129**, 252.
Lwoff, A., Dulbecco, R., Vogt, M., and Lwoff, M. (1955). *Virology* **1**, 128.
Macpherson, I., and Montagnier, L. (1964). *Virology* **23**, 291.
Magee, W. E., Sheek, M. R., and Sagik, B. P. (1958). *Proc. Soc. Exp. Biol. Med.* **99**, 390.
Mondal, S., and Heidelberger, C. (1970). *Proc. Nat. Acad. Sci. U. S.* **65**, 219.
Neuman, R. E., and McCoy, T. A. (1958). *Proc. Soc. Exp. Biol. Med.* **98**, 303.
Paul, J. (1970). "Cell and Tissue Culture," 4th ed. Williams & Wilkins, Baltimore, Maryland.
Pluznik, D. H., and Sachs, L. (1965). *J. Cell. Comp. Physiol.* **66**, 319.
Puck, T. T., and Kao, F. T. (1967). *Proc. Nat. Acad. Sci. U. S.* **58**, 1227.
Puck, T. T., and Marcus, P. I. (1955). *Proc. Nat. Acad. Sci. U. S.* **41**, 432.
Puck, T. T., Marcus, P. I., and Cieciura, S. J. (1956). *J. Exp. Med.* **103**, 273.
Rappaport, C., and Howze, G. B. (1966). *Proc. Soc. Exp. Biol. Med.* **121**, 1010.
Rinaldini, L. M. J. (1958). *Int. Rev. Cytol.* **7**, 587.
Robb, J. A. (1970). *Science* **170**, 857.
Robinson, W., Metcalf, D., and Bradley, T. R. (1967). *J. Cell. Physiol.* **69**, 83.
Sanders, F. K., and Smith, J. D. (1970). *Nature* (*London*) **227**, 513.
Sanford, K. K., Earle, W. R., and Likely, G. D. (1948). *J. Nat. Cancer Inst.* **9**, 229.
Schenck, D. M., and Moskowitz, M. (1958). *Proc. Soc. Exp. Biol. Med.* **99**, 30.
Schindler, R. (1964). *Exp. Cell Res.* **34**, 595.
Walters, R. A., Hutson, J. Y., and Burchill, B. R. (1970). *J. Cell. Physiol.* **76**, 85.
Wildy, P., and Stoker, M. (1958). *Nature* (*London*) **181**, 1407.

Chapter 3

Cell Fusion and Its Application to Studies on the Regulation of the Cell Cycle[1]

POTU N. RAO[2] AND ROBERT T. JOHNSON

The Eleanor Roosevelt Institute for Cancer Research, and Department of Biophysics and Genetics, University of Colorado Medical Center, Denver Colorado, and Department of Zoology, University of Cambridge, Cambridge, England

[1] This study was supported by U. S. Public Health Grant No. 5 PO1 HDO2080 from the National Institute of Child Health and Human Development.

[2] *Present address:* Department of Developmental Therapeutics, The University of Texas M. D. Anderson Hospital and Tumor Institute, Houston, Texas.

I.   Introduction

The ability to combine genetic material from two different sources to produce a single functional unit has long been the aim of investigators of higher eukaryotic cells. It is well known that macromolecular DNA can be taken up by mammalian cells and incorporated into the nucleus (Ledoux, 1965; Hill, 1966; Hill and Spurna, 1968; Schell, 1968; Robbins and Taylor, 1968; Ayad and Fox, 1968). While whole chromosomes may be taken up, there is no evidence that they are ever genetically active (Whang-Peng et al., 1967; Yosida and Sekiguchi, 1968). Because of the possibility of DNA being taken up by cells, attempts have been made to transform mammalian cells in vitro (Szybalska and Szybalska, 1962; Ledoux, 1965; Ayad and Fox, 1968). At present, however, it is clear that the techniques of transformation and transduction, which are widely used in the study of bacterial genetics, have been of little use in the case of mammalian cells. It is for this reason that the techniques of spontaneous and virus-induced cell fusion have been so extensively adopted in recent years. The cell fusion technique made it possible to obtain hybrids between cells of different genotypes thereby bypassing the process of fertilization. It has also provided an opportunity to study the nucleocytoplasmic interactions in mammalian cells in ways which are similar to the investigations so elegantly performed by the technique of nuclear transplantation in large cells like Protozoa or amphibian oocytes (Prescott and Goldstein, 1967; Goldstein and Prescott, 1967; Graham et al., 1966; Graham, 1966; Gurdon, 1967; De Terra, 1967; Ord, 1969).

Although mononucleate cells are the rule in higher plants and animals, the multinucleate condition does exist in some tissues (see Johnson and Rao, 1971 for extensive bibliography). Not only can a high proportion of multinucleate cells be seen, for example, in the endosperm of plants (Jungers, 1931), or in liver parenchymal tissue of animals including man (Carriere, 1969), but there is increasing evidence that mononucleate cells are extensively coupled electrically *in vivo* and *in vitro* (Loewenstein, 1966; Furshpan and Potter, 1968; O'Lague *et al.*, 1970). Communication between these apparently "isolated" cells occurs via the tight junctions, which exhibit a complex structure and through which large molecules can freely pass (Loewenstein, 1966). In a culture dish, cells may form extensive bridges among themselves, and, in some instances, DNA has been found in such bridges (Bendich *et al.*, 1967). Recently metabolic cooperation has been demonstrated between cells in tissue culture (Subak-Sharpe, 1969). These observations on functional, though not structural multinucleation in the strict sense, raise important questions concerning the regulation and coordination of events such as DNA synthesis and mitosis among cells in a given tissue (Johnson and Rao, 1971; Cone, 1969).

The purpose of this chapter is: (1) to present the technical aspects of virus-induced cell fusion, including procedures for the propagation of Sendai virus and for the fusion of cells in suspension or in monolayers by ultraviolet-inactivated Sendai virus and (2) to indicate the application of the cell fusion technique to studies on the regulation of the two major events of the cell cycle—DNA synthesis and the initiation of mitosis. The phenomenon of premature chromosome condensation (PCC) that results from the fusion of interphase cells with those in mitosis and its implications on the process of chromosome formation will be discussed. However, this chapter does not include those areas of cell fusion which deal with somatic cell genetics, differentiation and the regulation of phenotypic expression, or the induction of malignancy because these subjects have been dealt with at length in a number of excellent reviews and symposia. Information on these topics can be obtained in the following works: Harris (1968, 1970), Wistar Institute Symposium Monograph No. 9 on Heterospecific Genome Interactions (1969), Ephrussi (1970), Migeon and Childs (1970), and Watkins (1971a,b).

II. Types of Cell Fusion

A. Spontaneous Cell Fusion

Multinucleate cells originate from mononucleate cells in one of the two ways, first, due to the failure of cytokinesis (cytoplasmic division) immediately following karyokinesis (nuclear division) and, second, due to the fusion of two or more adjacent cells into one single cytoplasmic unit. The ability of tissue culture cells to fuse and to form multinucleate cells has long been recognized. Lewis (1927) has reviewed the early literature describing this phenomenon. Barski *et al.* (1960, 1961) were the first to describe the spontaneous hybridization of cells from two different tissue culture lines. This observation was confirmed and extended by Sorieul and Ephrussi (1961), Ephrussi and Sorieul (1962), Gershon and Sachs (1963), and Ephrussi *et al.* (1964) for a variety of mouse cell lines. Following the work of Harris and Watkins (1965), who showed that ultraviolet-inactivated Sendai virus could be used to fuse cells from different animal species, Ephrussi and Weiss (1965) described the spontaneous hybridization of cells from two different species. This work and its developments have been reviewed by Ephrussi (1965) and Ephrussi and Weiss (1965). The detection and isolation of the extremely rare spontaneous hybrid cells was made easier by the introduction of a selective method for cell lines with different drug-resistant properties (Littlefield, 1964b). The two genetic components of the hybrid cell were able to complement one another's deficiencies. Hence, in a defined medium that could not support the growth of either of the parental types, the hybrids were able to proliferate. This technique has been extended by Davidson and Ephrussi (1965, 1970), and has proved particularly useful in genetic analysis (Weiss and Green, 1967).

B. Virus-Induced Cell Fusion

Cell fusion is one of the results of infection by both DNA and RNA viruses. For example, members of the Herpes group, pox viruses, many myxoviruses, SV40, and visna virus produce multinucleation of infected cells (Roizman, 1962). The literature of the 19th century is abundant with examples of polykaryocytes from pathological lesions, and as Harris *et al.* (1966) point out, "it is very probable that some of the inflammatory lesions in which multinucleate cells were observed in the last century were caused by viruses, although, of course, the viral etiology of these conditions was not recognized until very much later."

With the development of tissue culture methodology the cell-fusing ability of viruses could be observed by time-lapse cinemicrography, and this enabled correlations to be made with the lesions found in the tissues of infected animals. For example, the work of Taniguchi *et al.* (1954) on the formation of giant cells by measles virus under *in vivo* conditions has been confirmed by the *in vitro* studies of Enders and Peebles (1954). Other myxoviruses, notably mumps (Henle *et al.*, 1954), respiratory syncytial virus (Morris *et al.*, 1956), the parainfluenza viruses (Okada *et al.*, 1957; Marston, 1958; Lepine *et al.*, 1959), and Newcastle disease virus (Johnson and Scott, 1964; Kohn, 1965) fuse tissue culture cells. The work of Roizman (1962) showed that herpes virus also fuses cells *in vitro.* The paramyxoviruses are able to fuse cells with high frequencies and have a wide host range. For this reason parainfluenza I (Sendai or hemagglutinating virus of Japan-HVJ) has been chosen by most investigators for cell fusion.

The use of virus-mediated cell fusion in somatic cell genetics was suggested by Ephrussi and Sorieul (1962), and the development of this possibility into a semiquantitative exercise has been due mainly to the work of Okada and his associates. Fukai *et al.* (1955) found that Ehrlich ascites tumor cells were agglutinated by a strain of parainfluenza I virus, and Okada *et al.* (1957) showed that this virus (HVJ) not only agglutinated the tumor cells but induced their fusion to form polykaryocytes both *in vivo* and *in vitro.* Ultraviolet irradiation abolished the infectivity but did not reduce the fusion ability of the virus (Okada, 1958). During the next few years the *in vivo* fusion technique was perfected (Okada, 1958; Okada, 1962a,b; Okada and Tadokoro, 1962; Murayama and Okada, 1965; Okada *et al.*, 1966). It has also been demonstrated that many types of cells from different species were susceptible to the action of the virus, although not all cells fused as readily (Okada and Tadokoro, 1962). In 1961, Okada showed that different strains of murine tumor cells could be fused together. Using cells from different species Harris and Watkins (1965), Harris (1965), and Okada and Murayama (1965a) demonstrated that multinucleated cells could be produced by means of this technique.

The technique of virus-induced fusion permitted the production and isolation of viable hybrid clones, and these hybrids are identical to those formed by spontaneous fusion in all respects (Yerganian and Nell, 1966; Coon and Weiss, 1969; Kao *et al.*, 1969a). Cell fusion by means of Sendai virus (Klebe *et al.*, 1970a; Coon and Weiss, 1969; Davidson, 1969a) has two main advantages over spontaneous fusion: (1) it vastly increases the numbers of hybrids, and (2) it permits fusion to occur between cells that either would not fuse at all or that fused at frequencies which would preclude experimentation.

C. Lysolecithin-Induced Cell Fusion

Poole *et al.* (1970) described the lysolecithin-mediated fusion between tissue culture cells and hen erythrocytes, and estimated that up to 20% of the cells were involved in fusion. The stability of the fused cells was increased by adding defatted serum to the mixture. It seems likely that lysolecithin will replace Sendai virus as the agent used for fusing cells as soon as the heterokaryons can be permanently stabilized against its lytic action. This will be a significant advance since it will ensure the absence of viral particles from the fused cells, and will also remove the necessity for growing Sendai virus that can fuse well. It should be noted, however, that lysolecithin, at any concentration, did not fuse baby hamster kidney cells (Elsbach *et al.*, 1969).

III.   Viruses and Cell Fusion

All of the Newcastle disease group of myxoviruses (Watson, 1966) have the ability to fuse cells. However, not all strains of these viruses induce fusion to the same extent (for example, see Okada and Hosokawa, 1961) and care must therefore be taken in selecting which strain to use. Sendai virus, which is probably synonymous with hemagglutinating virus of Japan, and which is also called parainfluenza I, is most widely used. The strains that have been found to give good fusion are ESW5 of Sendai (grown by Dr. J. F. Watkins, Oxford University) and the Z strain of HVJ (grown by Dr. Y. Okada, Osaka University). There are undoubtedly other strains which fuse well. Sendai virus has recently become commercially available from the American Type Culture Collection (ATCC No. VR105) and from Microbiological Associates, Bethesda, Maryland.

It should again be stressed that other paramyxoviruses are equally good at fusing cells. Of these, SV5 (parainfluenza 5) and Newcastle disease virus (Kohn, 1965) have been used for the routine production of multinucleate cells.

A.   Sendai Virus and Its Propagation

The routine growth and assay of this virus are straightforward and more detailed accounts for the procedures of growing this virus in hens' eggs, and its subsequent collection, concentration, and assay are given by Watkins (1971a,b).

The first requirement is to obtain a supply of good, fertile hens' eggs, preferably from a flock which has not been extensively treated with antibiotics. The eggs are routinely incubated at 37°C for 9 to 10 days before they are inoculated with virus. A small hole is made in the blunt end of each egg above the air-sac membrane, and 0.01–0.1 hemagglutinating units (HAU) of virus are inoculated into the allantoic sac by means of a syringe. Care must be taken to point the needle away from the center of the egg, and to ensure that penetration of the air-sac membrane does not exceed ¼ inch. The inoculum of infective virus may be diluted in allantoic fluid or in Hank's basal salt solution (BSS) without glucose to the desired concentration.

After sealing the hole with either paraffin wax or Scotch tape the eggs are replaced in the incubator in an upright position with the egg standing on its pointed end. There should be adequate humidity in the incubator, and the temperature may be reduced to 36°C. The eggs are thus incubated for 3 days, but it is common practice to candle the eggs every day to remove those in which the embryos are dead. Those eggs which have died during the first day of incubation should be discarded since the virus titer in these will be very low. In some laboratories the remaining eggs are reinoculated after 48 hours with double the dose of the virus used previously. This step often boosts the yield of virus.

The eggs from which virus is to be collected are placed at 4°C for 12 hours to ensure that there will be little or no bleeding during the collection of allantoic fluid. This fluid, which should be clear and faintly yellow is harvested by means of a blunt sterile Pasteur pipette. Between 3 and 10 ml of fluid is generally obtained per egg. If the allantoic fluid contains a great number of erythrocytes it should be discarded since the virus adsorbs to them, but if there is only a slight contamination then the allantoic fluid can be incubated for 30 minutes at 37°C to elute the virus.

When all the infected allantoic fluid has been harvested it is pooled, centrifuged at 1700 rpm for 10 minutes to remove any large debris, and a sample is removed for hemagglutination assay. This is done using a Salk pattern hemagglutination tray. Each of a number of cups on the tray is filled with 0.5 ml of phosphate buffered saline (PBS)–saline A of Dulbecco and Vogt (1954). Sequential doubling dilutions of the virus are made, and finally one drop (0.05 ml) of a 4% (v/v) solution of sheep, guinea pig, or chick erythrocytes in PBS is added to each cup. The end point is read off as the last reservoir in which complete hemagglutination has occurred after 1 hour at room temperature, and the concentration of virus in this last cup is equal to 1 HAU. Generally, the pooled allantoic fluid gives a titer of between 500 and 5000 HAU/ml.

The allantoic fluid is cleared by centrifugation at 2–4000g at 4°C, and the virus is concentrated as a pellet by centrifugation at 30,000g for 30 minutes in a Beckman 21 rotor, using stainless steel tubes. It is important not to spin too hard or for too long or else it becomes very difficult to disperse the pellet. The pelleted virus is resuspended in 2 ml of Hank's BSS without glucose (pH 7.2), and the hemagglutination titer is determined using 0.1 ml of the concentrate. The titer is generally 2 to 4 $\times$ 10^4 HAU/ml. An antibiotic stock solution is added to the virus to a final concentration of 250 units of penicillin G sodium, 250 μg streptomycin sulfate, 250 μg neomycin sulfate, and 2.5 units bacitracin (Sigma Chemical

Co., St. Louis, Missouri) per milliliter of virus solution. The virus can now be diluted with Hank's BSS without glucose to the required concentrations (e.g., 2000 HAU/ml) or left at a high concentration. In either case it is rapidly frozen in 1- or 2-ml quantities in glass ampoules or plastic tubes (Falcon Plastics, Los Angeles, California), by means of dry ice/ethanol, and stored at −70°C. For storage in liquid nitrogen the virus should be placed in sealed glass ampoules. The hemagglutination titer and the fusion ability of the virus does not decrease for periods of up to 9 months when stored at −70°C.

B.   Methods of Virus Inactivation

Since it is important to eliminate the complexity of viral infection, it is essential to inactivate the viral genome. This can be done by ultraviolet light or by β-propiolactone.

1.   Ultraviolet Light

Okada (1962a,b) first demonstrated that Sendai virus could be genetically inactivated by ultraviolet light while retaining its ability to fuse cells. This led to the widespread adoption of this method (Harris and Watkins, 1965; Harris et al., 1966; Coon and Weiss, 1969; Kao et al., 1969b). A 1- or 2-ml solution of the concentrated virus (2 to 8×10^4 HAU/ml) is placed either in a shallow watch glass or a plastic petri dish. It is exposed to ultraviolet light from a germicidal tube (Philips 15 watt, 18 inch, Type T.U.V. or GE G15T8) at a distance calculated to yield approximately 3000 $erg/cm^2/sec$. Exposures exceeding 5 minutes are likely to reduce the fusion capacity of the virus. The virus suspension should be gently shaken or pipetted at intervals of 1 minute to ensure maximum inactivation. After this treatment the virus can be diluted to give the desired concentration for immediate use or stored at −70°C for future use.

This procedure for virus inactivation is simple and rapid, and for most experimental purposes it is completely satisfactory. The one main disadvantage rests in the fact that although the infectivity of the virus is reduced by more than 10^6 there is still a small amount of residual infectivity (Harris and Watkins, 1965; Harris et al., 1966). This may, in part, be reduced by reinactivating the virus for 1 minute with ultraviolet light just before use.

2.   β-Propiolactone Inactivation

This method was first described by Neff and Enders (1968) and has been used by Klebe et al. (1970a,b), Baranska and Koprowski (1970), and Graham (1971). β-Propiolactone (β-prone, Fellows Testagar, Detroit, Michigan) is an alkylating agent and reacts with the guanosine of the viral RNA (Roberts and Warwick, 1963). It causes complete inactivation of the viral genome with no significant loss of fusion ability (Neff and Enders, 1968; Klebe et al., 1970a). The procedure for inactivation is as follows: A 10% solution of β-propiolactone in double-glass-distilled water is prepared immediately before use. This solution is diluted with sodium bicarbonate (1.68 gm NaHCO$_3$, 0.5 ml of 0.4% phenol red in 100 ml of isotonic NaCl) to give a 2% stock solution. All dilutions are rapidly carried out at 4°C. The concentrate of Sendai virus (10–20,000 HAU/ml) in PBS, supplemented with 0.5 to 1% bovine serum albumin, is treated with the β-propiolactone solution so that the final concentration of β-propiolactone does not exceed 0.13% (Neff and Enders, 1968). A concentration of 0.05% has been found suitable by Klebe et al. (1970a). The solution is shaken for 10 minutes at 4°C in tightly stoppered vessels to ensure complete mixing of the components, and then shaken at 37°C for 2 hours to inactivate the virus. The mixture is finally kept at 4°C for 12 hours to produce complete hydrolysis of the remaining β-propiolactone. When the virus is completely inactivated, it is diluted with Hank's BSS without glucose to give the required working concentration, then antibiotics are added, and the virus is rapidly frozen and stored at −70°C. It may be thus stored without appreciable loss of fusion activity for up to 5 months (Klebe et al., 1970a).

The advantages of β-propiolactone inactivation lie in the fact that it provides complete genetic inactivation of the virus. Consequently, this eliminates the doubt that results obtained with cell fusion studies are not spuriously complicated by any cytopathic effects of virus infection (Neff and Enders, 1968; Nichols, 1970).

C. The Mechanism of Cell Fusion

Cell fusion is the result of an interaction between the membranes of the virus and the cells involved. Both the viral and the cell surfaces are negatively charged (Bachtold et al., 1957) and it has been calculated that the rate of adsorption of virus onto cells is less than expected from Brownian movement (Cohen, 1963). This is probably due to the electrostatic repulsion that exists between them in a nonionic environment. The presence of cations in the medium in suitable concentrations would neutralize these repulsive forces by forming an electrical double layer or by binding to the negative charges. Consequently, adsorption of virus is increased by adding cations to the medium (Bachtold et al., 1957; Okada and Murayama, 1965b) so that the virus can approach closer to the cell for short range attractions to operate. At this stage electrostatic bonds are formed between the amino groups of the virus and the carboxyl groups of the cell surface (Levine and Sagik, 1956). This is followed by the attachment of the viral hemagglutinin spikes to the sialic acid receptor sites of the cell membrane (Cohen, 1963). At the sites of virus attachment the cell membrane becomes invaginated (Howe and Morgan, 1969), and, at this stage, the viral membrane probably fuses with the cell membrane. Radioisotope labeling of the viral and cell membrane phospholipids has shown that there is an exchange of free lipid between cell and virus (Hoyle, 1962), and this may set up a new lipid equilibrium which promotes membrane coalescence (Hoyle, 1962; Poole et al., 1970). At this stage the membrane of the virus starts to break down, and the nucleocapsid is liberated into the cell (Howe and Morgan, 1969). In the normal course of infection by low multiplicities of virus the cell membrane reforms, but at high multiplicities, such as those used in cell fusion, cytoplasmic or viral bridges are formed between cells (Schneeberger and Harris, 1966; Hosaka and Koshi, 1968) leading to a rapid coalescence of the cells which are thus joined.

D. Fusion Factor

Much work has been done to find out whether paramyxoviruses possess a distinct fusion factor. Available evidence suggests that the fusion factor most closely resembles the agent which is responsible for hemolytic activity, but it differs from it in several ways (Okada and Tadokoro, 1962; Kohn, 1965; Kohn and Klibansky, 1967; Zhdanov and Bukrinskaya, 1962; Yun-De and Gorbunova, 1962; Neurath, 1963, 1964; Neurath and Sokol, 1962). Fusion by paramyxoviruses depends on an intact lipoprotein membrane. At present it appears that if the fusion factor is a separate entity, it is predominantly lipid in composition (Kohn, 1965; Kohn and Klibansky, 1967; Poole et al., 1970; Lucy, 1970). Recently temperature-sensitive mutants of respiratory syncytial virus have been isolated (Gharpure et al., 1969), and one of them has lost the ability to fuse cells. It is possible that this represents a double mutation but, in any case, investigation of this virus may reveal the nature of the fusion factor.

It has been known for some time that extracts from tuberculosis bacilli promote the fusion of cells and that the active substance appears to be a lipopolysaccharide (Lewis, 1927; Sabin, 1932; Racadot and Frederic, 1955). Lysolecithin, which has been implicated in the hemolytic activity of paramyxoviruses (Rebel et al., 1962), has been shown to fuse cells in some cases (Poole et al., 1970), but not in others (Elsbach et al., 1969). Although the evidence is still incomplete, we may conclude that the ability of paramyxoviruses to fuse cells largely depends on the lipids they contain in their outer membranes.

IV. Methods of Cell Fusion

A. Fusion in Suspension

This is the most widely used and simple means of generating fused cells. The technique was first described by Okada (1958), and has been extensively used by many others (Harris et al., 1966; Coon and Weiss, 1969). The general procedure is as follows: The required number of cells are washed in cold (4°C) PBS (solution A of Dulbecco and Vogt, 1954) pH 7.2, then taken up in a small volume (0.5–1.0 ml) of Hank's BSS without glucose, pH 7.2, at 4°C. In general the smaller the volume in which the cells and the virus are resuspended the greater is the degree of fusion because the chances of the virus particles coming into contact with the cells are greatly improved. The cells may be thoroughly mixed using a vortex mixer before addition of virus. Serum is omitted from the fusion mixture because it has been our experience, and also that of Davidson's (1969a) that fusion is reduced in its presence. The cells may be added to precooled glass T-tubes (Harris et al., 1966), although we have found that small plastic tubes are adequate. About 0.5 ml of inactivated virus with the desired amount of hemagglutinating activity, in Hank's BSS without glucose, is added. This step must be carried out at 4°C. This results in an immediate and visible agglutination of the cells. For most purposes we have found that it is not necessary to maintain the fusion mixture at 4°C for 15 minutes unless there is a problem of poor agglutination. It is essential, however, to begin the fusion process with precooled solutions since viral agglutination occurs much faster at 4°C than at room temperature. The virus–cell mixture is next transferred to 37°C where the actual fusion begins (Okada, 1962b). We find that there is no need to shake the tubes during this incubation period, which usually lasts for 30 minutes. When the fusion is completed the cells are centrifuged at low speed, the supernatant removed, and medium containing serum is added. Much of the excess virus can be removed either by washing the centrifuged cells thoroughly with medium containing serum, or by layering the cells on a column of medium containing up to 30% serum (Coon and Weiss, 1969). The removal of excess virus has been found to increase the survival of hybrid cells (Davidson, 1969a).

The extent and type of cell fusion that can be achieved in suspension

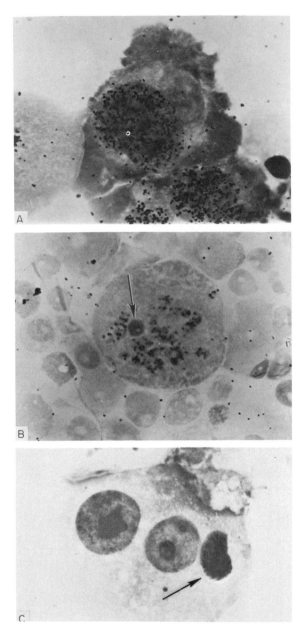

Fig. 1. Heterokaryons produced by the use of ultraviolet-inactivated Sendai virus. (A) HeLa/*Xenopus laevis* heterokaryon. The HeLa nucleus is labeled and the *Xenopus* nucleus is not. (B) HeLa/mosquito heterokaryon. The HeLa cell was in mitosis at the time of fusion. The HeLa chromosomes were prelabeled with thymidine-³H. The arrow indicates the unlabeled mosquito nucleus. Note the prominent nucleolus within the mosquito nucleus. (C) A multinucleate cell containing HeLa and bull sperm nuclei. The highly condensed sperm nucleus is indicated by the arrow. (From Johnson *et al.*, 1970.)

is controllable to a considerable degree. The actual numbers and ratios of cells used, the concentration of the virus, the volume of the fusion mixture, the lengths of time at 4° and 37°C, which are required for adequate agglutination and fusion, must be arrived at largely by trial and error. Account must be taken of the use to which the fused cells will be put. For example, if production and prolonged growth of hybrid cells is desired then one should consider the following points. (1) It will be necessary to optimize fusion conditions so that binucleate cells will be more numerous, since cells containing three or more nuclei often experience difficulty in passing through mitosis (Oftebro and Wolf, 1967). (2) Similarly, it may be necessary to fuse only those cells which are in interphase at the time of fusion since this eliminates the possibility of premature chromosome condensation caused by fusion between mitotic and interphase cells (Johnson and Rao, 1970; Sandberg et al., 1970; Klebe et al., 1970a). (3) It may ultimately be necessary to fuse populations of cells which have been synchronized in the same phase of the cell cycle so as to ensure maximum survival of the hybrids (Coon, 1967, Rao and Johnson, 1972). (4) Sendai virus should be completely inactivated by β-propiolactone. (5) Excess virus is removed after fusion. (6) If the fusion capacity of the two parental cell types differs markedly, the proportions must be carefully adjusted to obtain the maximum number of hybrids (Coon and Weiss, 1969). The fact that certain cells fuse preferentially with their own type should be taken into account (Mukherjee et al., 1970). (7) The less virus used, the lower will be the cytopathic effects, and therefore, the greater will be the survival of the hybrid cells.

If the aim of the cell fusion experiment is not the production of hybrid clones, then the experimental conditions need not be so rigorous. With higher concentrations of virus one can obtain a high degree of fusion and usually a variety of cell types can be induced to fuse (Fig. 1). Thus, it becomes possible to fuse lymphocytes or leukocytes (Harris et al., 1966; Miggiano et al., 1969), neurons (Jacobson, 1968), myoblasts and myotubes (Carlsson et al., 1970), insect cells, and spermatozoa (Johnson et al., 1970; Zepp et al., 1971). In most cases the differentiated cells do not require any special treatment before fusion with the exception of ova and spermatozoa. They will be discussed in a later section.

B. Fusion in Monolayer Cultures

This procedure which was first described by Kohn (1965) is becoming increasingly popular. It consists essentially of plating a mixture of the desired cell types at a required density and allowing them to settle and attach before virus is added to induce fusion. Under monolayer conditions contact between cells is usually restricted to two dimensions rather than three and it has been suggested that it is easier to manipulate the extent of fusion by this method than by fusion between rounded cells in suspension. By varying the cell density and proportions of the two types of cells it is possible to achieve reproducible hybridization frequencies by this method (Klebe *et al.*, 1970a).

The details of the procedure are as follows: A desired number of cells of both the parental types are mixed and plated either in plastic flasks or petri dishes. They are allowed to settle and attach to the bottom of the dishes at 37°C for up to 24 hours. The cells should not be confluent or too sparse, if effective and controlled fusion is to occur. Once the cells are finally attached, the medium is removed and the flasks or dishes are washed three times with cold (4°C) PBS to remove serum which hinders cell fusion. Virus is added to 0.2–1.0 ml of cold Hank's BSS without glucose, and the virus is allowed to adsorb onto the cells at 4°C for 15 minutes. The cells are then washed twice with cold PBS to remove excess virus, and incubated either with PBS or Hank's BSS without glucose, for 10 minutes at 37°C. This step greatly increases the yield of hybrid cells (Davidson, 1969a). Finally, the medium containing serum is added. The medium may be changed again after 2 hours.

This basic procedure can be modified in a variety of ways. For example, only one cell type is plated to produce a monolayer to which virus is later added and allowed to settle over the cells. Over these two layers, cells of the other parental type are plated forming a sandwich of the virus particles between the two cell types. Using this procedure Davidson (1969a) was able to obtain high frequencies of hybrid cells.

V.   Preparation of Some Specialized Cells for Fusion

A.   Ova and Blastomeres

The preparation and fusion of ova and blastomeres, either with one another or with somatic cells, has been described in detail by Graham (1969) and Baranska and Koprowski (1970). In brief, it has been found that β-propiolactone-inactivated Sendai virus was preferable to ultraviolet-inactivated virus, and that an extremely limited exposure to the virus and a rapid sequestering of the virus by serum, are essential for the survival of these cells (Graham, 1971).

B. Spermatozoa

The last few years have seen many attempts to induce fusion between spermatozoa and somatic cells, and there has been a certain amount of success (Coon; Koprowski; Pontecorvo; Pearson; Szollosi—all personal communications; Johnson et al., 1970). It has been our experience that spermatozoa are difficult to fuse with somatic cells, and we attribute this to the specialized nature of the membrane of the spermatozoan (Lung, 1968). In addition, it is often difficult to determine whether spermatozoa are lying on the somatic cells or are actually inside. If they are inside then it must be determined whether they are within vesicles or not. Because of the specialized nature of the spermatozoa surface we adopted a chemical treatment which "softens" the surface. This enabled us to fuse spermatozoa and somatic cells with a limited success (Fig. 1C).

Bovine spermatozoa were used within 1 week of ejaculation. They were washed once in Ham's F12 medium and supplemented with 10% fetal calf serum (Ham, 1965), at 37°C. The spermatozoa were then centrifuged at 200 rpm for 10 minutes in a clinical centrifuge, and after the supernatant had been removed they were washed again, this time in phosphate-buffered saline (pH 7.2). After centrifugation and removal of the supernatant, the sperm were treated with a solution of 0.25% trypsin and 0.01% EDTA for 60 minutes at 37°C. The total volume of fluid did not exceed 1 ml during this stage. The solution was next thoroughly agitated by means of a vortex mixer for 3 minutes before centrifuging at 2000 rpm. The pellet of spermatozoa was resuspended in cold (4°C) PBS at pH 7.2, and the spermatozoa were counted by means of a hemocytometer. Dilutions were made in cold PBS for the fusion mixture. Examination of these spermatozoa by phase contrast microscopy revealed that 80% of them had lost their tails but only 50% of them had lost the acrosome. At this stage of the treatment there was no contamination of the spermatozoa by any other types of cells.

In general, 10^7 spermatozoa were mixed with 10^6 somatic cells (e.g., HeLa) in a volume of 0.5 ml, using 500 HAU of ultraviolet-inactivated Sendai virus. Agglutination was allowed to proceed for 30 minutes at 4°C, and the tubes containing the fusion mixture were then transferred to 37°C and incubated for at least 1 hour. This procedure resulted in fusion between spermatozoa and HeLa cells, although the incidence of fusion was very low. The highest incidence of fusion that we observed was 14% (i.e., the percentage of multinucleate cells containing at least one spermatozoan nucleus). The spermatozoan nucleus, when inside a somatic cell, can be easily detected because of its size, shape, distribution of chromatin, and staining properties. It generally appears elongated and darkly stained (Fig. 1C). We have used a 10% Giemsa–1% May Grüne-wald stain (E. Gurr, London) in phosphate buffer (pH 6.8).

Preliminary experiments were carried out to assess whether the sperma-tozoa nucleus could be reactivated in heterokaryons (F. T. Kao, R. T. Johnson, and P. N. Rao, unpublished results). Fusion was carried out in the above manner between 10^7 spermatozoa and 10^6 auxotrophic mutants of a Chinese hamster ovary cell line, which require glycine (Kao and Puck, 1968). In one experiment heterokaryons were placed in F12 medium containing glycine for 48 hours, after which the medium was re-placed by F12 lacking glycine, while in another experiment they were directly plated with F12 medium without glycine. No colonies appeared in either of the experiments indicating that the spermatozoa nucleus had not supplemented the genetic deficiency of the somatic cell line. If the reactivation of the sperm nucleus is to occur, it might be essential to use a nondividing somatic cell as the partner in the heterokaryons in much the same way as in the X-irradiated A9-chick erythrocyte system of Harris et al. (1969).

C.   Plant Cells

Power et al. (1970) have recently described a method for the spon-taneous fusion of plant protoplasts from different species. Although the spontaneous fusion between plant and animal cells does not yet seem feasible, there is no reason why such fusions should not be attempted by paramyxoviruses or lysolecithin.

VI.   Methods for the Selection and Isolation of Hybrids

The use of hybrid cells created by spontaneous or virus-induced cell fusion between two different genotypes would be extremely limited if there were no efficient means of isolating the hybrids. One satisfactory solution to this problem has been to make use of cells with genetic markers. In this system both the parental cell types are eliminated under selection pressures leaving only the hybrid cells to form colonies. The selection pressures are based on (a) drug resistance, (b) nutritional re-quirements, and (c) differential sensitivities to chemical, physical, or biological agents. When selection pressure is placed on both parents, this may be termed as the double-selection method as opposed to the single-selection method when such pressures are applied only to one of the parents.

A. Double-Selection Method

1. DRUG RESISTANCE

The selective system developed by Littlefield has been used with little or no modification by a number of workers for the isolation of hybrid cells (Littlefield, 1963, 1964a,b, 1966; Davidson and Ephrussi, 1965; Weiss and Ephrussi, 1966; Scaletta *et al.*, 1967; Migeon, 1968; Migeon and Miller, 1968; Matsuya *et al.*, 1968; Siniscalo *et al.*, 1969; Miggiano *et al.*, 1969; Boone and Ruddle, 1969). In this method two cell lines with the following properties are fused. Parent A: (1) Lacks hypoxanthine guanine phosphoribosyltransferase (HGPRT) and is unable to utilize preformed purines. Hence it is resistant to the analog 8-azaguanine. The *de novo* pathway for purine synthesis by which the cell survives can be interrupted by the addition of aminopterin to the medium. (2) The enzyme thymidine kinase is present in these cells. Parent B: (1) Lacks thymidine kinase which is necessary for the phosphorylation of thymidine monophosphate to the triphosphate. The lack of this enzyme affects the conversion of bromodeoxyuridine (BUdR) to thymidine triphosphate and hence the growth of these cells is not affected by the presence of BUdR in the medium. The *de novo* pathway can be blocked by aminopterin. (2) It has a functional HGPRT.

The fused cells are plated in regular medium which, at the end of 48 hours, is replaced by the selective medium containing aminopterin, glycine, hypoxanthine, and thymidine. This would result in the elimination of both the parents because they are unable to utilize hypoxanthine or thymidine due to the lack of HGPRT or thymidine kinase. However, in the hybrids the lack of one enzyme in one of the cell components is compensated by its presence in the other. Thus the hybrid cells survive and grow into colonies while the parental cells are unable to do so.

2. NUTRITIONAL REQUIREMENTS

The nutritional requirements of certain cell lines have been used against them in selecting hybrids between such cells (Krooth and Weinberg, 1960; Krooth, 1964; Tedesco and Mellman, 1967). The auxotrophic mutants of Chinese hamster cells developed by Kao and Puck (1967) represent another example where this type of a selection system has been employed. For instance, the fusion between two glycine-requiring mutants of Chinese hamster cells and the plating of the fusion mixture in medium without glycine resulted in the survival of only the hybrid colonies due to genetic complementation (Kao *et al.*, 1969a,b).

3. Differential Sensitivity to Temperature

The temperature requirements of cells from different animal species are bound to vary because of differences in their body temperatures. For example, the chick fibroblasts can tolerate an ambient temperature of 45°C while mosquito cells and the cells from the gonads of rainbow trout cannot tolerate even 37°C which is optimal for most mammalian cells in culture (Brown, 1963; Rao and Engleberg, 1965; Buckley, 1969; Zepp *et al.*, 1971). As recently suggested by Steplewski and Koprowski (1970), the differential temperature sensitivities of cell lines could be exploited as a means for the isolation of hybrid cells from their parental types.

B.   Single-Selection Method

Differentiated cells such as avian erythrocytes (Harris, 1967), neurons (Jacobson, 1968), human leukocytes (Miggiano *et al.*, 1969), or Ehrlich ascites tumor cells (Harris *et al.*, 1966), which are not capable of growth in culture, are used as one of the parents in this system. Usually the other parent is a regular tissue culture line which could be eliminated by any one of the selection pressures while selecting for the hybrids.

C.   Selection on the Basis of Clonal Morphology

It is not always possible to find cell lines that possess either drug resistance or nutritional requirements. Under such circumstances it is sometimes possible to make use of differences in the clonal morphology and the rate of growth of the parents and the hybrids as reported by Coon and Weiss (1969) in rat liver cells, and Klein *et al.* (1971), Bregula *et al.* (1971), and Weiner *et al.* (1971) for mouse–mouse hybrids.

VII.   Preparation of Fused Cells for Microscopic Examination

A.   Fixation, Staining, and Radioautography

It is easier to score well-stretched cells more accurately with regard to the multiplicity of their nuclei than those which are more rounded. Tissue culture cells growing as a monolayer in a culture dish present an ideal condition for such examination. However, the difficulty that one would

FIG. 2. Cytocentrifuge. The photograph was kindly provided by the Shandon Elliot Co., England who are the manufacturers of this equipment.

come across in obtaining well-stretched cells soon after fusion has been overcome by the use of a cytocentrifuge developed by Shandon Elliot Co., of England (Fig. 2). This device makes it possible to produce a monolayer of cells on a regular microscope slide using centrifugal force and one can handle a maximum of 12 samples at a time. About $5\text{--}8 \times 10^4$ cells taken in 0.3–0.8 ml of medium gives an ideal spread localized into a spot 6.0 mm in diameter with little or no overlapping of cells. The cells were centrifuged at 600 rpm for 9 minutes and air-dried for 1 minute before fixing in a 3:1 mixture of absolute ethanol and glacial acetic acid for 5 minutes. The cells were stained by placing a drop of 2% solution of acetoorcein over which a cover slip was mounted. The slides were made permanent in the following manner. The cells were allowed to take the stain for about 10 minutes during which as much stain was placed by the edges of the cover slip as was needed to prevent

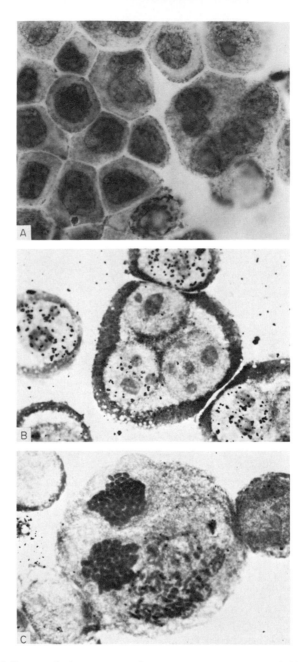

Fig. 3. Cell spreads by cytocentrifuge. (A) A microscopic field showing the even spreading and flattening of the cells. Note the clearance between the cell boundaries. (B) A trinucleate cell formed by the fusion between one labeled and two unlabeled cells. (C) This trinucleate cell was blocked in mitosis by Colcemid treatment. Because of the lack of mitotic spindles the chromosomes from each nucleus tend to form a separate entity. Note that one of the chromosome groups is labeled.

drying. At the end of this period excess stain was removed by blotting with blotting paper and the slides were placed on a smooth block of dry ice with cover slips facing up. The freezing of the fluid between the slide and the cover slip facilitates the separation of the two with the help of a razor blade. The slide and the cover slip were separated and the excess stain was removed by dipping them first in a 1:1 mixture of normal butyl alcohol:glacial acetic acid and then giving two changes of normal butyl alcohol. After drying, the cover slip was mounted onto the slide with a drop of Permount. In these preparations the cytoplasmic boundaries of the cell becomes clearly defined because the cytoplasm is also lightly stained (Fig. 3A).

The chromosome preparations were made using standard techniques (Tjio and Puck, 1958). They were fixed as before, hydrolyzed in $1 N$ HCl at 60°C for 6 minutes, and stained with crystal violet.

For radioautography the cells (on the slides) were given three 10-minute extractions with cold 5% trichloracetic acid (TCA) while rinsing them with distilled water in between the extractions. After the final wash, the slides were dried and processed for radioautography according to the procedure previously described (Rao and Engleberg, 1965).

B. Determination of Fusion Index

The simplest way of ascertaining the degree of cell fusion is by scoring the frequency of all multi- (bi, tri, tetra, etc.) nucleate cells and expressing it as a percentage of the total. This gives a measure of the extent of cell fusion but it does not indicate the intensity of fusion. An alternative method is the determination of fusion index (F.I.) as suggested by Okada and Tadokoro (1962). The fusion index is essentially a measure of the mean number of nuclei per cell and may be arrived at by the following equation:

$$F.I. = \frac{Nm}{Mc} - \frac{Np}{Cp}$$

where $F.I.$ = fusion index; Nm = total number of nuclei in all the multi-nucleate cells in a given sample after fusion; Mc = total number of multi-nucleate cells in the same sample; Np = total number of nuclei in a sample of the parent (control) culture before fusion; and Cp = total number of cells in the same sample.

It is important to remember that all tissue culture lines contain some multinucleate cells and their frequencies may range up to 5% or even more in some cases. In a suspension culture of HeLa cells in exponential growth the multinucleate cells, most of which are binucleate, constitute approximately 4% of the total population.

C.   Terminology

The following terms have been employed to describe the various types of multinucleate cells resulting from the fusion between cells of the same or different genotypes. *Homokaryon*—a multinucleate cell containing genetically identical nuclei; *heterokaryon*—a multinucleate cell containing genetically different nuclei; *hybrid*—a mononucleate cell whose nucleus contains two or more different genetic components; *homosynkaryon*—a mononucleate cell derived from a homokaryon by the fusion of genetically similar nuclei. It is most probable that the nuclei fuse together after entering synchronous mitosis. The fusion of nuclear membranes during interphase may also occur (Harris and Watkins, 1965). *Heterosynkaryon*—a mononucleate cell derived from a heterokaryon by the fusion of genetically different nuclei, this is synonymous with hybrid; *homophasic cell*—a multinucleate cell made up of genetically similar nuclei all of which were in the same phase of the cell cycle at the time of fusion; *heterophasic cell*—a multinucleate cell made up of genetically similar nuclei which were in different phases of the cell cycle at the time of fusion.

VIII.   Synchronization of HeLa Cells for Fusion

Since the basic aim of this investigation is the study of nucleo–cytoplasmic interactions in the hybrids, derived by the fusion of cells from different phases of the cell cycle, it is thus essential to obtain highly synchronized populations of cells in the G_1, S, G_2 phases, and in mitosis. Such synchronized populations of HeLa cells were obtained by the application of the excess thymidine (2.5 mM), double-block technique (Rao and Engleberg, 1966). S and G_2 cells were harvested at 1 hour and 6 hours, respectively, after the reversal of the second thymidine block. Mitotic cells were obtained either by Colcemid (6.74 $\times$ 10^{-7} M) or nitrous oxide (at 5.4 atm) treatments (Rao, 1968). The mitotic block produced by nitrous oxide is reversible thus allowing the cells to complete mitosis while the effects of Colcemid treatment are essentially irreversible in HeLa cells. Large quantities of mitotic or G_1 cells could be obtained from suspension cultures by the N_2O method using the apparatus shown in Fig. 4. The purity of mitotic cells in such a suspension culture ranges between 85 to 90%. However the purity could be increased to 99% by making the following changes. After the excess thymidine block was released the cells were resuspended in fresh medium and plated in 60-

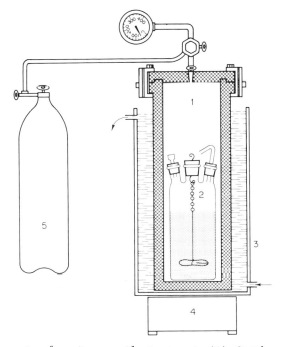

FIG. 4. Apparatus for nitrous oxide treatment. (1) Stainless steel pressure chamber with a removable top. (2) Cell chamber in which HeLa cells are grown in a suspension culture. (3) Water jacket through which 37°C water is circulated. The pressure chamber is immersed in the water jacket. (4) Magnetic stirrer. (5) Nitrous oxide tank.

mm plastic dishes (2.5–3×10^6 cells/dish). After incubating the dishes for 3.5 hours the floating cells were removed by changing the medium and then subjected to N_2O treatment for 9 hours. The rounded mitotic cells were shaken loose and collected. Highly synchronized populations of G_1 cells were obtained within 2 to 3 hours after placing the mitotic cells into suspension again. The mitotic cells can be stored at 4°C for 6 to 8 hours without loss of their ability to divide. The protocol for obtaining synchronized populations for various fusions is shown in Table I.

IX. Application of Cell Fusion to the Study of Events Related to Cell Cycle

The technique of virus-induced cell fusion has become a valuable tool in the study of a variety of problems which include the studies on gene

TABLE I

PROTOCOL FOR OBTAINING SYNCHRONIZED POPULATIONS OF HeLa CELLS
FOR THE THREE TYPES OF FUSION EXPERIMENTS

Time (hr)	Spinner A for prelabeled S and G_2 populations	Spinner B for G_2 population	Spinner C for G_1 population	Spinner D for G_1 population
0	Start Sp A; add thymidine-^{3}H (0.05 μCi/ml)	—	—	—
24	Place excess thymidine block	Start Sp B; place excess thymidine block	—	—
30	—	—	Start Sp C; place excess thymidine block	—
41	Release 1st TdR block; add ^{3}H-TdR (0.1 μCi/ml)	Release 1st TdR block	—	Start Sp D; place excess TdR block
50.5	Place 2nd TdR block	Place 2nd TdR block	—	—
51	—	—	Release TdR block; place cells in plastic culture dishes; incubate at 37°C	—
55	—	—	Transfer these dishes into N_2O chamber	—
62	—	—	—	Release TdR block; place cells in plastic culture dishes; incubate at 37°C
64	—	—	Release N_2O block; place cells in a spinner	—
65	Divide the cell suspension into three spinners A1, A2, and A3; release 2nd TdR block in A1	Release 2nd TdR block	—	—
66	—	—	Fuse cells from Sp A1 with those from Sp C to give G_1/S^{*a} fusion *Expt. II*	Transfer these dishes into N_2O chamber

TABLE I (*Continued*)

Time (hr)	Spinner A for prelabeled S and G_2 populations	Spinner B for G_2 population	Spinner C for G_1 population	Spinner D for G_1 population
70	Release 2nd TdR block in spinners A2 and A3	—	—	—
71	—	Fuse cells from Sp A2 with those from Sp B to give a S*/G_2 fusion *Expt. I*	—	—
75	—	—	—	Release N_2O block; place cells in a spinner
77	—	—	—	Fuse cells from Sp A3 with those from Sp D to give G_1/G_2* fusion *Expt. III*

a * Indicates prelabeled population. (From Rao and Johnson, 1970a.)

function and differentiation (Defendi *et al.*, 1964; Harris and Watkins, 1965; Harris, 1967; Bolund *et al.*, 1969; Davidson, 1969b); inheritance of neoplastic growth (Lawrence *et al.*, 1965; Ephrussi, 1970; Weiner *et al.*, 1971); complementation analysis (Kao *et al.*, 1969b); and chromosome mapping and linkage studies (Migeon, 1968; Kao and Puck, 1970). We applied this technique to answer some questions regarding the regulation of DNA synthesis and mitosis which will be discussed exclusively in the following pages.

A. Regulation of DNA Synthesis

As mentioned in the previous section, the availability of cells in different phases of the cell cycle with a high degree of synchrony made it possible to ask the question: How will the initiation and progression of DNA synthesis be affected when cells of the same or different stages are brought together by fusion? To answer this question two fusion experiments were performed where S cells were fused separately with G_1 and G_2 cells. The S cells were lightly prelabeled with thymidine-^{3}H and G_1 and G_2 cells were unlabeled. Following the fusion the cells were plated in a number of 35-mm plastic dishes to which thymidine-^{3}H (0.05 μCi/ml;

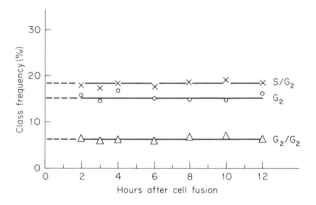

FIG. 5. The frequency of homo- and heterophasic binucleate cells in the S/G_2 fusion as a function of time after continuous labeling with thymidine-^{3}H. The S nucleus was prelabeled. The mononucleate G_2 class ($\bigcirc$—$\bigcirc$) is presented for comparison. X—X, S/G_2; $\triangle$—$\triangle$, G_2/G_2. (From Rao and Johnson, 1970a.)

6.7 Ci/mM) was added. Cells from these dishes were trypsinized and processed for examination at regular intervals. The frequencies of the three classes of binucleate cells which contain L/L, U/U, and L/U (L = labeled; U = unlabeled) nuclei were scored and plotted as a function of time. If there is any induction of DNA synthesis in the G_2 nuclei of the heterophasic cells then the frequency of the L/U class should decrease leading to a proportionate increase in the frequency of the L/L class. The definition of various classes and the procedures for calculating their frequencies have been described previously (Rao and Johnson, 1970a). The fact that the proportion of different classes in the S/G_2 fusion remained constant indicated that there was no induction of DNA synthesis in the G_2 nucleus (Fig. 5). The DNA synthesis in the S nucleus continued without any interruption even in the presence of G_2 component (Fig. 6).

A similar study with the G_1/S fusion revealed that the frequency of the L/U class decreased sharply following fusion (Fig. 7). This is the result of the induction of DNA synthesis in the G_1 nucleus under the influence of the S component in the fused cell. The DNA synthesis in the mononucleate G_1 cells or homophasic cells (G_1/G_1) did not commence until 8 hours later. The conclusions from these two experiments are: (1) There are some factors present in the S phase cell that can induce DNA synthesis in a G_1 nucleus but not in a G_2 nucleus following fusion. This also indicates that the DNA of the G_1 cell is available for replication while that of G_2 is not. (2) The absence of DNA synthesis in a G_1 cell is due to the lack of proper initiation factors rather than the presence of

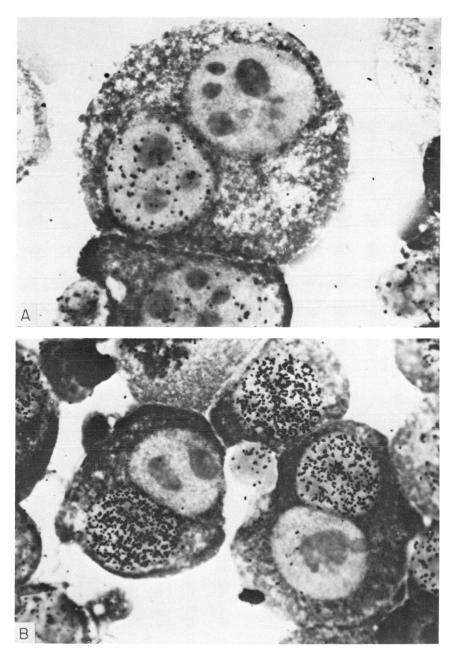

Fig. 6. (A) Heterophasic S/G$_2$ binucleate cell at $t = 0$ hours after fusion. The S nucleus was prelabeled with thymidine-^{3}H. (From Rao and Johnson, 1970a.) (B) Heterophasic S/G$_2$ binucleate cells at $t = 3$ hours after fusion. The S nucleus was prelabeled. The increase in the intensity of labeling of the S nuclei is an indication of continued DNA synthesis in these nuclei. The G$_2$ nuclei remained unlabeled.

101

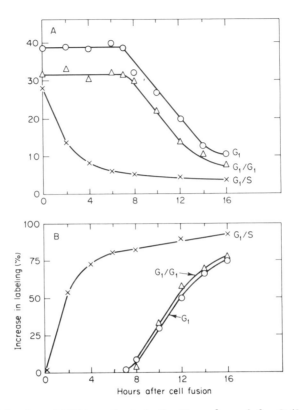

FIG. 7. Induction of DNA synthesis in the G_1 nucleus of the G_1/S fusion. (A) Class frequencies of G_1, G_1/G_1, and G_1/S as a function of time after fusion and continuous incubation with thymidine-^{3}H. The S nuclei were prelabeled. (B) Rate of induction of DNA synthesis in the G_1 nucleus of the G_1/S fused cells. (From Rao and Johnson, 1970a.)

any inhibitors. (3) There are no inhibitors of DNA synthesis in the S nucleus of G_2/S fused cell. (4) The presence of a G_2 component in a G_2/G_1 fused cell does not prevent the initiation and completion of DNA synthesis by the G_1 nucleus (Rao and Johnson, 1970a). This again confirms the absence of any inhibitors of DNA synthesis in the G_2 cells.

B.   Initiation of Mitosis

The effect of cell fusion on the initiation of mitosis was studied by fusing G_2/S^*; G_2^*/G_1 and S^*/G_1 phased populations ($*$ indicates that the nuclei were prelabeled with thymidine-^{3}H). In each of these experiments the fused cells were resuspended in fresh medium containing

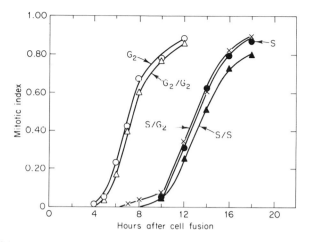

FIG. 8. Mitotic accumulation functions for the mono- and binucleate S and G_2 classes are compared with the heterophasic S/G_2 binucleate class. Note that the pattern of mitotic accumulation in S/G_2 is similar to that of the S parent. (From Rao and Johnson, 1970a.)

Colcemid ($6.74 \times 10^{-7}\ M$), and plated in a number of dishes. At regular intervals dishes were trypsinized, and the cells were processed for microscopic examination. Accumulation of cells in mitosis was plotted as a function of time following fusion. In the G_2/S^* and G_2^*/G_1 fusions the pattern of mitotic accumulation of the heterophasic binucleate cells (i.e.,

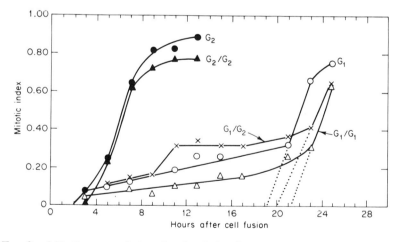

FIG. 9. Mitotic accumulations for the G_1/G_2 fusion, showing the interval between the parental G_1 and G_2 types. The mitotic accumulation function for the heterophasic G_1/G_2 is similar to the G_1 parent. (From Rao and Johnson, 1970a.)

cells containing one labeled nucleus and the other unlabeled) was very similar to that of the slow parent except that the hybrids reached mitosis slightly earlier than the homophasic cells of the same parent (Figs. 8 and 9). However, in the case of the S^*/G_1 fusion the hybrid cells reached mitosis much earlier than the homophasic cells of the G_1 parent (Fig. 10). This is due to the induction of DNA synthesis in the G_1 nuclei of the hybrid cells (Fig. 7). In these G_1 nuclei the pre-DNA synthetic period is essentially eliminated or considerably reduced. Another fact that emerges from these experiments is that the mitotic synchrony in a multinucleate cell is more common than asynchrony. The increase of asynchronous mitoses among the multinucleate cells in G_2/S and G_1/G_2 fusions is approximately 1% whereas it is over 10% among the G_2/S hybrids (Johnson and Rao, 1970). The asynchronous entry of one nucleus into mitosis has a profound effect on the structure of the lagging nucleus of that cell. A set of normal metaphase chromosomes associated with "pulverized chromosomes" was most commonly seen among G_2/S hybrids treated with Colcemid (Fig. 11). Later studies, which we will be discussing in detail in the next section, reveal that there are some factors present in a mitotic cell that can induce this phenomenon in an interphase nucleus.

From the above experiments we may conclude that: (1) In hybrids resulting from the fusion between cells of different phases the component in the earlier phase of the cell cycle becomes a limiting factor in the progression of this hybrid cell to mitosis. Consequently the nucleus of the advanced cell usually does not enter mitosis until the lagging nucleus

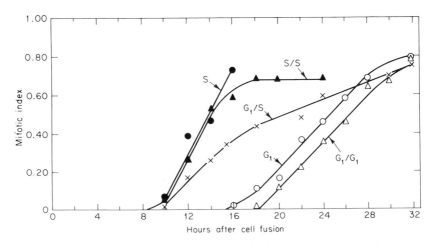

FIG. 10. Mitotic accumulation from the G_1/S fusion, showing the time interval between the parental S and G_1 classes and the intermediate nature of the heterophasic G_1/S binucleate cells. (From Rao and Johnson, 1970a.)

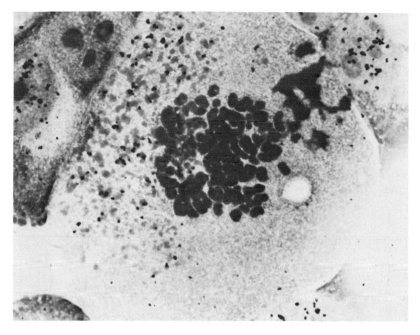

Fig. 11. Premature chromosome condensation in a heterophasic cell formed by the fusion of S and G_2 cells. The S cells were prelabeled with thymidine-³H before cell fusion while the G_2 cells were not. The initiation of mitosis in the G_2 nucleus resulted in the premature condensation of the S chromatin which could be identified by the silver grains in the autoradiograph. (From Johnson and Rao, 1970.)

goes through its normal course at the end of which they enter mitosis synchronously. (2) In the fused cells the G_2 component imparted some advantage with regard to an early onset of mitosis to the G_1 or S nucleus. The mitotic inducing effect of the G_2 cells was even more obvious with an increase in the dosage of G_2 component in the fused cells as shown in Fig. 12 (Rao and Johnson, 1970a). (3) The fact that the nuclei of the multinucleate cells become synchronous within one generation time following fusion suggests that there are some factors operating in this direction. The induction of DNA synthesis and the initiation of mitosis are probably the two main forces operating in order to bring about synchrony in multinucleate systems.

C. Premature Chromosome Condensation (PCC)

The observation of "pulverized chromosomes" in the G_2/S hybrids has led to an experiment in which a random population of HeLa cells were fused with cells blocked in metaphase by Colcemid treatment. The re-

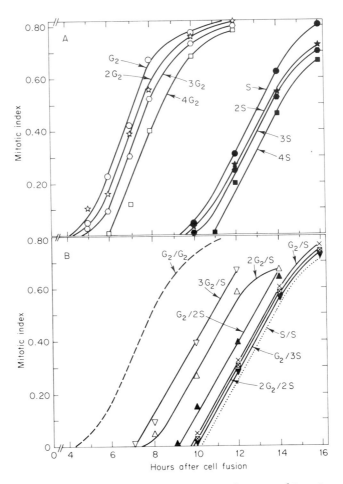

FIG. 12. (A) Mitotic accumulation functions of mono-, bi-, tri-, and tetra-nucleate homophasic cells of the S/G₂ fusion. The S nuclei were prelabeled. The greater the number of nuclei in a cell the slower was the onset of mitosis. (B) Dosage effect of the G_2 or S component on the rate of mitotic accumulation in the heterophasic multinucleate cells of the S/G₂ fusion. The homophasic binucleate G_2/G_2 and S/S are represented by dashed and dotted lines, respectively. (From Rao and Johnson, 1970a.)

sults of fusion between a mitotic and a random population of cells were variable. "Pulverized chromosomes" were observed in some hybrids while in others long slender chromosomes with one or two chromatids were seen (Fig. 13). Sandberg and his associates, working with Chinese hamster cells, made similar observations when they fused interphase cells with those in mitosis (Kato and Sandberg, 1968a,b,c; Takagi et al., 1969;

Sandberg *et al.*, 1970). The phenomenon of chromosome formation from interphase nuclei following fusion between mitotic and interphase cells has been designated as premature chromosome condensation or PCC (Johnson and Rao, 1970). The observation of such a phenomenon in G_2/S and mitotic/interphase fusions suggests that a mitotic inducer is present in either late G_2 or in mitotic cells and that it is capable of pulling a lagging nucleus into mitosis in a dose-dependent manner (Johnson and Rao, 1970).

1. The Structure of Interphase Chromosomes

In the following experiments synchronous populations of G_1, S, and G_2 HeLa cells were fused with mitotic cells to find out the cause of variability in the morphology of the prematurely condensed chromosomes observed in mitotic/interphase fusions. When mitotic cells were fused with a G_1 or G_2 cell there was a rapid condensation of the interphase chromatin into chromosomal patterns. The G_1 chromatin condensed into chromosomes with single chromatids, as might be expected from the unreplicated amount of DNA present and the G_2 chromatin condensed into chromosomes with double chromatids (Fig. 13A,C). The S chromatin, on the other hand, usually condensed less completely to form a patchwork of large and small fragments interspersed with material that was hardly condensed at all (Fig. 13B). It is probable that the replicative state of the S phase chromatin is responsible for the pattern of its condensation.

The morphological changes observed in the induction of PCC in interphase nuclei parallel those seen in normal mitosis. A prophase-like pattern was the first indication of division, and this was rapidly followed by dissolution of the nuclear membrane and the greater degree of condensation of the chromatin into whole or "fragmented" chromosomes according to the position occupied by the affected cell in the life cycle at the time of fusion. There was no sign that the induction of chromosome condensation in these cells was accompanied by the appearance of a mitotic spindle. The ability of a mitotic cell to induce PCC in an interphase nucleus depends largely on the ratio of mitotic to interphase nuclei in the cell at the time of fusion (Johnson and Rao, 1970). The G_1 phase nuclei were most readily induced into PCC, whereas S and G_2 nuclei were not so easily induced.

Attempts have also been made to test whether the PCC-inducing factors present in mitotic HeLa cells would produce similar effects in tissue culture cells from other animal species following fusion. The induction of PCC in the cells of the Chinese hamster, *Xenopus laevis*, chick fibroblasts, and mosquito following fusion with mitotic HeLa cells indicates

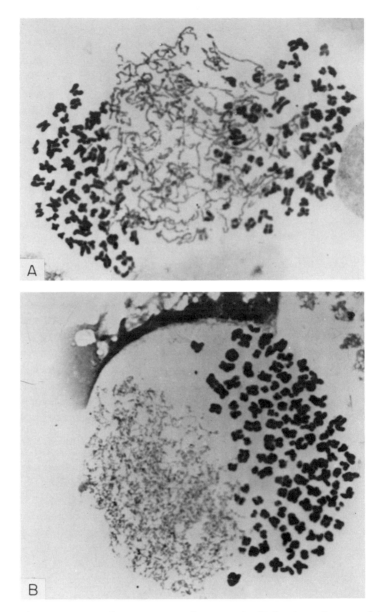

FIG. 13.   Premature chromosome condensation in HeLa interphase nuclei. (A) PCC of the G_1 nucleus in a heterophasic M/G_1 cell at 30 minutes after the addition of virus. Note the long, slender G_1 chromosomes with single chromatids. The more condensed chromosomes are from the mitotic cell blocked by Colcemid. (B) Premature condensation of the S chromatin at 45 minutes after the fusion between mitotic and S phase cells. The chromatin of the S nucleus presents a fragmented appear-

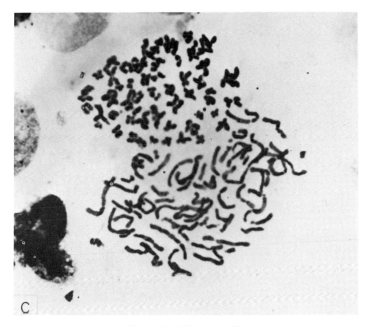

FIG. 13 (*Continued*)
ance. (C) PCC of the G_2 nucleus in an M/G_2 heterophasic cell. The more condensed chromosomes are from the mitotic cell treated with Colcemid for 19 hours. (From Johnson and Rao, 1970.)

that the PCC-inducing factors are common over a wide range of animal species (Johnson *et al.*, 1970). As in HeLa cells the structure of the PCC in these cell types varied according to the position of the affected cell in the cell cycle (Fig. 14).

2. THE FATE OF PCC

What will happen to the prematurely condensed chromosomes immediately following fusion? From a long range point of view, will they be retained in the hybrid cell or will they be lost? HeLa cells which were reversibly blocked in metaphase by nitrous oxide treatment were fused with G_1, S, or G_2 cells. During the first 5 hours after fusion, samples were taken at hourly intervals, fixed, and processed to study the organization of PCC when a functional mitotic spindle was present in the same cell. The PCC of the G_1 or G_2 types usually remained as a separate entity and did not appear to be associated with the mitotic spindle (Fig. 15). The PCC of the S type, on the other hand, appeared to be highly scattered around the mitotic chromosomes (Fig. 16A). However, the distribution of the prematurely condensed chromosomes between the daughter cells

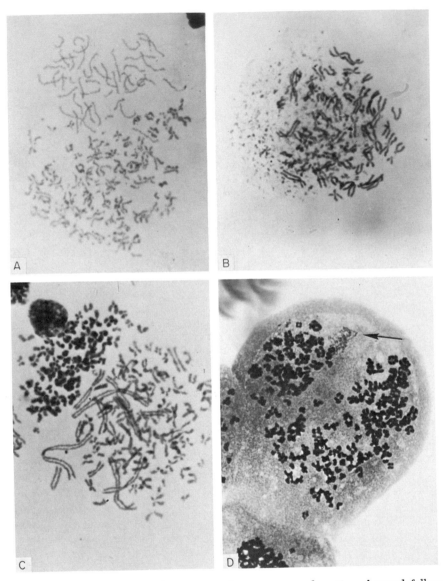

Fɪɢ. 14. Various types of premature chromosome condensation observed following fusion of tissue culture cells with mitotic HeLa cells. (A) Induction of PCC of
the *Xenopus* nucleus in the HeLa/*Xenopus* heterokaryon. Note the single stranded
G_1 chromosomes of the *Xenopus* as compared to the metaphase chromosomes of the
HeLa cell. In this preparation, 43 *Xenopus* chromosomes can be counted. The modal
chromosome number for the A6 *Xenopus* cell line is 46. (B) PCC of the S type of
the avian nucleus in a HeLa/embryonic chick fibroblast heterokaryon. (C) A HeLa/
Chinese hamster ovary cell heterokaryon showing the G_2 type of PCC of the CHO
nucleus. The metaphase chromosomes of the HeLa cell are more condensed than the
G_2 chromosomes of the CHO cell. (D) Induction of the G_1 type of PCC of the
mosquito nucleus (shown by arrow) in a HeLa/mosquito heterokaryon. (From
Johnson *et al.*, 1970.)

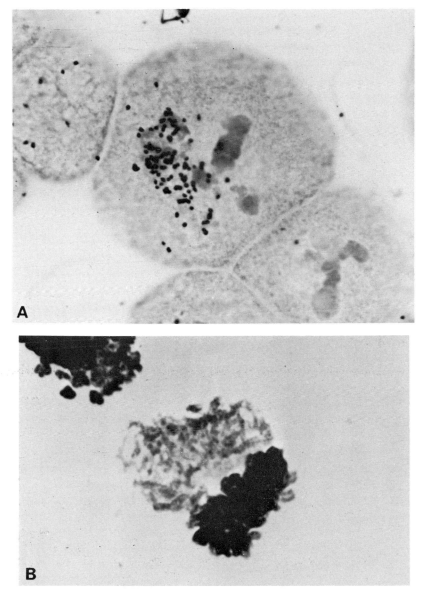

Fɪɢ. 15. Fate of PCC in HeLa cells immediately following fusion between an interphase cell and a mitotic cell capable of completing mitosis. (A) A radioautograph of a cell formed by the fusion of a mitotic cell with a G_1 cell which was prelabeled with thymidine-^{3}H. The unlabeled chromosomes of the mitotic cell are arranged in a discrete metaphase plate. The PCC which are covered with silver grains are not intimately associated with the metaphase plate. (B) A M/G_1 cell where the mitotic chromosomes form a metaphase plate while the lightly stained G_1 chromosomes (PCC) remain as a separate group.

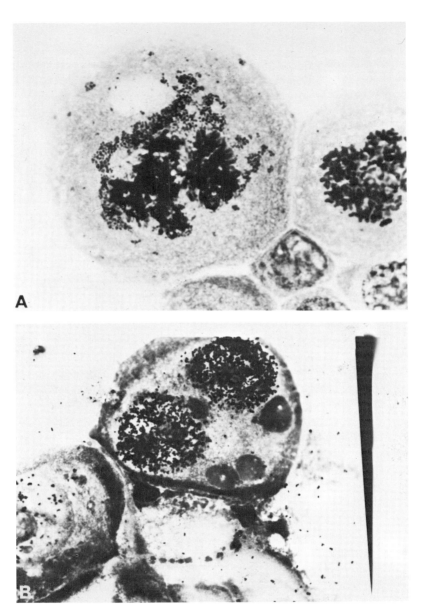

FIG. 16. M/S fused HeLa cells. S phase cells prelabeled with thymidine-^{3}H were fused with mitotic cells reversably blocked with N_2O. (A) In this cell the mitotic chromosomes are completing anaphase while the PCC of the S type are scattered around them. The PCC will probably be distributed randomly between the daughter nuclei. (B) A radioautograph of an M/S fused cell in which there was no induction of PCC of the S nuclei. The mitotic chromosomes were transformed into micronuclei under the influence of the interphase component. The presence of thymidine-^{3}H in the growth medium following fusion resulted in an increase in the number of grains on the S nuclei.

during the first mitosis after fusion is a random process. In mitotic/interphase cells, where the interphase nucleus failed to undergo PCC induction, the mitotic chromosomes were transformed into micronuclei under the influence of the interphase component (Fig. 16B).

The long-range effects of the PCC induction were studied in the auxotrophic mutants of the Chinese hamster cells (Rao and Johnson, 1972). These experiments were designed to study whether there is any difference in the rate of survival of hybrids between homophasic and heterophasic fusions involving two glycine-requiring mutants, glycine-A and glycine-B. The technique of selective detachment of the mitotic cells was employed for obtaining synchronous populations. The mitotic cells were harvested from a monolayer culture at the end of a 2-hour Colcemid (0.05 μg/ml) treatment, centrifuged, resuspended in fresh medium, and plated in plastic dishes. G_1, S, and G_2 populations were obtained by trypsinizing the cells at appropriate times after reversal of the Colcemid block. The protocol for various fusions is given in Table II. Twelve different fusions were made and twenty dishes were plated for each fusion with 1000 cells per dish. The colonies per dish were counted after 7 days and their average values for plating efficiency were expressed as the number of colonies that survived for every 1000 cells of each type plated (Table III). The absence of any colonies in the control where no virus was added to the mixture of the parental cell types indicated that there was no detectable amount of spontaneous fusion and that the selective technique was working perfectly. The rate of survival among the different homophasic fusions was very much alike but it was only about 30% of the plating efficiency of the fusion between the two random populations. This low plating efficiency among homophasic fusions is mainly due to the Colcemid treatment used for obtaining synchronized populations. The plating efficiencies of such populations obtained by the reversal of a 2-hour Colcemid block were significantly lower than that of untreated controls (Kato and Yosida, 1970). In general, the plating efficiencies are lower among the heterophasic fusions as compared to those of homophasic ones. Even among the heterophasic fusions the chances of recovering hybrids are better if one of the parental cell types was not in mitosis at the time of fusion. In other words, the induction of PCC in the fused cells results in lower plating efficiencies. This is generally what we would expect on the basis of the distribution of PCC we observed during mitosis immediately following fusion (Figs. 15 and 16). On the basis of these findings, we also expect that the rate of survival of hybrids in M/G_1 and M/G_2 fusions would be greater than that of M/S fusion. Indeed this has been borne out by the data presented in Table III.

We may summarize the various facts with regard to the survival of

TABLE II

PROTOCOL FOR THE COLLECTION OF G_1, S, G_2, AND M CELLS FROM GLY-A AND GLY-B MUTANTS FOR FUSION[a]

Phase in which cells are to be collected	Hours								
	0	2	4	6	7	9	11	13	15
Mutant-Glycine A									
Dish 1, 2[b] G_1	Add Colcemid	Collect M cells, reverse Colcemid block and plate[c]	Trypsinize and collect G_1 cells						
Dish 3, 4 S	Add Colcemid	Collect M cells, reverse Colcemid block and plate	↑	↑	↑	↑	Trypsinize and collect S cells		
Dish 5, 6, 7 G_2	Add Colcemid	Collect M cells, reverse Colcemid block and plate	↑	↑	↑	↑	↑	↑	Trypsinize and collect G_2 cells
Dish 8 M	↑	Add Colcemid	Collect mitotic cells and reverse Colcemid block						
Mutant-Glycine B									
Dish 1 G_1	Add Colcemid	Collect M cells, reverse Colcemid block and plate	Trypsinize and collect G_1 cells						
Dish 2 S	Add Colcemid	Collect M cells, reverse Colcemid block and plate	↑	↑	↑	↑	Trypsinize and collect S cells		

Dish	Phase							
Dish 3	G2	Add Colcemid	Collect M cells, reverse Colcemid block and plate; Add Colcemid	↑	↑	—	↑	↑ → Trypsinize and collect G2 cells
Dish 4	M	↑	Collect M cells, reverse Colcemid block and plate; Add Colcemid	↑	Add Colcemid	↑	↑	↑
Dish 5	G1	↑	↑	Collect mitotic cells and reverse Colcemid block	↑	↑	Trypsinize and collect G1 cells	
Dish 6	M	↑	↑	↑	Collect M cells and reverse Colcemid block; Add Colcemid	↑	Collect M cells, reverse Colcemid block and plate	
Dish 7	G1	↑	↑	↑	↑	↑	Collect M cells, reverse Colcemid block and plate	↑ → Trypsinize and collect G1 cells
Dish 8	S	↑	Collect M cells, reverse Colcemid block and plate; Add Colcemid	↑	↑	↑	↑	↑ → Trypsinize and collect S cells
Dish 9	M	↑	↑	↑	↑	↑	Add Colcemid	Collect M cells, reverse Colcemid block

Fuse (at first fusion point):
(A) M^a/M^b
(B) M^b/G_1^a
(C) G_1^a/G_1^b

Fuse (at second fusion point):
(D) S^a/S^b
(E) S^a/G_1^b
(F) S^a/M^b

Fuse (at final fusion point):
(G) G_2^a/G_2^b
(H) G_2^a/G_1^b
(I) G_2^a/S^b
(J) G_2^a/M^b

[a] From Rao and Johnson, 1972.

[b] 7×10^5 cells were plated in 100-mm plastic plates at 44 hours prior to the commencement of the experiment.

[c] Following the reversal of Colcemid block the cells were resuspended in 2 ml of medium and plated in 30-mm plastic dishes.

TABLE III

RATE OF SURVIVAL OF HYBRIDS IN HOMO- AND HETEROPHASIC FUSIONS BETWEEN
SYNCHRONIZED POPULATIONS OF GLY-A AND GLY-B MUTANTS

Type of fusion[a]	Number of colonies per 1000 cells of each parent
1 R^a/R^b	15.6
2 R_c^a/R_c^b	5.0
3 $R^a + R^b$ (w/o virus)	0
Homophasic fusions	
4 M^a/M^b	4.0
5 G_1^a/G_1^b	5.5
6 S^a/S^b	4.6
7 G_2^a/G_2^b	4.4
Heterophasic fusions	
8 M^a/G_1^b	2.0
9 M^a/S^b	0.62
10 M^a/G_2^b	1.56
11 S^a/G_1^b	3.2
12 S^a/G_2^b	2.7
13 G_1^a/G_2^b	2.5

[a] R = Random population; a = parent Gly-A; b = parent Gly-B; c = exposed to Colcemid (0.05 µg/ml) for 2 hours prior to fusion. (From Rao and Johnson, 1972.)

PCC as follows: (1) Induction of PCC reduces plating efficiency among the hybrid cells. (2) A hybrid cell with the PCC of G_1 or G_2 type has a better chance of survival than that with a S type of PCC. The rate of retention of G_1 or G_2 types of PCC by the hybrid cells seems to be significantly higher than that of S as evidenced by the higher plating efficiencies for the M/G_1 and M/G_2 fusions. (3) From the two preceding statements it is obvious that the interphase chromatin that had undergone premature condensation has not only been retained by the hybrid cell but also that it has not lost its genetic activity. (4) And last, the phenomenon of PCC induction may play a role in the elimination of chromosomes commonly observed among the human/mouse (Weiss and Green, 1967; Weiss *et al.*, 1968; Matsuya *et al.*, 1968; Migeon and Miller, 1968; Matsuya and Green, 1969) or human/hamster (Kao and Puck, 1970) heterokaryons.

3. VISUALIZATION OF CHROMOSOMES FROM DIFFERENTIATED CELLS

The technique of inducing premature chromosome condensation, for the first time, made it possible to see the structure of chromosomes of the interphase cells that are not ready to enter into mitosis. When such a technique is available it is natural to look for chromosomes of the differentiated cells some of which will never go into mitosis. In the following experiments we attempted to observe the structure of chromosomes in differentiated cells derived from a variety of animal species. HeLa cells blocked in mitosis with Colcemid were fused separately with chick erythrocytes, horse lymphocytes, and bovine sperm. The procedures for the preparation of these cells for fusion were given in a previous publication (Johnson et al., 1970). In all these fusions there was induction of PCC, but its frequency varied greatly depending upon the fusing ability of the cell type. The PCC of these three types of differentiated cells exhibited a chromosome structure similar to that of G_1 chromosomes (Fig. 17). This is indicative of the fact that these cells are held in the pre-DNA synthetic period due to differentiation. This method provides a new way of determining the phase of the cell cycle during which differentiation had occurred in a given cell type.

4. FACTORS THAT INFLUENCE PCC INDUCTION

The fact that PCC results from the fusion of an interphase cell with a mitotic cell and that the intensity of the effect depends on the number of mitotic cells present in the hybrid suggests that there are inducing factors present in the mitotic cell. Attempts have been made to induce PCC in a random population of HeLa cells by the addition of mitotic extract, but to no avail. Therefore, the effect of mitotic extract has been tested in a mitotic–interphase fused cell system (MIFCS) individually and also in combination with other chemicals that have been found to promote PCC. The procedures for the preparation of mitotic extract have been described in a separate publication (Rao and Johnson, 1971). The data shown in Fig. 18 indicate that the mitotic extract has the property of promoting PCC in this system.

Hybrid HeLa cells formed by the fusion of mitotic with interphase cells have been used as a test system to study the effects of various positively and negatively charged compounds on the induction of PCC of the interphase nuclei (Table IV). Among the various positively charged agents tested, spermine, putrescine, and Mg^{2+} were specific in promoting the PCC induction, while spermidine was unique in inhibiting this event. All the negatively charged compounds including estradiol-17β were uniformly inhibitory (Rao and Johnson, 1970b). The administration of cations to both cells and isolated nuclei results in a heterogeneous condensation of chromatin (Anderson and Norris, 1960; Philpot and Stanier, 1957; Robbins et al., 1970). In general, the condensation produced by cations does not lead to the formation of discrete chromosomes in interphase nuclei, although Anderson and Norris (1960) have reported that

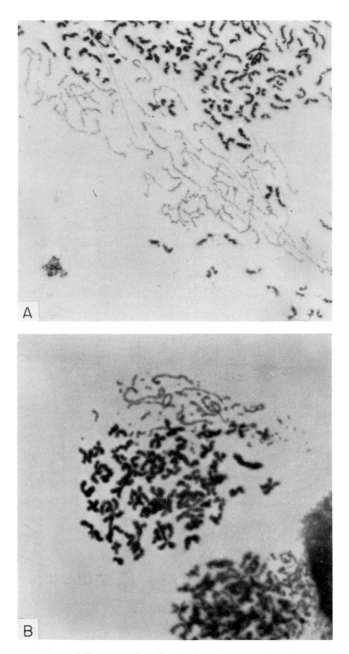

FIG. 17. PCC in differentiated cells. (A) Induction of PCC in a HeLa/horse lymphocyte heterokaryon. Note the single-stranded G_1 chromosomes of the horse lymphocyte in comparison with the metaphase HeLa. (B) G_1 type of PCC of the embryonic chick erythrocyte nucleus in a HeLa/chick RBC heterokaryon. (C) HeLa/

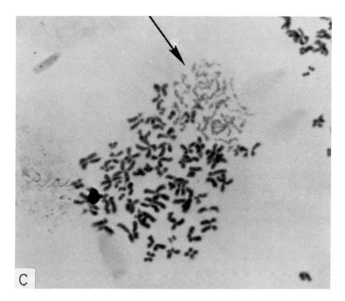

FIG. 17 (*Continued*)

bull's sperm heterokaryon showing the induction of PCC of the sperm nucleus. The chromosomes from the sperm nucleus (shown by the arrow) are of the G_1 type and are approximately 30 in number (Johnson *et al.*, 1970).

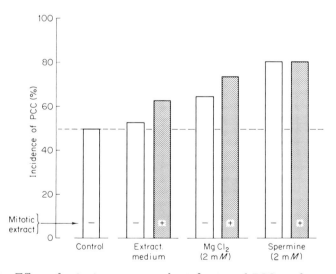

FIG. 18. Effect of mitotic extract on the induction of PCC in the mitotic-interphase fused cells. The presence or absence of the mitotic extract in the fusion mixture is indicated by + or −. The compound listed below each pair of columns indicates its presence in the fusion mixture. Extract medium = medium used for the homogenization of mitotic cells. (Rao and Johnson, 1971.)

TABLE IV

EFFECT OF POLYAMINES, DIVALENT CATIONS AND OTHER AGENTS ON THE INDUCTION
OF PCC IN THE MITOTIC-INTERPHASE FUSED CELL SYSTEM (MIFCS)

| | | Incidence of PCC (%) among hybrid cells[a] | | | |
| | | Binucleate | | Trinucleate | |
Treatment	Dose	1M:1I	%PCC relative to control	2M:1I	1M:2I
Control		50.0	100.0	79.4	12.2
Putrescine	$2 \times 10^{-3} M$	67.0	134.0	80.0	33.4
Spermine	$2 \times 10^{-3} M$	81.0	162.0	100.0	76.0
Cadaverine	$2 \times 10^{-3} M$	47.5	95.0	76.2	0
Spermidine	$2 \times 10^{-3} M$	14.0	28.0	14.4	0
MgCl$_2$	$2 \times 10^{-3} M$	65.0	130.0	95.0	40.0
CaCl$_2$	$2 \times 10^{-3} M$	42.5	85.0	70.0	45.0
Estradiol-17β	$8 \times 10^{-5} M$	14.0	28.0	14.4	0
Na$_2$HPO$_4$	$2 \times 10^{-2} M$	15.0	30.0	31.0	0
EDTA	0.1%	19.7	39.4	22.6	0
Cyclic AMP	$2 \times 10^{-3} M$	22.4	44.8	29.0	0
Heparin	100 units/ml	26.4	52.8	45.5	0

[a] M = Mitosis; I = interphase. (From Rao and Johnson, 1971.)

the number of condensed bodies may be correlated with the number of chromosomes known to exist in certain cells.

The striking differences in the effects of the four polyamines on the PCC-inducing system are probably explicable in terms of the degree of stereospecific binding to chromatin. In our system only putrescine and spermine are stimulatory, while spermidine is strongly inhibitory, as is cadaverine at $3 \times 10^{-3} M$ concentration. This suggests that only putrescine and spermine can bind to chromatin fibrils so as to form a discrete condensed chromosome. This is supported by the observation that the distance between the terminal primary amines of spermine is such that the molecule could bind to DNA easily. Recent X-ray diffraction work of Suwalsky *et al.* (1969) has shown that spermine binds to DNA highly stereospecifically. The bond distances in spermidine, however, cannot be made to fit so easily with the phosphate-bound oxygens of DNA, and our observations show that when spermidine is added to the PCC-inducing system, chromatin aggregation occurs but discrete, visible chromosomes are not produced.

Chromosomes have been shown to contain bound divalent metal cations (Steffenson, 1960, 1961; Cantor and Hearst, 1970) but their role in initiating or maintaining chromosome structure is still unknown. Our results indicate that magnesium rather than calcium is important for the organization of chromatin into prematurely condensed chromosomes. Some of our latest experiments indicate that proteins are also involved in the formation of visible chromosomes. On the basis of these observations we may conclude that the condensation of chromatin into chromosomes is initiated and maintained by Mg^{2+}, spermine, and some chromosome-specific proteins. Further work is necessary to clarify which proteins are involved and in what sequence these molecules interact.

X. Conclusions

On the basis of the data presented in this chapter we can draw some broad conclusions with regard to the regulation of DNA synthesis and initiation of mitosis in mammalian cells.

(1) As soon as a cell completes mitosis and enters G_1 phase, its DNA is in a state ready for replication. However, the replication does not commence until the necessary enzymes and other precursors become available. Once the replication of DNA in the chromosomes is completed it is not possible to initiate it again unless the daughter chromosomes are separated during mitosis.

(2) The inducers of DNA synthesis may have their origins in the nucleus but they are also present in the cytoplasm. As in the case of G_1/S fused cells these inducers from the S component can enter into the G_1 nucleus and initiate DNA synthesis. Further studies are necessary to determine the nature of these inducers.

(3) As with DNA synthesis, mitosis is also inducible. The fusion between mitotic and interphase cells would result in the induction of premature chromosome condensation of the interphase nucleus. The phenomenon of PCC appears very similar to the initiation of mitosis particularly with reference to condensation of chromatin into chromosomes and the dissolution of the nuclear membrane. In normal mitosis the dissolution of the nuclear envelope coincides with the appearance of the mitotic spindle. However, during PCC induction these two events are uncoupled suggesting that the formation of mitotic spindle is a separate event and the mechanisms necessary for chromosome condensation are different from those required for spindle development.

(4) The PCC inducers present in a mitotic HeLa cell are effective on cells from a variety of animal species. On this basis it is not unrealistic to think that the mitotic inducers, like PCC inducers, may be similar in a variety of cell types ranging from mosquito to man.

(5) The chemical nature of the mitotic inducers is yet to be ascertained. However, the data available at present indicate that positively charged agents like polyamines and divalent cations—spermine and Mg^{2+}, in particular—may play an important role in the initiation of mitosis. There are also indications that some chromosome-specific proteins may also be involved in the condensation of chromatin into discrete chromosomes.

ACKNOWLEDGMENTS

We are grateful to Dr. Theodore T. Puck for his critical reading of the manuscript and to Dr. C. F. Graham of Oxford University for allowing us to quote from his (then) unpublished manuscript on blastomere fusion. Our thanks are also due George Barela and David Peakman for their technical assistance. This is a joint contribution from the Department of Biophysics and Genetics (contribution No. 453), University of Colorado Medical Center, Denver, Colorado and the Department of Zoology, University of Cambridge, Cambridge, England.

REFERENCES

Anderson, N. G., and Norris, C. B. (1960). *Exp. Cell Res.* **19**, 605.

Ayad, S. R., and Fox, M. (1968). *Nature (London)* **220**, 35.

Bachtold, J. G., Bubel, H. C., and Gebhardt, L. P. (1957). *Virology* **4**, 582.

Baranska, W., and Koprowski, H. (1970). *J. Exp. Zool.* **174**, 1.

Barski, G., Sorieul, S., and Cornefert, F. (1960). *C. R. Acad. Sci. (Paris)* **251**, 1825.

Barski, G., Sorieul, S., and Cornefert, F. (1961). *J. Nat. Cancer Inst.* **26**, 1269.

Bendich, A., Vizoso, A. D., and Harris, R. G. (1967). *Proc. Nat. Acad. Sci. U. S.* **57**, 1029.

Bolund, L., Ringertz, N. R., and Harris, H. (1969). *J. Cell Res.* **4**, 71.

Boone, C. M., and Ruddle, F. H. (1969). *Wistar Inst. Symp. Monogr.* **9**, 77.

Bregula, U., Klein, G., and Harris, H. (1971). *J. Cell Sci.* **8**, 673.

Brown, A. (1963). *Virology* **21**, 362.

Buckley, S. M. (1969). *Proc. Soc. Exp. Biol. Med.* **131**, 625.

Cantor, K. P., and Hearst, J. E. (1970). *J. Mol. Biol.* **49**, 213.

Carlsson, S. A., Savage, R. E., and Ringertz, N. R. (1970). *Nature (London)* **228**, 869.

Carriere, R. (1969). *Int. Rev. Cytol.* **25**, 201.

Cohen, A. (1963). *In* "Mechanisms of Virus Infection" (W. Smith, ed.), pp. 153–190. Academic Press, New York.

Cone, C. D. (1969). *J. Theoret. Biol.* **22**, 365.

Coon, H. G. (1967). *J. Cell Biol.* **35**, 27a.

Coon, H. G., and Weiss, M. C. (1969). *Proc. Nat. Acad. Sci. U. S.* **62**, 852.

Davidson, R. (1969a). *Exp. Cell Res.* **55**, 424.

Davidson, R. (1969b). *Wistar Inst. Symp. Monogr.* **9**, 97.

Davidson, R. L., and Ephrussi, B. (1965). *Nature (London)* **205**, 1170.

Davidson, R. L., and Ephrussi, B. (1970). *Exp. Cell Res.* **61**, 222.

Defendi, V., Ephrussi, B., and Koprowski, H. (1964). *Nature (London)* **203**, 495.

De Terra, N. (1967). *Proc. Nat. Acad. Sci. U. S.* **57**, 607.

Dulbecco, R., and Vogt, M. (1954). *J. Exp. Med.* **99**, 167.

Elsbach, P., Homes, K. V., and Choppin, P. W. (1969). *Proc. Soc. Exp. Biol. Med.* **130**, 903.

Enders, J. F., and Peebles, T. C. (1954). *Proc. Soc. Exp. Biol. Med.* **86**, 277.

Ephrussi, B. (1965). *In* "Developmental and Metabolic Control Mechanisms and Neoplasia: 19th Annual Symposium Fundamental Cancer Research," pp. 486–502. Williams & Wilkins, Baltimore, Maryland.

Ephrussi, B. (1970). *In* "Genetic Concepts and Neoplasia," pp. 9–28. Williams & Wilkins, Baltimore, Maryland.

Ephrussi, B., and Sorieul, S. (1962). *In* "Approaches to the Genetic Analysis of Mammalian Cells" (D. J. Merchant and J. B. Neal, eds.), pp. 81–97. University of Michigan Press, Ann Arbor, Michigan.

Ephrussi, B., and Weiss, M. C. (1965). *Proc. Nat. Acad. Sci. U. S.* **53**, 1040.

Ephrussi, B., Scaletta, L. J., Stenchever, M. A., and Yoshida, M. C. (1964). *In* "Cytogenetics of Cells in Culture" (R. J. C. Harris, ed.), pp. 13–25. Academic Press, New York.

Fukai, K., Okada, V., and Suzuki, T. (1955). *Med. J. Osaka Univ.* **6**, 17.

Furshpan, E. J., and Potter, D. D. (1968). *In* "Current Topics of Developmental Biology" (A. A. Moscona and A. Monroy, eds.), Vol. III, pp. 95–127. Academic Press, New York.

Gershon, D., and Sachs, L. (1963). *Nature (London)* **198**, 912.

Gharpure, M. A., Wright, P. F., and Chanock, R. M. (1969). *J. Virol.* **3**, 414.

Goldstein, L., and Prescott, D. M. (1967). *In* "The Control of Nuclear Activity" (L. Goldstein, ed.), pp. 3–17. Prentice-Hall, New Jersey.

Graham, C. F. (1966). *J. Cell Sci.* **1**, 363.

Graham, C. F. (1969). *Wistar Inst. Symp. Monogr.* **9**, 19.

Graham, C. F. (1971). *In* "Karolinska Symposia No. 3—*In vitro* Methods in Cell Biology," in press.

Graham, C. F., Arms, K., and Gurdon, J. B. (1966). *Develop. Biol.* **14**, 349.

Gurdon, J. B. (1967). *In* "Ciba Foundation Symposium" (A. V. S. de Reuck and J. Knight, eds.), pp. 65–78. Little Brown, Boston, Massachusetts.

Ham, R. G. (1965). *Proc. Nat. Acad. Sci. U. S.* **53**, 288.

Harris, H. (1965). *Nature (London)* **206**, 583.

Harris, H. (1967). *J. Cell Sci.* **2**, 23.

Harris, H. (1968). "Nucleus and Cytoplasm," 2nd Ed. Clarendon Press, Oxford.

Harris, H. (1970). "Cell Fusion: The Dunham Lectures," Harvard Univ. Press, Cambridge, Massachusetts.

Harris, H., Sidebottom, E., Grace, D. M., and Bramwell, M. E. (1969). *J. Cell Sci.* **4**, 499.

Harris, H., and Watkins, J. F. (1965). *Nature (London)* **205**, 640.

Harris, H., Watkins, J. F., Ford, C. E., and Shoefl, G. I. (1966). *J. Cell Sci.* **1**, 1.

Henle, G., Deinhardt, F., and Girardi, A. (1954). *Proc. Soc. Exp. Biol. Med.* **87**, 386.

Hill, M. (1966). *Exp. Cell Res.* **41**, 253.

Hill, M., and Spurna, V. (1968). *Exp. Cell Res.* **50**, 208.

Hosaka, Y., and Koshi, Y. (1968). *Virology* **34**, 419.

Howe, C., and Morgan, C. (1969). *J. Virol.* **3**, 70.

Hoyle, L. (1962). *Cold Spring Harbor Symp. Quant. Biol.* **27**, 113.

Jacobson, C. O. (1968). *Exp. Cell Res.* **53**, 316.

Johnson, C. F., and Scott, A. D. (1964). *Proc. Soc. Exp. Biol. Med.* **115**, 281.

Johnson, R. T., and Rao, P. N. (1970). *Nature (London)* **226**, 717.

Johnson, R. T., and Rao, P. N. (1971). *Biol. Rev.* **46**, 97.

Johnson, R. T., Rao, P. N., and Hughes, S. D. (1970). *J. Cell. Physiol.* **76**, 151.

Jungers, V. (1931). *Cellule* **40**, 298.

Kao, F. T., and Puck, T. T. (1967). *Genetics* **55**, 513.

Kao, F. T., and Puck, T. T. (1968). *Proc. Nat. Acad. Sci. U. S.* **60**, 1275.

Kao, F. T., and Puck, T. T. (1970). *Nature (London)* **228**, 329.

Kao, F. T., Chasin, L., and Puck, T. T. (1969a). *Proc. Nat. Acad. Sci. U. S.* **69**, 1284.

Kao, F. T., Johnson, R. T., and Puck, T. T. (1969b). *Science* **164**, 312.

Kato, H., and Sandberg, A. A. (1968a). *J. Nat. Cancer Inst.* **40**, 165.

Kato, H., and Sandberg, A. A. (1968b). *J. Nat. Cancer Inst.* **41**, 1117.

Kato, H., and Sandberg, A. A. (1968c). *J. Nat. Cancer Inst.* **41**, 1125.

Kato, H., and Yosida, T. H. (1970). *Exp. Cell Res.* **60**, 459.

Klebe, R. J., Chen, T.-R., and Ruddle, F. H. (1970a). *J. Cell Biol.* **45**, 74.

Klebe, R. J., Chen, T.-R., and Ruddle, F. H. (1970b). *Proc. Nat. Acad. Sci. U. S.* **66**, 1220.

Klein, G., Bregula, U., Weiner, F., and Harris, H. (1971). *J. Cell Sci.* **8**, 659.

Kohn, A. (1965). *Virology* **26**, 228.

Kohn, A., and Klibansky, C. (1967). *Virology* **31**, 385.

Krooth, R. S. (1964). *Cold Spring Harbor Symp. Quant. Biol.* **29**, 189.

Krooth, R. S., and Weinberg, A. N. (1960). *Biochem. Biophys. Res. Commun.* 3, 518.

Lawrence, J., Scaletta, L. J., and Ephrussi, B. (1965). *Nature (London)* **205**, 1169.

Ledoux, L. (1965). *Progr. Nucleic Acid Res. Molec. Biol.* **4**, 231–267.

Lepine, P., Chany, C., Droz, B., and Robbe-Fossat, F. (1959). *Ann. N. Y. Acad. Sci.* **81**, 62.

Levine, S., and Sagik, B. P. (1956). *Virology* **2**, 57.

Lewis, M. R. (1927). *Amer. Rev. Tuberc.* **15**, 616.

Littlefield, J. W. (1963). *Proc. Nat. Acad. Sci. U. S.* **50**, 568.

Littlefield, J. W. (1964a). *Nature (London)* **203**, 1142.

Littlefield, J. W. (1964b). *Science* **145**, 709.

Littlefield, J. W. (1966). *Exp. Cell Res.* **41**, 190.

Loewenstein, W. R. (1966). *Ann. N. Y. Acad. Sci.* **137**, 441.

Lucy, J. A. (1970). *Nature (London)* **227**, 815.

Lung, B. (1968). *J. Ultrastr. Res.* **22**, 485.

Marston, R. Q. (1958). *Proc. Soc. Exp. Biol. Med.* **98**, 853.

Matsuya, Y., and Green, H. (1969). *Science* **163**, 697.

Matsuya, Y., Green, H., and Basilico, C. (1968). *Nature (London)* **220**, 1199.

Migeon, B. R. (1968). *Biochem. Genet.* **1**, 305.

Migeon, B. R., and Childs, B. (1970). *In* "Progress in Medical Genetics" (A. G. Steinberg and A. G. Bearn, eds.), Vol. VII, pp. 1–28. Grune & Stratton, New York.

Migeon, B. R., and Miller, C. S. (1968). *Science* **162**, 1005.

Miggiano, V., Nabholz, M., and Bodmer, W. (1969). *Wistar Inst. Symp. Monogr.* **9**, 61.

Morris, J. A., Blount, R. F., Jr., and Savage, R. E. (1956). *Proc. Soc. Exp. Biol. Med.* **92**, 544.

Mukherjee, A. B., Dev, V. G., and Miller, O. J. (1970). *J. Cell Biol.* **47**, 146a.

Murayama, F., and Okada, Y. (1965). *Biken J.* **8**, 103.

Neff, J. M., and Enders, J. F. (1968). *Proc. Soc. Exp. Biol. Med.* **127**, 260.

Neurath, A. R. (1963). *Acta Virol.* **7**, 490.

Neurath, A. R. (1964). *Acta Virol.* **8**, 143.
Neurath, A. R., and Sokol, F. (1962). *Acta Virol.* **6**, 66.
Nichols, W. W. (1970). *Ann. Rev. Microbiol.* **24**, 479.
Oftebro, R., and Wolf, I. (1967). *Exp. Cell Res.* **48**, 39.
Okada, Y. (1958). *Biken J.* **1**, 103.
Okada, Y. (1962a). *Exp. Cell Res.* **26**, 98.
Okada, Y. (1962b). *Exp. Cell Res.* **26**, 119.
Okada, Y., and Hosokawa, Y. (1961). *Biken J.* **4**, 217.
Okada, Y., and Murayama, F. (1965a). *Biken J.* **8**, 103.
Okada, Y., and Murayama, F. (1965b). *Exp. Cell Res.* **40**, 154.
Okada, Y., and Tadokoro, J. (1962). *Exp. Cell Res.* **26**, 108.
Okada, Y., Suzuki, T., and Hosaka, Y. (1957). *Med. J. Osaka Univ.* **7**, 709.
Okada, Y., Murayama, F., and Yamada, K. (1966). *Virology* **28**, 115.
O'Lague, P., Dalen, H., Rubin, H., and Tobias, C. (1970). *Science* **170**, 464.
Ord, M. J. (1969). *Nature (London)* **221**, 964.
Philpot, J. St. L., and Stanier, J. E. (1957). *Nature (London)* **179**, 102.
Poole, A. R., Howell, J. I., and Lucy, J. A. (1970). *Nature (London)* **227**, 810.
Power, J. B., Cummins, S. E., and Cocking, E. C. (1970). *Nature (London)* **225**, 1016.
Prescott, D. M., and Goldstein, L. (1967). *Science* **155**, 469.
Racadot, J., and Frederic, J. (1955). *C. R. Acad. Sci. (Paris)* **240**, 563.
Rao, P. N. (1968). *Science* **160**, 774.
Rao, P. N., and Engleberg, J. (1965). *Science* **148**, 1092.
Rao, P. N., and Engleberg, J. (1966). *In* "Cell Synchrony: Studies in Biosynthetic Regulation" (I. L. Cameron, and G. M. Padilla, eds.), pp. 332–352. Academic Press, New York.
Rao, P. N., and Johnson, R. T. (1970a). *Nature (London)* **225**, 159.
Rao, P. N., and Johnson, R. T. (1970b). *J. Cell Biol.* **47**, 167a.
Rao, P. N., and Johnson, R. T. (1971). *J. Cell Physiol.* **78**, 217.
Rao, P. N., and Johnson, R. T. (1972). *J. Cell Sci.* (in press).
Rebel, G., Fontanges, R., and Colobert, L. (1962). *Ann. Inst. Pasteur (Paris)* **102**, 137.
Robbins, A. B., and Taylor, D. M. (1968). *Nature (London)* **217**, 1228.
Robbins, E., Pederson, T., and Klein, P. (1970). *J. Cell Biol.* **44**, 400.
Roberts, J. J., and Warwick, G. P. (1963). *Biochem. Pharmacol.* **12**, 1441.
Roizman, B. (1962). *Cold Spring Harbor Symp. Quant. Biol.* **27**, 327.
Sabin, F. (1932). *Physiol. Rev.* **12**, 141.
Sandberg, A. A., Aya, T., Ikeuchi, I., and Weinfeld, H. (1970). *J. Nat. Cancer Inst.* **45**, 615.
Scaletta, L. J., Rushforth, N. B., and Ephrussi, B. (1967). *Genetics* **57**, 107.
Schell, P. L. (1968). *Biochim. Biophys. Acta* **166**, 156.
Schneeberger, E. E., and Harris, H. (1966). *J. Cell Sci.* **1**, 401.
Siniscalo, M., Klinger, H. P., Eagle, H., Koprowski, H., Fujimoto, W. Y., and Seegmiller, J. E. (1969). *Proc. Nat. Acad. Sci. U. S.* **62**, 793.
Sorieul, S., and Ephrussi, B. (1961). *Nature (London)* **190**, 653.
Steffenson, D. M. (1960). *In* "The Cell Nucleus" (J. S. Mitchel, ed.), pp. 216–221. Academic Press, New York.
Steffenson, D. M. (1961). *Int. Rev. Cytol.* **12**, 163.
Steplewski, Z., and Koprowski, H. (1970). *In* "Methods in Cancer Research" (Harris Busch, ed.), Vol. V, pp. 155–191. Academic Press, New York.

Subak-Sharpe, J. H. (1969). *In* "Ciba Foundation Symposium on Homeostatic Regulators" (G. E. W. Wolstenholme and J. Knight, eds.), pp. 276–288. Churchill, London.

Suwalsky, M., Traub, W., Shmneli, U., and Subirana, J. A. (1969). *J. Mol. Biol.* **42**, 363.

Szybalska, E. H., and Szybalska, W. (1962). *Proc. Nat. Acad. Sci. U. S.* **48**, 2926.

Takagi, N., Aya, T., Kato, H., and Sandberg, A. A. (1969). *J. Nat. Cancer Inst.* **43**, 335.

Taniguchi, T., Kamahora, J., Kato, S., and Hagawara, K. (1954). *Med. J. Osaka Univ.* **5**, 367.

Tedesco, T. A., and Mellman, W. J. (1967). *Proc. Nat. Acad. Sci. U. S.* **57**, 829.

Tjio, J. H., and Puck, T. T. (1958). *J. Exp. Med.* **108**, 259.

Watson, P. (1966). *J. Lab. Clin. Med.* **68**, 494.

Watkins, J. F. (1971a). *In* "Methods in Virology" (H. Koprowski and K. Maramorosch, eds.), Vol. V, in press. Academic Press, New York.

Watkins, J. F. (1971b). *In* "Perspectives in Virilogy" (M. Pollard, ed.), Vol. VII, in press, New York.

Weiner, F., Klein, G., and Harris, H. (1971). *J. Cell Sci.* **8**, 681.

Weiss, M. C., and Ephrussi, B. (1966). *Genetics* **54**, 1095.

Weiss, M. C., and Green, H. (1967). *Proc. Nat. Acad. Sci. U. S.* **58**, 1104.

Weiss, M. C., Ephrussi, B., and Scaletta, L. J. (1968). *Proc. Nat. Acad. Sci. U. S.* **59**, 1132.

Whang-Peng, J., Tjio, J. H., and Cason, J. C. (1967). *Proc. Soc. Exp. Biol. Med.* **125**, 260.

Yoshida, T. H., and Sekiguchi, T. (1968). *Mol. Gen. Genet.* **103**, 253.

Yerganian, G., and Nell, M. (1966). *Proc. Nat. Acad. Sci. U. S.* **55**, 1066.

Yun-De, H., and Gorbunova, A. S. (1962). *Acta Virol.* **6**, 193.

Zepp, H. D., Conover, J. H., Hirschhorn, K., and Hodes, H. L. (1971). *Nature New Biol.* (*London*) **229**, 119.

Zhdanov, V. M., and Bukrinskaya, A. G. (1962). *Acta Virol.* **6**, 105.

Chapter 4

Marsupial Cells in Vivo and in Vitro[1]

JACK D. THRASHER

Department of Anatomy, UCLA School of Medicine, Los Angeles, California

[1] Portions of this work were supported by Research Grant USPHS GM 14749-01, General Research Support Grant, UCLA, and from funds from the Cancer Research Coordinating Committee, University of California.

I. Introduction

The order Marsupiala consists of two major groups, one in the Western continent (families Didelphidae and Caenolestidae) and the other in Australiasia (families Dasyuridae, Notoryctidae, Peramelidae, Phalangeridae, Phascolomidae, and Macropodidae). Recent estimates are that close to 69 species comprise the Western group (Cabrera, 1957), while approximately 160 have been described in the Australian Islandic group (Sharman, 1954, 1961; Walker, 1968). The purpose of this chapter is to review those characteristics of marsupials which are pertinent to cellular biology. Other aspects of marsupial biology such as embryology (Mc-Crady, 1938; Hartman, 1952), parasites and diseases (Potkay, 1970; Thrasher et al., 1971a), and breeding (Farris, 1952; Barnes and Barthold, 1969; Thrasher, 1969b) will not be covered.

Marsupials are unique as subjects for experimental biology. The aspects which have received greatest attention have been the simple karyotypes and the "embryonal-fetal" nature of development. For example, Australiasian marsupials have a diploid number of chromosomes ranging from 10 to 28. The few Western marsupials that have been studied have either 7 or 11 pairs in the $2n$ condition. As an order, marsupials have a bimodal distribution of chromosome numbers at 12 and 22. Thus, the metatherians differ from eutherians by possessing fewer chromosomes, as well as by having a bimodal instead of a unimodal distribution (Sharman, 1961).

Developmentally, marsupials are best known for their unique mode of reproduction (McCrady, 1938; Hartman, 1952; Barnes and Barthold, 1969) and for the embryology of their blood-forming organs (Kalmutz, 1962; Block, 1964; Rowlands et al., 1964; Miller et al., 1965). For example, the Virginia opossum ovulates up to 56 ova of which as many as 32 are fertilized. The period of intrauterine development lasts 12.8 days and the young are born equivalent to an 8- to 10-week-old human embryo or 10-day-old mouse fetus (McCrady, 1938; Block, 1964). The embryos weigh 0.1 gm at birth. Their body weight increases by a factor of 25 during the first 25 days after birth. Between days 25 and 65 the body weight increases to 25 gm. The newborn Virginia opossum remains attached to the nipple until the 45th day, after which it leaves the pouch intermittently until the 65th day of life. The young begin eating solid food between days 60 to 100 and can be weaned at day 120 (Block, 1964).

The postnatal development of the blood-forming organs is even more striking than the growth rate. At birth, large lymphocytes are present only in the thymus gland. A serial population of lymphatic tissues occurs as follows: cervical lymph nodes, day 3 after birth; thoracic lymph nodes, days 4 and 5; gastrointestinal lymph nodes, days 8 and 9; and spleen, days 17 to 22. The only circulating blood cells at birth are nucleated eosinophilic megaloblasts from the yolk sac with a few neutrophilic myelocytes, metamyelocytes, and large lymphocytes interspersed (Block, 1964). The development of the immune response correlates with the undifferentiated nature of the lymphatic system. Antibody titers to injected antigens from *Salmonella typhi* flagellum and ϕX174 are first detectable on the eighth day after birth (Kalmutz, 1962; Rowlands *et al.*, 1964). Rejection of allograft skin transplants also begins on day 8 (LaPlante *et al.*, 1966). Finally, Miller *et al.* (1965) have demonstrated that thymectomy of opossum embryos between days 10 and 12 after birth slows the rate of appearance and decreases the number of small and medium lymphocytes in lymphatic tissues.

Various marsupials have been shown to have other unique features that make them invaluable as research tools. For example, Hunsacker (personal communication, 1970) has been able to raise wallaby pouch young in an incubator. Both *Didelphis* (Virginia opossum) and *Caluromys* (wooly opossum) have been superovulated. Currently, *in vitro* fertilization of these ova with epididymal spermatozoa is being undertaken (Fritz, personal communication, 1970). The genus *Marmosa* (Fig. 1) is mouse-size, is easy to breed, has a gestation period of less than 13 days, and lacks a pouch (Barnes, 1968; Barnes and Barthold, 1969;

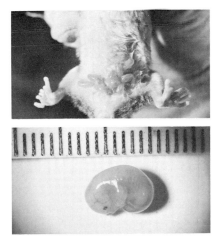

FIG. 1. The upper figure shows a newborn litter of *Marmosa mitis*. Note the absence of a pouch. In the lower figure is a newborn *Marmosa*. A newborn weighs about 60 mg and measures approximately 7–8 mm crown–rump. Note the poorly developed cranium, eyes, and hind limbs.

Thrasher, 1969b). Therefore, it is an "open window" for investigation of the effects of irradiation, mutagens, oncogenic viruses, etc., upon development. Finally, *Potorous tridactylus* (Tasmanian rat kangaroo) and *Protemnodon bicolor* (wallaby) have a 2n chromosome complement of 13♂ : 12♀ and 11♂ : 10♀, respectively. Two X chromosomes are present in the female and an XY_1Y_2 in the male (Sharman and Barber, 1952; Sharman, 1961; Walen and Brown, 1962; Moore and Gregory, 1963; Shaw and Krooth, 1964; Moore, 1965).

Several other aspects of marsupial biology could constitute this chapter as introductory remarks only. Unfortunately, I must limit myself to marsupial cells *in vivo* and *in vitro*. Therefore, the words of Moore and Uren (1966) in their paper on karyotype changes of cell lines from *Antechinus swainsonii* fully describe the purpose of this review:

> The large size and distinctive appearance of the normal chromosomes of this species [see Fig. 19] render a near diploid line of great potential value for cytogenetic and radiobiological studies. The near-diploid and near-tetraploid aneuploid lines of the same species will be of use for any investigations of the effect of alterations in chromosome number or total DNA on a test system, e.g., the induction of neoplastic change. These cells are easily cultured in commercially available media and can be cloned. It is anticipated that they will prove very useful in investigations projected by this laboratory.

II.   Karyotypes of Marsupials

Marsupials are karyotypically simple, but they also have a wide range in chromosome number and morphology. For convenience only, they will be divided into Australiasian and Western groups. Generic terminology will be strictly adhered to, but for those individuals not familiar with the complex taxonomy of marsupials, common names will be used parenthetically. Because considerable inconsistency exists in marsupial classification, the generic terminology of Sharman (1961) has been used.

A.   Karyotypes of Representative Australiasian Marsupials

1.   Families with an Even Number of Chromosomes

Sharman (1961) reviewed the literature and described the karyotypes of a variety of Australiasian marsupials. The families Dasyuridae (marsupial mice and rats, native cats, Tasmanian devil, and Tasmanian wolf), Peramelidae (bandicoots), and Phascolomidae (wombats) are characterized by six pairs of autosomes and an XX/XY set of sex chromosomes

TABLE I

CHROMOSOME NUMBERS FROM REPRESENTATIVE DASYURIDS,
PERAMELIDS, AND PHASCOLOMIDS[a]

Species	$2n$	Sex chromosomes	Reference
Dasyuridae			
Dasyurus hallucatus	14	XX/XY	Sharman (1961)
Dasyurus maculatus	14	XX/XY	Greenwood (1923); Koller (1936)
Dasyurus quoll	14	XX/XY	Drummond (1938)
Antechinus flavipes	14	XX/XY	Sharman (1961)
Antechinus swainsonii	14	XX/XY	Moore and Uren (1965)
Sminthopsis crassicaudata	14	XX/XY	Sharman (1961)
Sminthopsis macrura	14	XX/XY	Sharman (1961)
Myromecobius fasciatus	14	XX/XY	Sharman (1961)
Peramelidae (see Table II)			
Perameles bourganvillei	14	XX/XY	Sharman (1961)
Phascolomidae			
Lasiorhinus latifrons	14	XX/XY	Sharman (1961)
Phascolomis ursinus	14	XX/XY	Sharman (1961); Altman and Ellery (1925)

[a] Genus and species in this table are adapted after Sharman (1961).

(Table I). The Dasyuridae have twelve autosomes. *Sminthopsis crassicaudata*) (narrow-footed marsupial mouse) has five pairs of submetacentrics, while the largest autosome set is metacentric. The autosomes of the native cat (*Dasyurus hallucatus*) are submetacentrics. The karyotype of the Dasyurids can be represented by *Antechinus swainsonii* (broad-footed marsupial mouse). The complement can be arranged by decreasing length. Autosomes 1, 3, 4, and 5 are submetacentrics, while 2 and 6 can be classified as metacentrics. In this family, the X chromosomes are the smallest metacentrics of the complement, while the Y chromosome is visible as a small dot at metaphase (see Fig. 19).

The Peramelids (bandicoots) have karyotypes similar to the Dasyurids, except that the X and Y chromosomes are larger. Typical karyotypes of several species are given in Fig. 2. The chromosomes of the bandicoots, as in the case of *A. swainsonii*, can be arranged by a combination of size and position of the centromere. Jackson and Ellem (1968) have given a detailed description of the chromosomes of *Perameles nasuta* pouch young. According to these authors and Hayman and Martin (1965b), the karyotypes of all Peramelids are very similar with respect to chromosome size and placement of the centromere. In *P. nasuta*, for example (Fig. 3),

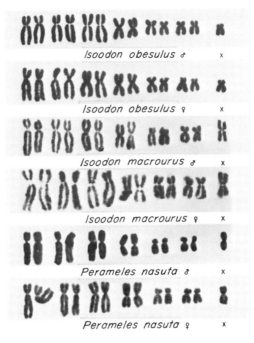

FIG. 2. Karyotypes of representative Peramelids (courtesy of Hayman and Martin, 1965a,b).

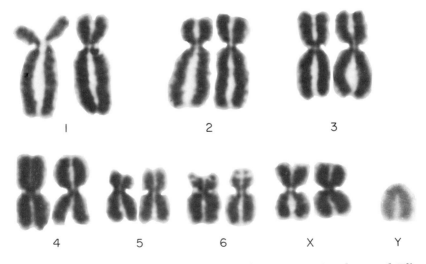

FIG. 3. Karyotype of *P. nasuta*, a Peramelid (courtesy of Jackson and Ellem, 1968).

chromosome 1 has a submedian centromere and measures about 32 μ in length. The second pair is also submetacentric and measures 27.4 μ. Chromosome 3 is almost 29 μ and has a more submedian centromere than 1 and 2. Chromosome 4 is smaller (about 21 μ) and is metacentric. Chromosomes 5 and 6 are the smallest of the autosomes (about 12 μ) and are slightly submetacentric with 6 being characterized by the presence of satellites. The X chromosomes (about 14 μ) are metacentrics, while the Y chromosome (about 9 μ) is the only telocentric of the complement. Sex chromosome mosaicism in Peramelids appears to be prevalent in adult tissues (Table II). Hayman and Martin (1965b) have suggested from their observations on five species that one X in females and the Y chromosome of males are lost during development. In those species examined, the female germ line is XX and the adult female somatic tissues are XO. The male germ line is XY and the male soma are XO. Apparently, this group of marsupials has solved the dosage compensa-

TABLE II

THE CHROMOSOME NUMBERS AND SEX CHROMOSOME CONSTITUTIONS IN DIFFERENT TISSUES OF *Isoodon* SPP., *Perameles* SPP., AND *Peroryctes longicauda*[a]

Species	Sex	Tissue	2n	Sex chromosomes
Isoodon macrourus		Thymus	13[b]	XO
		Bone marrow	13	XO
	♂ and ♀	Leukocytes	13	XO
		Spleen	13	XO
		Corneal epithelium	13	XO
Isoodon obesulus		Intestinal epithelium	13[b]	XO
	♂	Testes	14	XY
	♀	Ovarian tissue	14	XX
Perameles nasuta		Tissue culture (skin)	14[d]	XX and XY
		Corneal epithelium	14	XX and XY
	♂ and ♀	Intestinal epithelium	14[c]	XX and XY
		Bone marrow	13	XO
		Leukocytes	13	XO
Perameles gunni		Spleen	13	XO
	♂	Testes	14	XY
	♀	Ovarian tissue	14	XX
Peroryctes longicauda	♀	Corneal epithelium	14	XX
		Spleen	13	XO

[a] Courtesy of D. L. Hayman.
[b] Only *I. obesulus* examined.
[c] Only *P. nasuta* examined.
[d] Jackson and Ellem (1968) (pouch young of *P. nasuta*).

TABLE III

CHROMOSOME NUMBERS IN THE PHALANGERIDS AND MACROPODS[a]

Species	$2n$	Sex chromosomes	Reference
Phalangeridae			
Cercaertus concinnus	14	XX/XY	Sharman (1961)
Cercaertus lepidus	14	XX/XY	Sharman (1961)
Petaurus breviceps	22	XX/XY	Drummond (1933)
Trichosurus vulpecula	20	XX/XY	Altman and Ellery (1925); Koller (1936)
Trichosurus caninus	20	XX/XY	Sharman (1961)
Phascolarctos cinereus	16	XX/XY	Koller (1936); Greenwood (1923)
Pseudocherius peregrinus	20	XX/XY	Altman and Ellery (1925); Koller (1936)
Schoinobates volans	22	XX/XY	Agar (1923)
Macropodidae			
Lagorchestes hirsutus	22	XX/XY	Sharman (1961)
Lagostrophus fasciatus	24	XX/XY	Sharman (1961)
Macropus major fulignosus	16	XX/XY	Sharman (1961)
Macropus major major	16	XX/XY	Sharman (1961); Moore (1965)
Macropus major malanops	16	XX/XY	Sharman (1961)
Macropus major ocydromus	16	XX/XY	McIntosh and Sharman (1953)
Macropus major tasmaniensis	16	XX/XY	McIntosh and Sharman (1953)
Macropus robustus	16	XX/XY	Sharman (1961)
Macropus rufus	20	XX/XY	Sharman (1961); Moore (1965); McIntosh and Sharman (1953)
Onychogalea unguifer	20	XX/XY	Sharman (1961)
Petrogale penicillata pearsoni	22	XX/XY	Sharman (1961)
Petrogale rothschildi	22	XX/XY	Sharman (1961)
Protemnodon agilis	16	XX/XY	Sharman (1961)
Protemnodon bicolor	10 ♀, 11 ♂	XX/XY1Y2	Sharman (1961); Hayman and Martin (1965a)
Protemnodon dorsalis	16	XX/XY	Sharman (1954)
Protemnodon eugenii	16	XX/XY	Sharman (1954)
Protemnodon irma	16	XX/XY	Sharman (1961)
Protemnodon parryi	16	XX/XY	Matthey (1934)
Protemnodon rufogrisea	16	XX/XY	McIntosh and Sharman (1953)
Setonix brachyurus	22	XX/XY	Drummond (1933); Sharman (1954)
Thylogale billardierii	22	XX/XY	McIntosh and Sharman (1953)
Thylogale stigmatica	22	XX/XY	Sharman (1961)
Thylogale thetis	22	XX/XY	Sharman (1961)

TABLE III (*Continued*)

Species	2n	Sex chromosomes	Reference
Bettongia cuniculus	22	XX/XY	McIntosh and Sharman (1953)
Bettongia lesueur	22	XX/XY	Sharman (1961); Drummond (1933)
Bettongia penicillata	22	XX/XY	Sharman (1961)
Hypsiprymnodon moschatus	22	XX/XY	Sharman (1961)
Potorous tridactylus	12 ♀, 13 ♂	XX/XY1Y2	Moore (1965); McIntosh and Sharman (1953); Hayman and Martin (1965a); Bick and Jackson (1968); Sharman and Barber (1952); Shaw and Krooth (1964); Walen and Brown (1962)

[a] Genus and species in this table are adapted after Sharman (1961).

tion problem by the elimination, instead of inactivation, of the sex chromosomes.

The karyotypes of the family Phascolomidae (wombats) have been described for *Lasiorhinus latifrons* and *Phascolomis ursinus* (Sharman, 1961). Research of the literature has failed to uncover detailed description of the chromosomes of other species of this group. In general, the chromosomes are similar to those of the bandicoots in morphology and size. The complement consists of 12 autosomes with chromosomes 1 and 2 being large and submetacentrics, chromosome 3 a metacentric, chromosome 4 is smaller and submetacentric, and chromosomes 5 and 6 of the complement are the smallest metacentrics. The X chromosomes are similar to autosomes 5 and 6 but have an achromatic region near the centromere. The Y chromosome is the smallest and the only telocentric.

2. FAMILIES WITH VARYING NUMBERS OF CHROMOSOMES

The families Phalangeridae (phalangers, cuscuses, possums, and koala) and Macropodidae (kangaroos, tree kangaroos, rat kangaroos, wallabies, and wallaroos) have a wide variety of chromosome numbers (Sharman, 1961). The family Phalangeridae has considerable heterogeneity in both the number and morphology of chromosomes (Table III). For example, the two species of the genus *Cercaetus* (dormouse possum) have 14 chromosomes and a karyotype similar to the Dasyurids. The remaining genera, on the other hand, have 10 and 11 pairs of chromosomes in the diploid state. In the genera with the larger number of chromosomes *Petaurus breviceps* (gliding possum), all chromosomes are metacentric, while the chromosomes of *Trichosurus vulpecula* (bush-tail possum) are almost all acrocentric (Sharman, 1961). Again, as in the Phascolomidae, very few detailed studies on chromosome morphology have been reported.

The family Macropodidae consists of a large number of genera. The diploid number of chromosomes is fairly constant within a genus, but varies between genera. Table III summarizes the chromosome count on a large number of macropods. The genera *Macropus* (except *M. rufus*) and *Protemnodon* (except *P. bicolor*) have a total of 16 chromosomes in

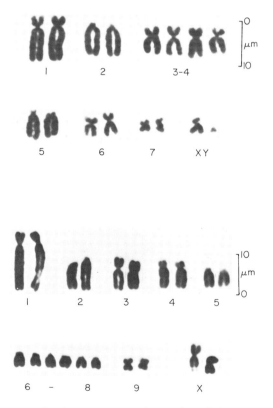

Fig. 4. Karyotype of *Macropus major* (upper) and *Macropus rufus* (lower), family Macropodidae (courtesy of Moore, 1969, personal communication).

the 2n condition. The remaining genera (except *Potorous tridactylus*) have a diploid complement of 20, 22, or 24 (Sharman, 1961).

In the genera with 16 chromosomes there is usually one pair of acrocentric autosomes, while the remaining autosomes are either submetacentric or subtelocentric. The short arms of the subtelocentrics are so minute that it is often difficult to classify them as either acrocentric or subtelocentric (Sharman, 1961; Moore, 1965). The karotypes of *M. major* (great gray kangaroo) and *M. rufus* (red kangaroo) are given in Fig. 4. Detailed analysis of log percentage length and short arm/long arm ratio values for these species have been described by Moore (1965). The chromosomes of *M. major* and *M. rufus* are easily grouped according to size and the position of the centromere. The morphology of the chromosomes of *M. major* are typical of the Macropods with 16 chromosomes. The largest chromosome (1) is submetacentric. Chromosomes 2 and 5 can be classified as either acrocentric or subterminal centromeres (subtelocentrics). Autosomes 3 and 4 are similar in size and cannot be distinguished from each other by either length or arm ratio. Chromsome 6 is

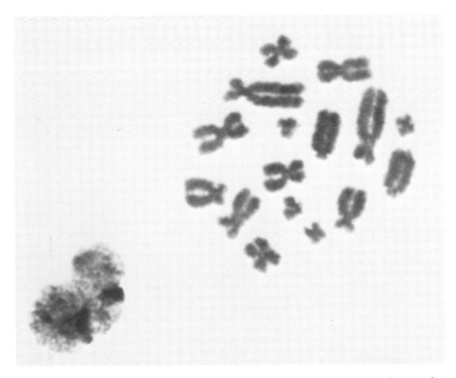

FIG. 5. Karyotype of *Macropus robustus* (courtesy of Taylor and Hunsaker, 1969).

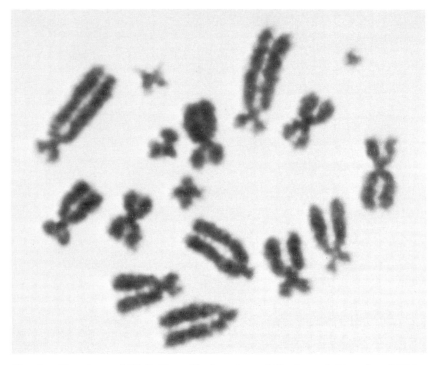

FIG. 6.   Karyotype of *Wallabia agilis* (courtesy of Taylor and Hunsaker, 1969).

the smallest submetacentric, while chromosome 7 is the smallest meta-centric. The X chromosome is metacentric and intermediate in size to chromosomes 6 and 7. The Y chromosome appears as a small dot at meta-phase. The karyotype of *M. rufus* demonstrates the increase in the num-ber of acrocentrics that occur in Macropods with larger numbers of chromosomes. Chromosomes 1, 3, and 4 are submetacentrics, while 2, 5, 6, 7, and 8 are acrocentrics. However, 2 and 5 may be classified as having subterminal centromeres. Chromosome 9 is the smallest metacentric. The X chromosome is metacentric with a satellite. Recently, Taylor and Hun-sacker (1969) have examined the karyotypes of *M. robustus* and *Wall-abia agilis*. They are similar to that of *M. major* and are shown in Figs. 5 and 6.

The karyotype of *Potorous tridactylus* (Tasmanian rat kangaroo) is given in Fig. 7. The diploid complement is 12 in the female and 13 in the male. Chromosomes and each pair of homologs and each unpaired chromosome can be readily identified. The chromosomes are large, rang-ing from about 2.5 to 10 μ. Chromosome 1 is the largest of the autosomes and has a near subterminal centromere. Chromosome 2 is almost as large

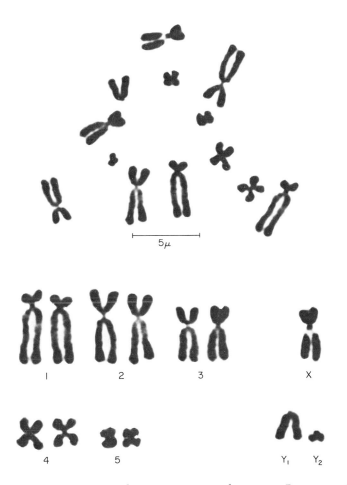

5μ

1	2	3	X

4	5	Y₁ Y₂

FIG. 7. Karyotype of a male Tasmanian rat kangaroo. *Potorous tridactylus* (courtesy of Shaw and Krooth, 1969).

as 1 but is submetacentric. Chromosome 3 is also submetacentric but is shorter than 1 and 2. Chromosomes 4 and 5 are small metacentrics with 4 being almost twice the size as 5. The sex chromosomes Y_1 and Y_2 are acrocentric. The X chromosome is submetacentric and is similar in size to chromosome 3. However, it is characterized by an acromatic gap adjacent to the centromere on the long arm.

Protemnodon bicolor (black-tailed wallaby) has the smallest number of chromosomes in the diploid state of any mammal reported to date (11♂ : 10♀). As in other marsupials with a low modal number of chro-

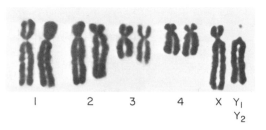

FIG. 8. Karyotype of the black-tailed wallaby, *Protemnodon bicolor* (courtesy of Moore and Uren, 1969).

mosomes, the autosomes can be easily distinguished by length and the position of the centromere (Fig. 8). Chromosome 1 is the largest with a submedian centromere. Autosome 2 is about the same, but the centromere is more subterminal. Chromosome 3 is about one-half the size and is submetacentric. Chromosome 4 is the smallest of the autosomes and is acrocentric. The X chromosome is nearly the same as autosomes 1 and 2 and has a subterminal centromere. However, it is identifiable by a large achromatic gap in the centromeric region. The Y chromosomes are acrocentric, with Y_1 the smaller and barely visible in most diploid cells and Y_2 the larger.

The numbering of the two Y chromosomes of *P. bicolor* and *P. tridactylus* should be clarified at this point. In general, the procedure is to give the largest chromosome of the complement the smallest number. Hence, Y_1 is the larger and Y_2 the smaller in *P. tridactylus*. However, the numbering sequence in *P. bicolor* is based upon the presumption that the smaller Y is the ancestral sex chromosome and the larger Y arose from an autosomal X translocation. Hence, Y_1 is reserved for the original male sex chromosome and Y_2 for the translocated autosome. It is suggested that the latter classification be perpetuated. The hypothesis of the origin of Y_2 is substantiated by the observations of Sharman and Barber (1952). The sex trivalent of *P. bicolor* consists of two sets of pairing segments with different properties and relationships. The first pairing segment is the large Y and most of the long arm of the X. These are euchromatic and condense in a manner that suggests they are autosomal in origin. The small Y pairs with the short arm of the X (allosomic portion) and both are heterochromatic. They remain condensed through mid-diplotene in a manner similar to the X and Y chromosomes of other marsupials.

B.   Karyotypes of Western Marsupials (Didelphidae)

Research of the literature has failed to uncover studies on the karyotypes of the family Caenolestidae (rat opossum). The Caenolestids consist of three genera and seven species. They are found in Western South America (Walker, 1968). Apparently very few specimens have been trapped and they are considered to be very rare. Therefore, it is not surprising that these shrewlike marsupials have not been studied.

The family Didelphidae (opossums) are most readily accessible marsupial for investigation in the United States. The Didelphids consists of 12 genera and about 69 species (Cabrera, 1957; Walker, 1968). Of these only *Didelphis marsupialis virginianus* (Virginia opossum) is native to North America. The remaining genera, such as *Marmosa, Philander,* and *Caluromys,* can be obtained through most wild animal dealers who import them directly from Mexico and Central, and South America. The chromosome numbers of several Didelphids are given in Table IV.

The karyotype of *Didelphis marsupialis virginianus* has been described by a number of investigators (Painter, 1922; Biggers *et al.,* 1965; Schneider and Rieke, 1967; Sinha, 1967a,b) and is shown in Fig. 9. The diploid number is 20 autosomes and a sex complement of XX or XY. The chromosomes are about three times larger than human chromosomes. Like other marsupials, they are particularly amenable to investigation with autoradiography and tritium-labeled precursors of proteins and nucleic acids (see below). The autosomes can be arranged into three groups according to their size. Chromosomes 1–3 are subtelocentric and

TABLE IV
CHROMOSOME NUMBERS IN DIDELPHIDS

Species	$2n$	Sex chromosomes	Reference
Didelphis azarae	22	XX/XY	Saez (1931)
Didelphis marsupialis aurita	22	XX/XY	Dreyfus and Campos (1941)
Didelphis marsupialis virginianus	22	XX/YY	Biggers *et al.* (1965); Painter (1922); Schneider and Rieke (1967); Sinha (1967a)
Lutreolina crassicaudata	22	XX/XY	Saez (1938)
Philander opossum	22	XX/XY	Perondini and Perondini (1966)
Caluromys derbianus	14	XX/XY	Biggers *et al.* (1965); Legator *et al.* (1966)
Marmosa mitis (robinsoni)	14	XX/XY	Wolf *et al.* (1970); Reig (1968)
Marmosa elegans	14	XX/XY	Wolf *et al.* (1970)
Marmosa mexicana	14	XX/XY	Biggers *et al.* (1965)

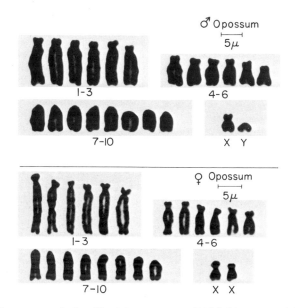

Fɪɢ. 9. Karyotype of the Virginia opossum, *Didelphis marsupialis virginianus* (courtesy of Schneider and Rieke).

are 12–15 μ in length. Chromosomes 4, 5, and 6 are also subtelocentric, but are about one-half the length of group 1. Chromosomes 7 to 10 consist of four pairs of acrocentrics measuring about 5 μ and becoming progressively smaller. The X chromosomes are identifiable as small submetacentrics, while the Y chromosome is the smallest acrocentric.

Philander opossum (four-eyed opossum) also has 20 autosomes and XX/XY chromosomes (Biggers *et al.*, 1965; Perondini and Perondini, 1966). However, unlike the Virginia opossum, the autosomes can be discriminated into only two groups (Biggers *et al.*, 1965). The autosomes are divisible into three pairs of large (about 10–12 μ) acrocentrics and seven pairs of medium-sized (about 6 μ) acrocentrics. The X chromosome is identifiable as a small acrocentric and the Y as the smallest acrocentric.

The simplest karyotypes of the Didelphidae have been found in *Caluromys derbianus* (woolly opossum) and *Marmosa* sp. (pouchless opossums). Figure 10 shows the karyotype of the woolly opossum, which consists of 14 chromosomes. The autosomes are divisible into three groups according to size and position of the centromere. Three pair form the largest autosomes (about 14 μ) and are submetacentric. One pair of metacentrics comprises the second group, while the third group consists

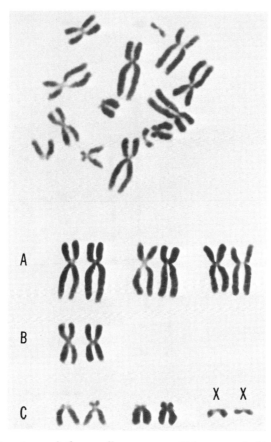

FIG. 10. Karyotype of the woolly opossum, *Caluromys derbianus* (courtesy of H. G. Wolf).

of two pairs of subtelocentrics. The X chromosomes are small acrocentrics, and the Y chromosome is identifiable as the smallest chromosome of the complement (Biggers *et al.*, 1965; Legator *et al.*, 1966; Wolf *et al.*, 1970).

Only three species of the genus *Marmosa* (*M. mitis* or *robinsonii*, *M. elegans*, and *M. mexicana*) have been examined cytologically (Biggers *et al.*, 1965; Reig, 1968; Wolf *et al.*, 1970). Apparently, the karyotype is similar in the three species and is very close to that of *Caluromys*. The chromosomes are divisible into three groups (Fig. 11). Group 1 consists of three pairs of large (about 11 μ) submetacentrics. The second group contains one pair of medium-sized metacentrics. The last group (3) contains two pairs of subtelocentrics. The X chromosome is the smallest metacentric, and the Y is the only acrocentric of the complement.

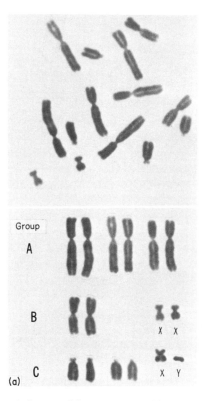

Fig. 11. Karyotype of the pouchless opossum, *Marmosa mitis* (a) and *Marmosa elegans* (b) (courtesy of H. G. Wolf).

III.   The Marsupial Cell Cycle

Very little information is available on the marsupial cell cycle, the duration of its phases, or the variability that may exist under a variety of experimental conditions. Such information is vital before a full appreciation of the mechanisms that regulate cell division in marsupial cells can be obtained. Knowledge of the cell cycle in a variety of marsupial cells, both *in vivo* and *in vitro*, is prerequisite to further studies on cell division in these animals. For this reason, the author has gone to some length to obtain information on the cell cycle in epithelia of *Didelphis marsupialis virginianus* and *Marmosa mitis* (Thrasher, 1969a, 1970; Thrasher *et al.*, 1971b). In addition, information on the cell cycle of *Didelphis* lymphocytes *in vitro* (Schneider, 1969, personal communication) and kidney

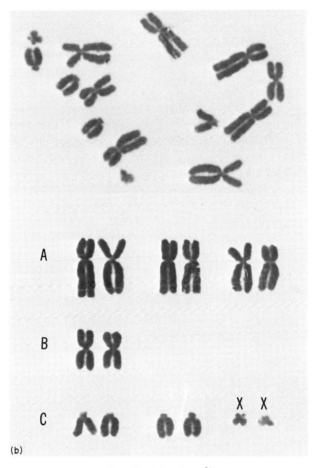

FIG. 11 (*Continued*)

cells from *P. tridactylus* (Whissell and Cleaver, personal communication, 1969) have been made available (Table V).

A. *Didelphis marsupialis virginianus*

The cell cycle of lymphocytes stimulated by PHA is short, lasting about 11 to 12 hours (Fig. 12). However, it is interesting to note that the S phase (about 9 hours) is more extended than those reported for eutherian mammals. The remainder of the cell cycle is occupied principally by G_2 (2.0 hours) and D (0.5 hour). G_1 cannot be accurately measured, but appears to be between 0 and 1 hour. The S phase *in vivo* is slightly longer, averaging 10 hours in the body and pyloric regions of the stomach

TABLE V

ESTIMATES FOR THE MEAN DURATION OF THE PHASES OF THE
CELL CYCLE OF VARIOUS MARSUPIAL CELL TYPES

Cell type	Cell cycle (hr)	S (hr)	D (hr)	G_2 (hr)	G_1 (hr)	Reference
Lymphocytes (Didelphis)	11–12	9.0	0.5	2.0	0–1	Schneider (1969, personal communication)
Stomach (body) (Didelphis)	—	10.0	2.5	1.5	—	Thrasher (1969a)
Stomach (pylorus) (Didelphis)	—	10.0	2.5	1.5	—	Thrasher (1969a)
Tongue (Marmosa)	44	12.8	2.5	1.0	27.7	Thrasher et al. (1971b)
Stomach (body) (Marmosa)	—	14.0	1.5	1.0	—	Thrasher et al. (1971b)
Duodenum (Marmosa)	20	8.5	2.0	0.5	9.0	Thrasher (1970)
Jejunum (Marmosa)	28	8.6	1.5	1.0	16.4	Thrasher (1970)
Ileum (Marmosa)	32	9.7	1.5	1.0	20.8	Thrasher (1970)
Colon (Marmosa)	40	11.3	1.5	1.0	26.7	Thrasher et al. (1971b)
Kidney (P. tridactylus)	28	14.5	—	4.0	9.5 (G_1 + D)	Whissell and Cleaver (1969, personal communication)

(Thrasher, 1969a). The duration of G_2 (1.5 hours) is about the same as
that of the lymphocytes *in vitro,* while the duration of D (2.5 hours) is
considerably longer. The major difference in the cell cycle between lymphocytes *in vitro* and the gastric mucosa *in vivo* lies in the average time
that a given cell spends in G_1.

B. *Marmosa mitis*

Although the cell cycle in *Didelphis* does not have unusual kinetic
parameters, the situation in *M. mitis* is quite different (Table V). *Marmosa* is probably the only small mammal in which a wide variety of
values for the S phase exist in the epithelia of intestinal tract. The values
obtained for the S phase do not conform to the constancy hypothesis
that has been derived from observations on intestinal epithelia of small
rodents (Lesher *et al.,* 1961; Cameron and Greulich, 1963; Cairnie *et al.,*
1965; Thrasher, 1967; Cleaver, 1967). The shortest time for the S phase

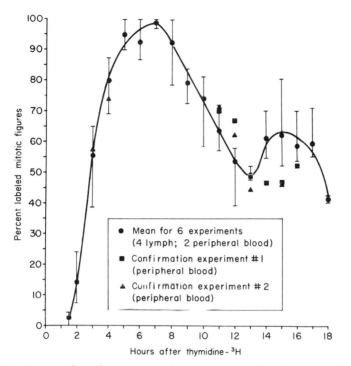

FIG. 12. Curves that demonstrate the percentage of labeled mitotic figures following a pulse label with ³H-Tdr in lymphocyte cultures from the Virginia opossum (courtesy of Schneider, 1969).

(8.5–9.7 hours) is in the small intestine of *Marmosa*. At the far ends of the intestinal tract, the values for the descending colon and tongue epithelia are 11.3 and 12.8 hours, respectively. The variability of the S phase duration is even more evident by the fact that it is too long to measure in the gastric mucosa, greater than 14 hours (Thrasher *et al.*, 1971b). There is a possibility that the differences between the S phases of the small intestine, tongue, and colon may be explained on the basis of the low body temperature of marsupials (about 34°C, Wislocki, 1933). However, the long S phase in the gastric mucosa does not support the hypothesis of a temperature effect. Thus, it may be possible that the cell cycle and S phase of intestinal epithelia of *Marmosa* may be independently regulated by genetic or transcriptional differences, local environmental conditions, intramural blood circulation, diurnal rhythms, or by yet unexplainable mechanisms. It should be pointed out at this time that these measurements have been made on wild trapped animals whose genetic and environmental backgrounds are unknown.

The latter point is particularly interesting in the light of recent observations made in this laboratory on *Marmosa mitis* (Barenfus *et al.*, unpublished observations, 1970; Thrasher *et al.*, 1971a). Wild trapped *M. mitis* and laboratory-reared domestic animals are easily stressed by undetermined factors under laboratory conditions. The stress is manifested in perforating gastric and duodenal ulcers as well as nonspecific hemorrhagic enteritis in about 50% of the animals. Therefore, it appears that *M. mitis* may be a good model to investigate the etiology of intestinal ulceration in relation to altered cell population kinetics. Finally, it is worth pointing out that the major variability in the cell cycle occurs in the duration of G_1. The average values obtained for D and G_2 are slightly longer than those times observed in small eutherian mammals (Cleaver, 1967; Cameron, 1971).

C. *Potorous tridactylus*

Recently, the cell cycle of kidney cells from *Potorous tridactylus* was measured *in vitro* (Whissell and Cleaver, personal communication, 1969). As in the cases of *Marmosa* and *Didelphis*, an extended S phase (14.5 hours) was observed. In addition, G_2 (about 4 hours) was considerably longer than corresponding measurements on other opossum cells. The generation time was 28 hours with G_1 and D occupying 9.5 hours of the total cell cycle. Information regarding these cells can be obtained from the National Tissue Culture Collection.

IV.   Tissue Culture of Marsupial Cells

Several investigators have reported the culture conditions for marsupial lymphocytes (Hayman and Martin, 1965a; Moore, 1965; Moore and Uren, 1965; Schneider and Rieke, 1967, 1968; Sinha, 1967a,b,c; Bick and Jackson, 1968; Burton, 1968a,b; Wolf *et al.*, 1970) and cell lines (Walen and Brown, 1962; Moore and Uren, 1966; Uren *et al.*, 1966; Moore and Radley, 1968; Jackson and Ellem, 1968; Bick and Brown, 1969). In addition, this laboratory has attempted to culture cells from the kidneys, spleen, and abdominal connective tissues of *Marmosa mitis*. It appears at the present time that marsupial cells can be easily cultured with slight modifications of standard methodologies and media. Therefore, only those aspects of culturing marsupial cells which appear to be of value will be described.

A. Lymphocytes

Peripheral blood and thoracic duct lymphocytes have been cultured by slight modifications of the method of Moorhead *et al.* (1960). Schneider and Rieke (1968) have defined the culture conditions for *Didelphis* lymphocytes. It appears that human, horse, and opossum sera elicit minimal cell division, while normal and heat-inactivated Lewis rat serum evokes a strong response. Apparently, both contaminating red blood cells and a rat serum complement are the limiting factors in successful culturing of *Didelphis* lymphocytes. The culture conditions for either peripheral blood or thoracic duct lymphocytes are as follows: small Erlenmeyer flasks inoculated with 3 to 5×10^6 of relatively pure lymphocytes (buffy coat is permissible) per milliliter of culture medium (Earle's MEM), containing 40 μg/ml of PHA, penicillin (100 units/ml), and heat-inactivated Lewis serum (the percentage was not defined, but presumably 10–15% concentration of serum should be adequate). Under these conditions, *Didelphis* lymphocytes enter DNA synthesis and mitosis in the largest numbers between 36 to 48 hours after initiation of the culture. The optimal time for obtaining colchicine-arrested metaphases is between 55 to 72 hours for other marsupials (Hayman and Martin, 1965a; Moore, 1965; Moore and Uren, 1965; Sinha, 1967a,b,c; Bick and Brown, 1969).

Observations on lymphocytes of *M. giganteus* (grey kangaroo) have shown that some marsupial lymphocytes may have an earlier response to the mitogenic agent phytohemagglutin (PHA) than do human lymphocytes (Burton, 1968a,b). Synthesis of DNA begins at 12 hours and the percentage of incorporated tritiated thymidine (^{3}H-Tdr) increases rapidly thereafter. Mitotic figures are first observed at 28 hours and reach significant numbers by 36 hours after PHA treatment (Burton, 1968b). Also, the mean histone content begins to decrease by 5 minutes and drops rapidly by 24 minutes after addition of PHA. The stainable histones remain at about 75 to 80% of the control values (Burton, 1968a). Similarly, early responses to PHA are detectable in changes in the rates of RNA and protein synthesis. Stimulated RNA synthesis, as measured by the increased incorporation of cytidine-^{3}H, begins as early as 12 minutes and cytoplasmic labeling is detectable by 18 minutes after initial introduction of PHA. Incorporation of phenylalanine-^{3}H increases rapidly after 30 minutes of incubation (Burton, 1968b).

Culture conditions for peripheral lymphocytes of *Marmosa mitis* have recently been described by Wolf *et al.* (1970). The buffy coat from six heparinized capillary tubes from peripheral blood obtained by centrifugation at 4200 rpm for 5 minutes in an International clinical centrifuge are combined. These are placed in 5 ml of medium in a 15-ml Falcon tissue culture flask and incubated at 38°C in an atmosphere of 95% air and 5% CO_2. The composition of the medium is 75% RPMI-1629 (GIBCO), 20% fetal calf serum (GIBCO), 2% L-glutamine (GIBCO), and 5% phytohemagglutinin (Wellcome). One-hundred units of procaine penicillin and 0.125 mg of streptomycin/ml of medium are added. It is interesting to note that the concentration of phytohemagglutinin is four times that recommended for human lymphocytes. Colcemid is added 9 hours before harvest. Under these conditions DNA synthesis begins by 27 hours after inception of the culture and most of the cells are in their first division by 43 hours.

In conclusion, it appears that *Didelphis, Macropus,* and *Marmosa* lymphocytes are excellent cell systems to investigate the molecular events of the prereplicative phase following stimulation by PHA. The major differences between marsupial and human lymphocytes with regard to responses to PHA are: (1) RNA synthesis and labeling in the cytoplasm occurs between 12 and 18 hours in human lymphocytes (Winter and Yoffey, 1965), while it is detectable in kangaroo lymphocytes as early as 12 minutes after addition of PHA; (2) DNA synthesis begins at 12 hours and maximum mitotic activity occurs between 28 and 36 hours after PHA stimulation. These times are about 12 hours earlier than similar observations on human lymphocytes (Killander and Rigler, 1965; Burton, 1968b); and (3) a greater concentration of PHA is required.

B.   Cell Lines

Fibroblastic cell lines have been established from *Antechinus swainsonii* (Moore and Uren, 1966), *Protemnodon bicolor* (Moore and Radley, 1968; Uren *et al.,* 1966), and *Potorous tridactylus* (Walen and Brown, 1962; Bick and Brown, 1969). The culture media and methods for establishing cell lines are slight modifications of standard techniques. Fibroblastic outgrowths from *A. swainsonii* and *P. bicolor* were initially obtained in plasma clots. Outgrowing cells were subcultured with a rubber policeman. Thereafter, subculturing was done by either a rubber policeman or 0.5% trypsin treatments on glass surfaces. The cells have been grown as monolayers in Eagle's MEM containing up to 25% fetal calf serum in 5% CO_2 gas phase. Giant cell feeder layers are necessary to obtain good plating efficiency of *A. swainsonii* (Moore and Uren, 1966), while pyruvate and serine were added to the culture medium for *P. bicolor* (Moore and Radley, 1968; Uren *et al.,* 1966). Cells from a male *P. tridactylus* were successfully cultured in Earle's salts solution supplemented with 2% lactalbumin hydrolyzate, 2% yeast hydrolyzate, and 10% fetal calf serum (Walen and Brown, 1962). More recently, Bick and

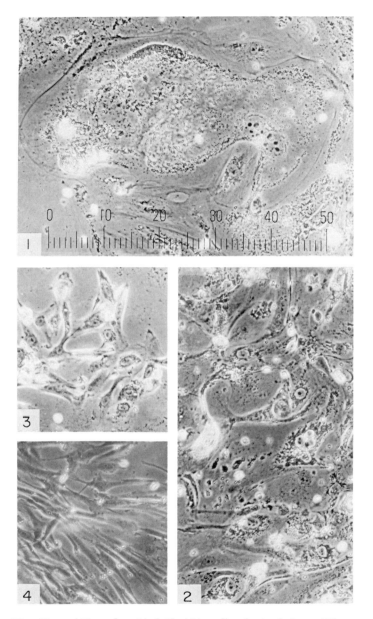

FIG. 13. Giant (1) and epitheloid (2) cells obtained from kidney explants from adult *M. mitis* after 7 days in culture (MEM, Hank's salts, 10% fetal calf serum, 37°C). Fibroblasts (3, 4) from abdominal connective tissue cultured in MEM, Earle's salts, 0.2 mM nonessential amino acids, 0.4 mM L-glutamine, 10% fetal calf serum, 37°C. Each division in (1) represents 10 μ.

Brown (1969) have cultured and established cell lines from kidney and testicular tissues of *P. tridactylus* in medium 199 containing 12% horse serum.

Cells from the kidney, spleen, and abdominal connective tissue of adult and newborn *Marmosa mitis* have been cultured in this laboratory. The use of enzymatic digestion (trypsin, pronase, collagenase) to obtain cells from tissue fragments resulted in irreparable damage to *Marmosa* cells. Therefore, monolayers were established by permitting small fragments (0.5 mm³) to first adhere to the surface of plastic 25-ml Falcon flasks in 1–2 ml of medium at 37°C in 5% CO_2. Figure 13 shows cells obtained under two different growth conditions. MEM (Hank's salts) supplemented with 10% fetal calf serum and containing 100 units of penicillin and streptomycin per milliliter permitted the growth of monolayers of kidney giant and epitheloid cells in 7 days. Similar observations were made on abdominal connective tissue fibroblasts. Excellent growth of fibroblasts was obtained in MEM (Eagle's salts), supplemented with 10% fetal calf serum, 0.4 mM of L-glutamine, and 0.2 mM nonessential amino acids. Subculturing was carried out by either a rubber policeman or a 5-minute treatment with 0.25% trypsin in Earle's salts. At the present time, an analysis of the cell cycle and chromosome replication is in progress at various temperatures to obtain information on optimal growth. Recently, Zee (1969) has announced the establishment of a fibroblastic cell line from *Marmosa* embryos under similar conditions.

V.   Chromosome Synthetic Activities and Cytogenetics of Karyotypically Simple Cells

Investigations of chromosome replication in marsupial cells has been limited. In general, it can be safely stated that studies of DNA, RNA, and protein synthesis in marsupial chromosomes have confirmed observations made on eutherian mammals as well as producing evidence that may question the hypothesis of X chromosome inactivation. In addition, the problems of transformation and increased gene frequencies of mammalian cells *in vitro* resulting from selection pressures may possibly be attacked through relatively simple karyotypes of marsupials.

A.   Chromosomal DNA, RNA, and Protein Synthesis and Ploidy in *Didelphis*

Schneider and Rieke (1967) and Sinha (1967a,b,c) have investigated nucleic acid and protein synthesis in lymphocyte chromosomes of *D. mar-*

supialis virginianus. According to their observations, the autosomes begin DNA synthesis synchronously and terminate asynchronously in both sexes (Schneider and Rieke, 1967). The chromosomes have similar labeling patterns when exposed continuously to ³H-Tdr *in vitro.* At the beginning of the S phase they proceed from predominantly moderate to heavy incorporation of the isotope. The autosomes are heavily labeled by hour 3 of S and terminated DNA synthesis by 1 hour before the end of S. The rate of incorporation of ³H-Tdr decreased in the autosomes throughout the S phase. Thus, the autosomes of *Didelphis* do not fall into early and late replicating chromosomes. This does not exclude the possibility that replicating units of individual chromosomes may undergo early or late DNA synthesis. However, Sinha (1967b) has demonstrated that excessive hyperchromatization and late replicating chromosomes may occur to some extent in *Didelphis* (Fig. 14). More information is needed to clarify this point.

With regard to the sex chromosomes Schneider and Rieke (1967) have shown that one female X chromosome synthesizes DNA throughout the S phase, while the other begins about 3.5 hours later. The two X chromosomes, however, terminate DNA synthesis synchronously (Fig. 15) about 1 hour later than the autosomes. On the other hand, demonstrating the lack of detailed knowledge on *Didelphis* chromosomes, Sinha (1967b) has suggested that some autosomes may terminate replication later than the X. The male X chromosome synthesizes DNA throughout S, while the

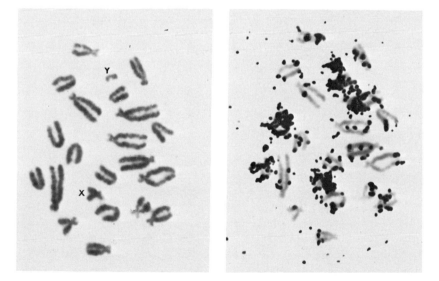

Fig. 14. Metaphase lymphocytes of a male *Didelphis* before autoradiography (left) and after autoradiography (right). The cells were exposed to ³H-Tdr and treated with colchicine 4½ hours before harvest (courtesy of Sinha).

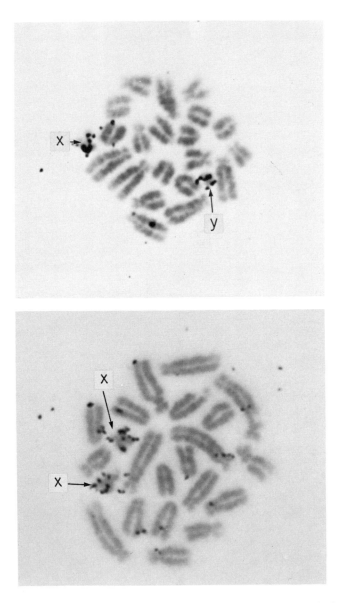

FIG. 15. Terminal DNA synthesis in the sex chromosomes of the Virginia opossum labeled with ³H-Tdr near the end of the S phase (courtesy of Schneider and Rieke, 1969).

Y has late replicating characteristics. As in the female, both sex chromosomes of the male terminate DNA synthesis synchronously and possibly later than the autosomes (Fig. 15).

Nuclear protein synthesis in *Didelphis* lymphocytes is similar to that reported for human leukocytes (Cave, 1966). Protein synthesis increases sharply during the first half of S, decreases in the later half of S, and increases, thereafter, through G_2. In addition, recent evidence suggests that both X chromosomes may undergo RNA synthesis (Schneider, personal communication, 1969). This latter observation may be significant with respect to the Lyon hypothesis of sex chromosome inactivation (Lyon, 1966). In addition, the question of sex chromosome functions in marsupials may be interesting from the point of view that anomalies do occur, e.g., mosaicism in Peramelids, and Y_1Y_2 of *P. bicolor* and *P. tridactylus*.

Another interesting phenomenon of *Didelphis* chromosomes has been reported by Sinha (1967c). Lymphocytes from a phenotypically normal male opossum showed tetraploid and near haploid cells amid a majority of diploid leukocytes in short-term culture (Figs. 16 and 17). At present, it appears that proliferating bone marrow cells may have fused producing tetraploids. The "hybrid-cell types" may subsequently have given rise to near-haploid daughter cells by a process of double-reduction division. However, the probability of endomitosis, nondisjunction, or aberrant mitotic spindles cannot be ruled out. A possible solution for which mechanism may be applicable could be obtained from the formation of parabiotic opossum "embryos" between opposite sexes. This is particularly appealing since it is accepted that newborn opossums are immuno-

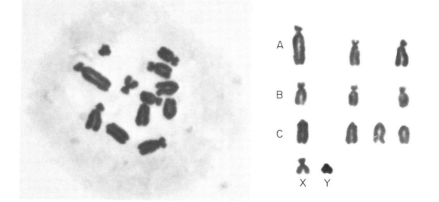

FIG. 16. Near-haploid metaphase obtained after short-term culture of lymphocytes from a phenotypically normal male Virginia opossum (courtesy of Sinha).

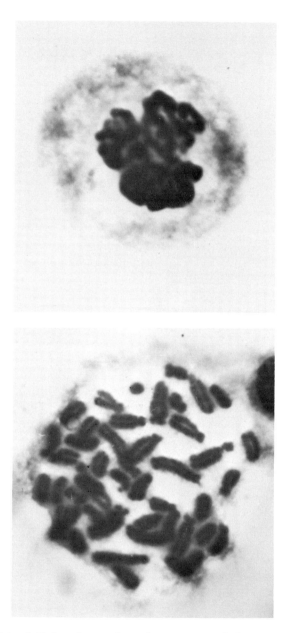

FIG. 17. Tetraploid lymphocyte from the same animal shown in Fig. 16 (courtesy of Sinha, 1969).

logically incompetent until the eighth day of life. Thus, parabiosis of opossum embryos may be a very fruitful area in marsupial biology to investigate.

B. DNA Synthesis in Chromosomes of *Protemnodon bicolor* and *Potorous tridactylus*

Moore and Uren (1965) have also demonstrated that the autosomes of a male wallaby, *P. bicolor*, complete DNA synthesis at about the same time. It is now apparent that the smaller the number of chromosomes the more likely that such a phenomenon will be observed. However, DNA synthesis within each chromosome does follow a rigid pattern. In addition, while there is a strong tendency toward synchrony of DNA synthesis between homologs, some asynchrony between them was observed.

Observations on autosomal labeling patterns have permitted the division of individual chromosomes into several replicating regions (Fig. 18). The replicating units (replicons) of a particular chromosome varied with respect to the time at which DNA synthesis was completed. For example, chromosome 1 replicated in the following sequence: L3 and L5, followed by S1, S3, L1, and L4 with L6, S2, and L2 completing synthesis last, where L and S signify long and short arms, respectively.

The sex chromosomes showed marked asynchrony with respect to the autosomes. The small Y_1 chromosome replicates early, possibly before the other chromosomes of the complement begin. The short arm of the X synthesizes DNA along its entire length late in the S period. This pattern

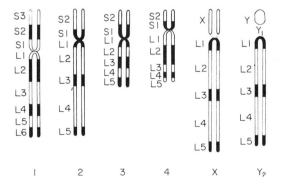

FIG. 18. Diagrammatic representation of the chromosomes of *P. bicolor* ♂ with a division into regions which vary with respect to the time at which DNA synthesis is completed. L and S specify long and short arms (courtesy of Moore and Uren, 1969).

is different from that observed in eutherian mammals (Bader *et al.*, 1963; Moorhead and Defendi, 1962; Morishima *et al.*, 1962). The early completion of DNA synthesis in Y_1 may result from either its small size or its positive heteropycnosis. In addition, it is different from that of *Didelphis* where the X synthesizes DNA throughout S and the Y is late replicating (Schneider and Rieke, 1967).

DNA synthesis in chromosomes of female *P. tridactylus* and *P. bicolor* has also been reported (Hayman and Martin, 1965a). The short arm (allosomic portion) of one of the X chromosomes replicates late. The autosomes and the long arm of the X chromosomes synthesize DNA synchronously and throughout the S phase. Thus, it appears that the females of these two species have autosomal replication similar to the males. However, more information is needed to fully evaluate early versus late synthesis of the X chromosomes.

C.   Evolution of Karyotypes of Cells from *Antechinus swainsonii* in *Vitro*

Investigations on the changes in the karyotypes of cell lines of *A. swainsonii* have demonstrated the cytogenetic value of karyotypically simple mammalian cells (Moore and Uren, 1966). Development *in vitro* of aneuploid cell lines of the marsupial mouse follows a course of events similar to that described in mouse fibroblasts (Hsu *et al.*, 1961). The primary transformation is from a diploid to a tetraploid population of cells. For example, in one cell line the modal number of chromosomes shifted from 14 to 28 as early as the 47th passage. The completeness of this transformation indicates either all diploid cells transform or the fitness of tetraploid cells is greater than diploids. The fitness may be a function of viability, division rate, and/or changes in beneficial gene frequencies.

The changes in the karyotype and possibly gene frequencies were further evident by specific alterations in ploidy that occurred when cells from the original tetraploids were cloned. The modal number of centromeres of three cloned lines shifted from 28 to 20, 21 and 25. Other profound changes occurred in three chromosomes. In a male cell line, the Y chromosome was absent in about 20% of the cells and chromosome 5 had abnormal morphology suggesting that a combination of translocation, deletion, and possibly endoreplication had taken place (Fig. 19). Furthermore, female cell lines tended to lose part or all of chromosome 3. Thus, the changes in ploidy and chromosome morphology could possibly result from changes that take place in the tissue culture environment.

In conclusion, the data from cell lines of *A. swainsonii* raise the question of whether or not diploid cells can proliferate indefinitely in tissue culture. If not, aneuploid changes *in vitro* may be causally related to environmental selection pressures. These pressures could, therefore, greatly influence or bring about changes in gene frequencies on cells whose life cycles are shortened *in vitro* over the *in vivo* conditions. Rapid proliferation in culture, thus, probably requires profound readjustment in the genetic complement of the cell.

D. Evolution of Karyotypes and Changes in DNA Synthetic Patterns in Cells from *Protemnodon bicolor in Vitro*

The wallaby *P. bicolor* has a very simple karyotype, 11♂, 10♀ (Fig. 20) which also permits investigation on karyotype changes *in vitro* (Uren *et al.*, 1966; Moore and Radley, 1968). The alterations in the morphology and DNA synthetic patterns of wallaby fibroblasts have been examined in detail between the 55th and 80th passage *in vitro*. The modal centromeric number remained at 10 in 90% of the cells, with the remaining 10% ranging from 11 to 40.

Approximately 90% of the cells contain a distinctive marker chromosome, M1. It is submetacentric with three achromatic gaps on its short arm. At the 55th passage, two basic karyotypes appeared which comprised about 40% of the cells. Both cell lines (named type I and type II) contained M1 chromosomes and are shown in Fig. 20. However, the two cell lines have different alterations in the karyotypes: type I has two large submetacentrics (M2 chromosomes) and type II has one M2 and an extra chromosome 1.

A comparison between individual chromosomes of the major aneuploid karyotypes (types I and II) and the diploid karyotype of fibroblasts by means of densitometry, measurements of length and arm ratio, and examination of late replication units with ^{3}H-Tdr was undertaken. These measurements showed that: (1) the aneuploid chromosome 1 is identical to the diploid chromosome 1 in all three analyses; (2) aneuploid chromosomes 3, 4, and Y were different from the diploid chromosomes by having an increase in the amount of late replicating regions, although densitometric and size measurements could not reveal a significant difference; and (3) the origin of the marker chromosomes (M1, M2) could not be determined.

In conclusion, it appears, as observed in *A. swainsonii* fibroblasts, that the breakdown of the normal cell cycle of diploid cells *in vitro* and the emergence of well-adapted aneuploid cell lines may be causally related. In addition, the change in the DNA synthetic patterns in aneuploid cells to late replicating units may facilitate the development of cell lines. The

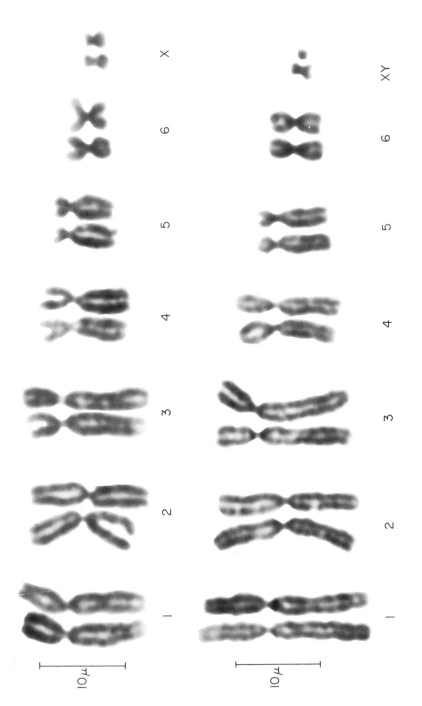

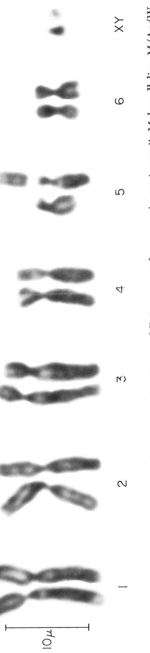

FIG. 19. Karyotype of normal female (*top*) and male (*middle*) marsupial mouse, *A. swainsonii*. Male cell line M/An/W₁ (*bottom*) that has had a normal chromosome 5 replaced by an abnormal chromosome with an achromatic gap at the 31st passage (courtesy of Moore and Uren, 1969).

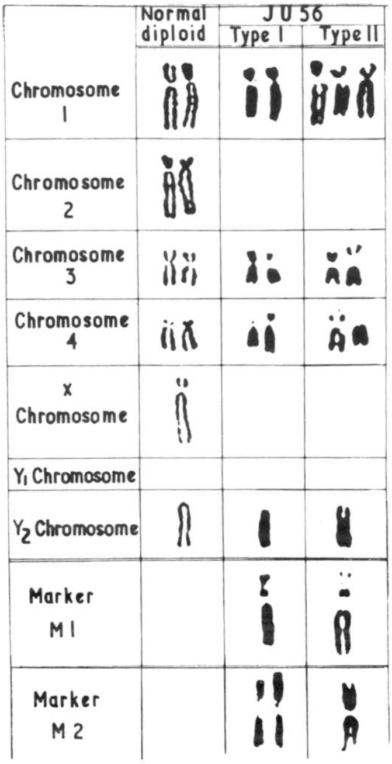

	Normal diploid	JU 56 Type I	JU 56 Type II
Chromosome 1			
Chromosome 2			
Chromosome 3			
Chromosome 4			
X Chromosome			
Y₁ Chromosome			
Y₂ Chromosome			
Marker M1			
Marker M2			

FIG. 20.   The two basic karyotypes (I and II) found at the 55th passage of fibroblasts established from *P. bicolor* compared to the normal diploid complement (courtesy of Uren, Moore, and van den Brenk, 1969).

exact role of these changes is not fully understood at the present time. However, the change in ploidy, the increase in the number of late replicating regions, and the overt differences between chromosomes of diploid and aneuploid cells may indicate that an evolution of effective gene frequencies has taken place. The increase in ploidy (i.e., development of marker chromosomes in the wallaby or tetraploidy in the marsupial mouse) may be the first step toward the development of gene duplications required for fitness in the *in vitro* environment. The subsequent increase in late replicating regions of various chromosomes (e.g., aneuploid chromosomes 3, 4, and Y) may be one of the steps toward masking of unnecessary genes. Further adjustments can be made through translocation or deletions of particular portions of chromosomes. No doubt these mechanisms must work in concert in order to obtain optimal gene activity for homeostasis in the new environment. It would appear at the present time that analysis of karyotype changes in marsupial cells *in vitro* may lead to profound readjustment of our knowledge concerning mammalian evolution and cellular differentiation. It is not too unreasonable to suggest that the role of endoreplication, endomitosis, translocation, deletion, nondisjunction, and gene duplication in mammalian differentiation may be easily studied with karyotypically simple marsupial cells by alterations of the *in vitro* environment. Finally, analysis of alterations in the karyotype of marsupial cells should begin with the first day *in vitro*. Furthermore, attempts should be made to correlate these changes with specific enzyme pathways and proteins.

E.   Diploidy, Contact Inhibition, and Temperature Effects in a Cell Line from *Potorous tridactylus*

In contrast to the changes in ploidy observed in *A. swainsonii* and *P. bicolor*, 96% of kidney epithelial cells from *P. tridactylus* have retained diploidy as well as contact inhibition after 4 years and 4 months (420 passages) *in vitro*. The retention of these characteristics has been attributed to frequent subculturing and changing of medium (Bick and Brown, 1969). However, it would be useful to have data on the relative DNA content of chromosomes as done on cell lines from *P. bicolor* (Radley, 1966).

It is interesting to note that the kidney cells from *P. tridactylus* survive extended exposures to low temperatures. These cells have been maintained at 27°C for several months. Subsequent subculturing at 37°C revealed that no changes had occurred in chromosome morphology as a result of the low temperature treatment. In addition, cultures have been kept at 27°C for at least 14 months with a monthly change of medium. During this period of time very little or no division took place and the population density decreased. When these cultures were raised to 30°–31°C cell proliferation began. Similarly, in this laboratory fibroblasts from *M. mitis* have been maintained at 25°C for 72 hours without adverse changes on cell morphology (Thrasher, unpublished observations, 1969). Thus, it appears that marsupial cells may be synchronized by temperature shock.

F.   Chromosome Breakage by Heliotrine in *Potorous tridactylus* Leukocytes

The radiomimetic effects of pyrrolizidine alkaloid heliotrine on the chromosomes of *P. tridactylus* have been reported (Bick and Jackson, 1968). The data again emphasize the advantages of a simple karyotype to analyze the effects of mutagenic agents on chromosomes. Heliotrin had the following effects upon *P. tridactylus* chromosomes: (1) despiralization of chromosomes in 10% of the cases; (2) suppression of mitosis in concentrations above $2 \times 10^{-7} M$; (3) chromatid interchange at the centromeric regions; (4) formation of centric and acentric parts; and (5) total isochromatid deletions.

VI. Concluding Remarks

An attempt has been made to summarize the current information that has been obtained on marsupial cells *in vivo* and *in vitro* in order to demonstrate their potentiality as cytogenetic tools for the cell biologist. A variable cell cycle and simple karyotypes are the predominant attributes of marsupial cells. In addition, marsupial cells are easily cultured by slight modifications of standard tissue culture media. Preliminary observations indicate that cells *in vitro* are resistant to prolonged exposures to low temperatures. Thus, they appear to be excellent models for cell synchrony by temperature shocks as well as for investigating the evolution of chromosomes under *in vitro* culture conditions.

It seems plausible to predict at the present time that marsupials will be used for developmental biology, *in vitro* fertilization and culturing of embryos, and for investigating the ontogeny of the immune system. The limiting factors for establishing marsupials as routine laboratory animals appear to be lack of knowledge on behavior, food requirements, and reproductive physiology.

ACKNOWLEDGMENTS

I wish to thank Mrs. Z. Trirogoff for her skilled technical assistance and to personally acknowledge the contributions of Drs. Hayman, Jackson, Shaw, Hunsaker, Moore, Schneider, Sinha, and Wolf.

REFERENCES

Agar, W. E. (1923). *Quart. J. Microsc. Sci.* **67**, 183.
Altmann, S. C. A., and Ellery, M. E. W. (1925). *Quart. J. Microsc. Sci.* **69**, 463.
Bader, S., Miller, O. J., and Mukherjee, B. B. (1963). *Exp. Cell Res.* **31**, 100.
Barnes, R. D. (1968). *Lab. Animal Care* **18**, 251.
Barnes, R. D., and Barthold, S. W. (1969). *J. Reprod. Fert. Suppl.* **6**, 477.
Bick, Y. A. E., and Brown, J. K. (1969). *Cytobios* **2**, 123.
Bick, Y. A. E., and Jackson, W. D. (1968). *Aust. J. Biol. Sci.* **21**, 469.
Biggers, J. D., Fritz, H. I., Hare, W. C. D., and McFeeley, R. A. (1965). *Science* **148**, 1602.
Block, M. W. (1964). *Ergeb. Anat. Entwicklungsgesch.* **37**, 237.
Burton, D. W. (1968a). *Exp. Cell Res.* **49**, 300.
Burton, D. W. (1968b). *Exp. Cell Res.* **53**, 329.

Cabrera, A. (1957). *Rev. Mus. Argent. Cienc. Natur. Bernadino Rivadavia* No. 4, pt. 1.

Cairnie, A. B., Lamerton, L. F., and Steel, G. G. (1965). *Exp. Cell Res.* 38, 528.

Cameron, I. L. (1971). *In* "Cellular and Molecular Renewal in the Mammalian Body" (I. L. Cameron and J. D. Thrasher, eds.). Academic Press, New York.

Cameron, I. L., and Greulich, R. C. (1963). *J. Cell Biol.* 18, 31.

Cave, M. D. (1966). *J. Cell Biol.* 29, 209.

Cleaver, J. E. (1967). "Thymidine Metabolism and Cell Kinetics," 259 pp. Wiley, New York.

Dreyfus, A., and Campos, J. E. (1941). *Univ. Sao Paulo Fac. Fil. Cienc. Letras Biol. Ger.* 3, 3.

Drummond, F. H. (1933). *Quart. J. Microsc. Sci.* 76, 1.

Drummond, F. H. (1938). *Cytologia* 8, 343.

Farris, E. J. (ed.). (1952). "The Care and Breeding of Laboratory Animals," p. 256. Wiley, New York.

Greenwood, A. W. (1923). *Quart. J. Microsc. Sci.* 67, 203.

Hartman, C. G. (1952). "Possums," 174 pp. University of Texas Press, Austin, Texas.

Hayman, D. L., and Martin, P. G. (1965a). *Cytogenetics* 4, 209.

Hayman, D. L., and Martin, P. G. (1965b). *Genetics* 52, 1201.

Hsu, T. C., Billen, D., and Levan, A. (1961). *J. Nat. Cancer Inst.* 27, 515.

Jackson, L. G., and Ellem, K. A. O. (1968). *Cytogenetics* 7, 183.

Kalmutz, S. E. (1962). *Nature (London)* 193, 851.

Killander, D., and Rigler, R., Jr. (1965). *Exp. Cell Res.* 39, 701.

Koller, P. C. (1936). *J. Genet.* 32, 451.

LaPlante, E. S., Burrell, R. G., Watne, A. L., and Zimmerman, B. (1966). *Surg. Forum* 17, 200.

Legator, M., Jacobson, C., Perry, M., and Dolimpo, D. (1966). *Life Sci.* 5, 397.

Lesher, S., Fry, R. J. M., and Kohn, H. I. (1961). *Exp. Cell Res.* 24, 334.

Lyon, M. F. (1966). *In* "Advances in Teratology" (Woollam, D. H. M., ed.), Vol. 1, p. 25. Academic Press, New York.

McCrady, E., Jr. (1938). "The Embryology of the Opossum," 233 pp. Wistar Institute of Anatomy and Biology, Philadelphia. (The American Anatomical Memoirs, No. 16.)

McIntosh, A. J., and Sharman, G. B. (1953). *J. Morphol.* 93, 509.

Matthey, R. (1934). *C. R. Soc. Biol. (Paris)* 117, 406.

Miller, J. A. F. P., Block, M., Rowlands, D. T., and Kind, P. (1965). *Proc. Soc. Exp. Biol. Med.* 118, 916.

Moore, R. (1965). *Cytogenetics* 4, 145.

Moore, R., and Gregory, G. (1963). *Nature (London)* 200, 234.

Moore, R., and Radley, J. M. (1968). *Exp. Cell Res.* 49, 638.

Moore, R., and Uren, J. (1965). *Exp. Cell Res.* 38, 341.

Moore, R., and Uren, J. (1966). *Exp. Cell Res.* 44, 273.

Moorhead, P. S., and Defendi, V. J. (1962). *J. Cell Biol.* 15, 390.

Moorhead, P. S., Nowell, P. C., Mellman, W. J., Battips, D. M., and Hungerford, D. A. (1960). *Exp. Cell. Res.* 20, 613.

Morishima, A., Grumbach, M. M., and Taylor, J. H. (1962). *Proc. Nat. Acad. Sci.* 48, 756.

Painter, T. S. (1922). *J. Exp. Zool.* 35, 13.

Perondini, A. L. P., and Perondini, D. R. (1966). *Cytogenetics* 5, 28.

Potkay, S. (1970). *Lab. Animal Care* 20, 502.

Radley, J. M. (1966). *Exp. Cell. Res.* **41**, 217.

Reig, O. A. (1968). *Experientia* **24**, 185.

Rowlands, D. T., LaVia, M. F., and Block, M. H. (1964). *J. Immunol.* **93**, 157.

Saez, F. A. (1931). *Amer. Natur.* **65**, 287.

Saez, F. A. (1938). *Rev. Soc. Argent. Biol.* **14**, 156.

Schneider, L. K., and Rieke, W. O. (1967). *J. Cell Biol.* **33**, 497.

Schneider, L. K., and Rieke, W. O. (1968). *Cytogenetics* **7**, 1.

Sinha, A. K. (1967a). *Experientia* **23**, 671.

Sinha, A. K. (1967b). *Experientia* **23**, 889.

Sinha, A. K. (1967c). *Exp. Cell Res.* **47**, 443.

Sharman, G. B. (1961). *Aust. J. Zool.* **9**, 38.

Sharman, G. B. (1954). *West Aust. Natur.* **4**, 159.

Sharman, G. B., and Barber, H. N. (1952). *Heredity* **6**, 345.

Shaw, M. W., and Krooth, R. S. (1964). *Cytogenetics* **3**, 19.

Taylor, K., and Hunsacker, D. (1969). Personal communication.

Thrasher, J. D. (1967). *Anat. Rec.* **157**, 621.

Thrasher, J. D. (1969a). *Exp. Cell Res.* **57**, 442.

Thrasher, J. D. (1969b). *Lab. Animal Care* **19**, 67.

Thrasher, J. D. (1970). *Experientia* **26**, 871.

Thrasher, J. D., Barenfus, M., Rich, S., and Shupe, D. (1971a). *Lab. Animal Science* **21**:526.

Thrasher, J. D., Berg, N., and Hauber, E. (1971b). *Cell Tissue Kinet.* **4**, 185.

Uren, J., Moore, R., and van den Brenk, H. (1966). *Exp. Cell Res.* **43**, 677.

Walen, K. H., and Brown, S. W. (1962). *Nature* (*London*) **194**, 406.

Walker, E. P. (1968). "Mammals of the World," 2nd Ed., Vol. 1, 644 pp. Johns Hopkins Press, Baltimore, Maryland.

Winter, G. C. G., and Yoffey, J. M. (1965). *Nature* (*London*) **208**, 1018.

Wislocki, G. B. (1933). *Quart. Rev.* **8**, 385.

Wolf, H. G., Klein, A. K., and Foin, A. T. (1970). *Cytobios* **6**, 97.

Zee, Y. C. (1969). "Marsupial Newsletter," Vol. 1, No. 2, p. 2. Radiobiology Laboratory, University of California, Davis, California.

Chapter 5

Nuclear Envelope Isolation

I. B. ZBARSKY

Institute of Developmental Biology, Academy of Sciences of the USSR, Moscow, USSR

I. Introduction

Cell nuclei can be now isolated from many different sources and fractionated into subnuclear components (Zbarsky, 1963; Busch, 1967;

Roodyn, 1969), and isolation and purification of substructural entities such as chromatin and nucleoli from these nuclei is now possible (Frenster *et al.*, 1963; Busch *et al.*, 1965; Zbarsky and Yermolaeva, 1968). However, among the major membraneous cellular structures only the nuclear membranes were not isolated and studied biochemically until quite recently.

It is quite apparent that the nuclear envelope deserves thorough investigation, which is impossible without the isolation of the nuclear envelope itself and/or its membranes.

The nuclear envelope consists of two nuclear membranes, the inner and the outer one. The outer nuclear membrane is directly connected to the endoplasmic reticulum and is usually regarded as a part of it. There is a space between the two nuclear membranes, called the perinuclear space. A remarkable feature of the nuclear envelope is the presence of rather large nuclear pores (400–1000 Å in diameter). At the boundaries of the pores the two nuclear membranes are connected and form an area of enhanced mechanical stability (Gall, 1964; Franke, 1970; Yoo and Bayley, 1967). The structure of the nuclear envelope (Fig. 1) is very similar in different eukaryotic cells (Franke, 1970).

Isolation of the nuclear envelope from individual amphibian oocyte nuclei was accomplished by Callan and Tomlin in 1950 using a microsurgical technique. Only during the last few years have methods been developed which allow one to obtain nuclear membranes in larger scale and to learn something about their chemical composition and their biochemical and biophysical properties (Zbarsky *et al.*, 1967, 1969; Zbarsky, 1969; Kashnig and Kasper, 1969; Berezney *et al.*, 1970; Franke *et al.*, 1970).

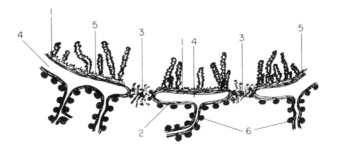

Fig. 1. Schematic structure of the nuclear envelope fragment. 1, Inner nuclear membrane; 2, outer nuclear membrane; 3, pores and "pore complex"; 4, perinuclear space; 5, chromatin adjacent to the inner nuclear membrane; 6, endoplasmic reticulum with attached ribosomes.

II. Isolation of the Nuclear Envelope from Individual Cells

The first method for the isolation of nuclear envelopes was described by Callan and Tomlin (1950), working with oocytes of a newt and a toad. The gigantic nuclei, extracted from the oocytes by puncture, were dissected and the nuclear envelope separated from the nuclear content. This method permitted one to obtain the material only from individual nuclei and, therefore, could not be used for biochemical investigations, but proved valuable for electron microscopy. Later this method was repeated and used by other authors (Gall, 1954, 1964, 1967; Franke and Scheer, 1970) for different animal species.

The details of the Callan and Tomlin method follow. A part of the ovary from a newt, *Triturus cristatus*, or a toad, *Xenopus laevis*, is excised. The excised fragments are immersed in liquid paraffin or placed in dry cups and covered by glass. In liquid paraffin, the oocyte is still surrounded by body cavity fluid and does not touch the paraffin, being, at the same time, protected from the air. These ovarian fragments remain in their native state for more than 24 hours at a temperature of 3–6°C.

A piece of *Triturus* ovary containing a few oocytes is rinsed a few times with distilled water. Then it is placed in a glass vessel under a binocular microscope at low magnification (×16). The place of attachment of a large transparent oocyte (the nucleus being visible inside) is seized with watchmakers forceps, and the follicular envelope surrounding the oocyte is punctured by a sharp needle with a shallow, almost tangential prick. Slight pressure (by means of the needle) is then applied, and the nucleus is extruded with the cytoplasm. The nucleus is then removed by a pipette with an orifice (smoothed in a small gas flame) a little wider than the diameter of the nucleus.

In the case of *Xenopus*, the oocytes of which are not transparent, the nucleus lies just below the surface of the pigmented pole. The follicular envelope in this species is much stronger than in *Triturus*, and if it is pricked near the nucleus the nucleus often drops from the oocyte free from cytoplasm.

The isolated nuclei are put in distilled water, where they are allowed to stretch until they are about 1½ times the initial diameter. Then the nuclei are punctured, and the nuclear envelopes are stretched over grids, fixed, and examined by electron microscopy.

If the nuclei are left in water, they continue to distend until they finally burst, and the nuclear contents, including nucleoli, liberated into the surrounding solution. In this case, the nuclear membranes are suitable for examination, but they are quite distended.

The method is applicable to other animal species and has been used by several authors with some modifications. Gall (1967) successfully used this method for another newt, *Triturus viridescens*, a frog, *Rana pipiens*, and a starfish, *Henricia sanguinolenta*. Instead of distilled water he uses a solution composed of 5 parts 0.1 M KCl and 1 part 0.1 M NaCl. Oocytes are broken with forceps and the nucleus sucked in and out of a pipette a few times to remove the yolky contamination. The isolated nucleus is flattened against the supporting Formvar film by drawing off the liquid with filter paper and then washed in an additional drop of the liquid.

Franke and Scheer (1970) used the same KCl–NaCl medium with the addition of 2 to 10 mM MgCl$_2$ or CaCl$_2$. They also used a medium consisting of 0.2 M sucrose, 2% gum arabic, 0.02 M tris buffer (pH 7.2), and 4 mM n-octanol (Franke, 1966) or a pure hypotonic sucrose solution (0.1 or 0.05 M) containing 0.02 M tris buffer, pH 7.2.

III.   Isolation of the Nuclear Envelope by Cell Nuclei Fractionation

A.   Fractionation of Cell Nuclei

The old view that cell nuclei consist almost exclusively of nucleohistone was challenged by Stedman and Stedman (1943), who described a large amount of an acidic protein "chromosomin" in cell nuclei from different tissues, and by Mirsky and Pollister (1946), who found a nonhistone protein in the deoxyribonucleoprotein fraction extracted from rat liver cell nuclei with 1 M NaCl. In both cases, the nuclei were isolated in acidic media, and, therefore, a reasonable fractionation of nuclear material was impossible.

The first data on a systematic fractionation of nuclei isolated in natural media were presented by Zbarsky and Debov (1948, 1951) and Zbarsky (1950). Nuclei isolated from different tissues with dilute neutral sodium citrate (Zbarsky and Perevoshchikova, 1948, 1951) were exhaustively extracted with 1 M NaCl to obtain a "nucleoprotein fraction." The extraction of the residue with a 0.01–0.05 N NaOH solution yielded a fraction of "acidic protein." The residue after alkaline extraction consisted of an extremely insoluble fibrous material called "residual protein."

A few years later, Soudek and Beneš (1955) showed that the residual protein fraction originated from the nuclear envelope. The method was then improved in our laboratory; preliminary extraction of the globulin fraction was added to the procedure, and the cytological characteristics of each fraction were described (Georgiev, 1958; Zbarsky and Georgiev, 1959a,b). It was found that the globulin fraction extracted with 0.14 M NaCl corresponded to the nuclear sap, the deoxyribonucleoprotein fraction to chromatin, and the acidic protein fraction to nucleoli and "residual chromosomes." The origin of the residual protein fraction from the nuclear envelope was also confirmed (Georgiev et al., 1960). The data on fractionation and cytological characteristics of nuclei were summarized by Zbarsky (1963).

B. Nuclear Residual Protein Fraction

A more detailed study of this fraction has proved its origin from the nuclear envelope. The residual fraction may be obtained after the extraction of purified nuclei with dilute (0.01–0.05 N) alkali. This fraction can be used for characterization of the nuclear envelope structural proteins. Electron microscopy reveals that the fraction is membraneous in appearance, and amino acid analysis demonstrates the very hydrophobic character of its constituent amino acids and a high proline content (Zbarsky et al., 1962a,b; Zbarsky, 1969; Soudek and Nečas, 1963; Steele and Busch, 1964). This fraction contains approximately 3–5% of the total nuclear protein in liver and other normal tissues studied. In neoplastic tissues the nuclear residual protein fraction is more abundant, has a different amino acid composition, and, in addition to the membraneous structures, also contains amorphous lamellar material (Zbarsky et al., 1962a,b).

C. Other Fractions Enriched with Nuclear Membranes

It is clear from the fractionation procedure described above that the extraction of isolated nuclei with dilute alkali yields in residue a denatured fraction containing essentially nuclear membranes (the residual protein fraction). However, a fraction enriched in nuclear membranes (acidic protein plus residual protein fractions) can also be obtained after extraction of the isolated nuclei with salt solutions of high ionic strength (e.g., 1–2 M NaCl).

This principle was used to extract thymus nuclei by Conover (1970b) and Betel (1972), and a fraction rich in nuclear membranes with cytochrome oxidase activity about tenfold higher than in initial nuclei was obtained. Essentially similar results were obtained with rat liver nuclei in our laboratory.

Conover (1970b) used isolated calf thymus nuclei treated with Triton and purified by high-density sucrose centrifugation. The nuclei (1.5–2.0 gm protein) were then treated for 10 minutes at 30°C with DNase (20 mg) and polyethylene sulfonate (100 mg) dissolved in 50 ml of 0.25 M sucrose solution containing 0.003 M CaCl$_2$, 0.01 M MgCl$_2$, and 0.01 M tris buffer, pH 7.2. The nuclear suspension was then diluted with 3 volumes of cold 0.25 M sucrose–0.003 M CaCl$_2$ solution and sedimented by centrifugation at 800g for 10 minutes.

The nuclear pellet is suspended in 80 ml of a solution containing 0.15 M NaCl, 0.02 M tris buffer (pH 7.2), and 0.002 M EDTA by means of a Dounce homogenizer and extracted at 4°C for 20 minutes with stirring. The pellet after centrifugation of this suspension at 12,000g for 10 minutes is now resuspended and extracted in the same manner with 80 ml of 1.0 M NaCl–0.02 M tris buffer, pH 7.2. The fraction enriched in nuclear membranes is pelleted at 40,000g for 20 minutes, suspended in 0.25 M sucrose–0.03 M CaCl$_2$, and again centrifuged at 40,000g for 20 minutes and resuspended in 0.25 M sucrose–0.003 M CaCl$_2$.

The principle of salt extraction with high ionic strength was used by several authors for isolation of purified nuclear membranes (see Section V).

IV. Isolation of the Nuclear Envelope and Nuclear Membranes by Disintegration of Nuclei and Gradient Centrifugation

Cell disintegration and gradient centrifugation were used by several authors for the isolation of different types of biological membranes. In our laboratory we described two procedures for the isolation of the nuclear envelope and of the two nuclear membranes from rat liver and ascites tumor nuclei (Zbarsky *et al.*, 1967, 1969; Perevoshchikova *et al.*, 1968). A similar method for the isolation of the nuclear envelope from onion roots and, later, from mouse liver was independently suggested by Franke (1966, 1967).

A. Isolation of Nuclei from Rat Liver and Ascites Tumors

The nuclei from rat liver are isolated by a modification of the Chauveau method (Chauveau et al., 1956). We usually employ a two-step procedure described by us (Georgiev et al., 1960) or the modification of a method by Di Girolamo et al. (1964). The nuclei should be essentially free from impurities. The critical point is the absence of extranuclear membraneous admixtures such as endoplasmic reticulum, mitochondria, and plasma membranes, which may give significant contamination after the isolation of the membraneous material.

Rats weighing 150–250 gm are sacrificed by decapitation. The livers are removed in a cold room as soon as possible, cut with scissors into small pieces, and rinsed one or two times with cold solution of 0.25 M sucrose, 0.003 M $CaCl_2$, and 0.01 M tris buffer, pH 7.5. The pieces of liver are homogenized in 10 volumes of the same solution in a glass Potter–Elvejhem homogenizer with a Teflon pestle for 2–3 minutes. The homogenate is filtered through nylon (160 mesh) or through four and then eight layers of gauze and centrifuged at 1000g for 10 minutes.

The crude nuclear pellet is suspended in 10 volumes of 2.3 M sucrose solution containing 0.0015 M $CaCl_2$ and 0.01 M tris buffer, pH 7.5, and gently homogenized in a glass Potter–Elvejhem homogenizer with a Teflon pestle. The nuclei are sedimented from this suspension by centrifugation at 35,000g for 30 minutes. The pellet of purified nuclei is washed by resuspension in 3 to 4 volumes of 0.25 M sucrose–0.003 M $CaCl_2$ solution and centrifuged at 400g for 5 minutes.

The purity of isolated nuclei is controlled by light microscopy (staining with methylene blue or methyl green pyronin) and phase-contrast microscopy. Nuclear preparations are systematically checked by electron microscopy of ultrathin sections, enzymatic tests (for glucose-6-phosphatase and succinate dehydrogenase), and chemical analysis.

The nuclear preparations are essentially free from cytoplasmic contamination and contain both the inner and the outer nuclear membranes (Fig. 2).

The nuclei from tumors are more difficult to obtain in the pure state. Reasonably clean nuclei can be isolated in media containing low concentration of nonionic detergent. The method of Fisher and Harris (1962) in which a 0.1% solution of Tween 80 is used proved to be satisfactory for isolation of nuclei, while preserving both nuclear membranes (Fig. 3), from mouse Ehrlich ascites carcinoma cells and rat Zajdela ascites hepatoma.

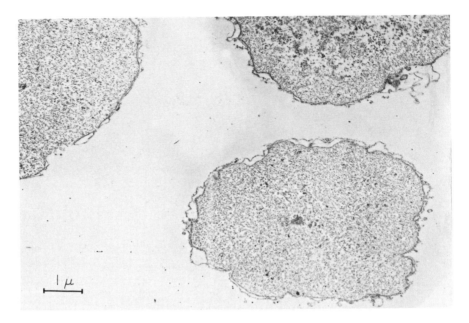

FIG. 2.   Ultrathin section of isolated rat liver nuclei. Sections are stained with lead hydroxide. Both membranes of the nuclear envelope are clearly seen.

The ascitic fluid is collected from mice weighing about 20 gm on the 7th day after inoculation and placed in a cooled beaker containing 1 ml of 5% sodium citrate for each 20 ml of ascitic fluid. The cells are collected by centrifugation at 2500g for 10 minutes and washed twice by suspension in 4 volumes of 0.075 M K_2HPO_4 solution and centrifugation at 2500g for 10 minutes. The pellet of washed cells is resuspended in 3 volumes of cooled 0.1% (v/v) solution of Tween-80 in distilled water and homogenized in a Waring blendor with nylon knives at 5000 rpm for 10 minutes. Nuclei are sedimented by centrifugation at 2500g for 10 minutes and washed by suspension in an equal volume of 0.1% Tween-80 solution and recentrifuged.

The nuclei from Zajdela ascites hepatoma can be isolated by the same method but a 10-minute homogenization results in disruption not only of the cells but also the nuclei; therefore, they are homogenized for 5 minutes at 2500 rpm and centrifuged at 2500g 5 minutes. The nuclei at the end of the isolation are washed twice with 0.1% Tween-80 solution. Depending upon the material and equipment the best procedure for any other type of cell can be found empirically.

In general, other methods for isolation of the nuclei can also be used

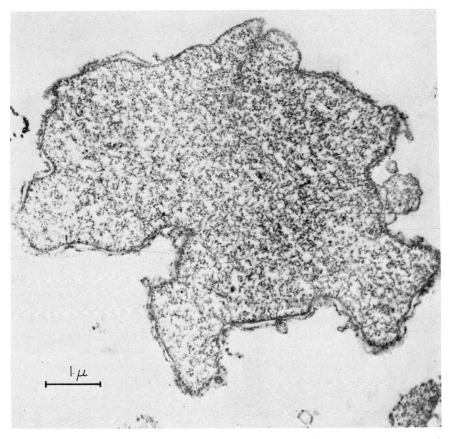

FIG. 3. Ultrathin section of an Ehrlich ascites cell nucleus. Sections are stained with lead hydroxide. Both nuclear membranes are preserved.

if the nuclear preparations retain both nuclear membranes and are sufficiently pure.

Franke (1967) used a slightly modified method of Kuehl (1964). Small pieces of mouse liver were incubated from 5 to 15 hours in a 100-fold volume of isolation medium containing 4% gum arabic, 0.1 M sucrose, 0.004 M n-octanol, and 0.02 M tris buffer, pH 7.5. In contrast to the procedure of Kuehl, however, this medium did not contain Ca^{2+} ions. After the incubation, the pieces of liver were homogenized in fresh medium in a Waring blendor for 10 to 20 seconds and filtered through nylon and flannelette. After centrifugation at 220g for 8 minutes, the nuclear pellet was resuspended in 0.775 ml of isolation medium to which 6 ml of superconcentrated sucrose solution (245 gm sucrose in 100 ml of water) were added. The suspension was layered over a similar super-concentrated sucrose solution (a modified discontinuous sucrose gradient according to Birnstiel et al., 1962) and centrifuged in the SW-25 rotor of a Spinco L50 ultracentrifuge—first for 10 minutes at 10,000g and then for 45 minutes at 72,000g. The nuclei are concentrated at the interface of the two sucrose layers. They are collected, diluted in 3 volumes of initial isolation medium, and resedimented by centrifugation at 500g for 10 minutes.

B.   Isolation of the Nuclear Envelope

According to our method (Zbarsky et al., 1967, 1969; Perevoshchikova et al., 1968) the nuclear envelopes from rat liver and ascites tumors can be isolated by osmotic shock of purified nuclei followed by gradient centrifugation in a discontinuous sucrose gradient. This method results in isolation of whole nuclear envelope fragments and preserves both nuclear membranes and pore complexes. The method is apparently suitable for cell nuclei from different tissues.

Isolated nuclei are washed initially with 5 volumes of 0.25 M sucrose solution containing 0.003 M $CaCl_2$ and then with 5 volumes of 0.25 M sucrose solution without calcium. The washing is performed each time by suspension and centrifugation for 5 minutes at 400g. Then 4 volumes of 0.02 M phosphate buffer, pH 7.2, are added to 1 volume of nuclei, and the suspension is shaken in a cold room at approximately 4°C for 30 minutes. This treatment produces not only osmotic shock but also detachment of the chromatin adjacent to the inner nuclear membrane (Whitfield and Perris, 1968).

The suspension is then centrifuged at 400g for 17 minutes. The supernatant is stored, and the pellet containing intact nuclei and chromatin clumps is slightly homogenized (3 to 4 strokes) in a loosely fitted glass Potter–Elvejhem homogenizer in 1 volume of the same phosphate buffer and again centrifuged at 400g for 7 minutes. The pellet is discarded, and the pooled supernatants are thoroughly mixed with 2 volumes of 69% sucrose solution. Aliquots (7.5 ml) of this suspension are transferred to centrifuge tubes of the Spinco No. 40 angle rotor. This suspension is overlayered with the sucrose solutions with densities (d_4^{20}) 1.19, 1.18, 1.16, 1.14, and 1.09, 1.0 ml of each forming a discontinuous sucrose gradient, and centrifuged for 75 minutes at 36,000 rpm (80,000g, r_{av}) in the Spinco angle rotor No. 40. A small pellet of whole nuclei and nuclear fragments is discarded. The membraneous material is concentrated in the sucrose layer $d = 1.19$ or at the interface of the layer $d = 1.19$ and the initial suspension.

Sometimes a faint membraneous nebula appears at the interface of sucrose layers with $d = 1.16$ and 1.14. This material probably contains a small portion of outer nuclear membranes (see below, Section IV,C), detached in the course of the isolation procedure, and it is usually discarded.

The material from the layer $d = 1.19$ is collected with a pipette or a syringe equipped with a needle, suspended in 10 volumes of distilled water, and sedimented by centrifugation at 23,000g for 30 minutes. The crude pellet of the nuclear envelope fragments is purified by flotation in the above-mentioned discontinuous sucrose gradient and centrifugation in the SW-39 rotor. The pellet is suspended in a small volume of distilled water, and the suspension is transferred to centrifuge tubes of the SW-39 rotor (0.8 ml of the suspension into each tube); then 2 volumes (1.6 ml) of 69% sucrose solution are added and thoroughly mixed. The above-mentioned discontinuous sucrose gradient is layered over this suspension using 0.5 ml of each sucrose solution. The gradient is centrifuged in an SW-39 rotor at 105,000g (36,000 rpm) for 75 minutes.

The material from the layer $d = 1.19$ is again collected, diluted with 10 volumes of distilled water, and centrifuged for 30 minutes at 23,000g. Sucrose is removed from the pellet by suspension two to three times in approximately 10 volumes of cold distilled water and centrifugation at 23,000g for 30 minutes. The washed pellet of nuclear envelope fragments may be used for biochemical studies and electron microscopy. The isolation procedure is illustrated by the scheme shown in Fig. 4.

Electron microscopy of positively contrasted preparations of purified isolated nuclear envelope clearly demonstrates the nuclear pores and their annuli with a diameter of 500 to 1000 Å. One square micron of the surface contains approximately 30 pores (Fig. 5). Ultrathin sections of purified pieces of isolated nuclear envelope also reveal both nuclear membranes and pores (Fig. 6).

For the isolation of nuclear envelope from ascites tumor nuclei purified in Tween-80–containing media, 1 volume of 0.1 M phosphate buffer, pH 7.2, is added to 5 volumes of wet nuclei. The suspension is shaken in the cold for 90 minutes and centrifuged at 1100g for 15 minutes. The pellet is discarded and the supernatant mixed with 2 volumes of 69% sucrose solution. Otherwise the procedure is similar to those described above for the isolation of the nuclear envelope from rat liver.

A rather similar method was described by Franke (1966, 1967). The pellet of nuclei isolated according to the method of Kuehl (1964), without calcium (Section IV,A), is suspended in a few drops of isolation medium and diluted with 4 ml of distilled water or 0.02 M sucrose solution. This treatment leads to swelling of nuclei, rupture of the nuclear

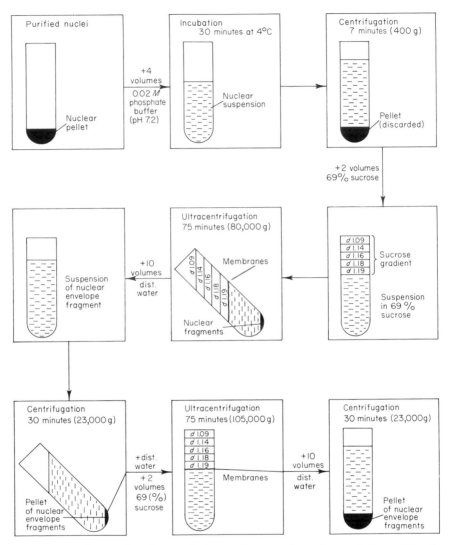

FIG. 4. Scheme of isolation of the nuclear envelope from rat liver (Zbarsky *et al.*, 1967, 1969; Perevoshchikova *et al.*, 1968).

envelope, and detachment of nucleoli and most of the chromatin from the envelopes. This suspension is then gently sonicated (1–3 seconds at power setting 1 of the Branson sonifier S-125).

For electron microscopy the author recommends immediate fixation in 4 ml of 2% OsO_4 dissolved in Veronal-acetate buffer, pH 7.2. After fixation for 30 minutes to 1 hour the undamaged nuclei and nucleoli are

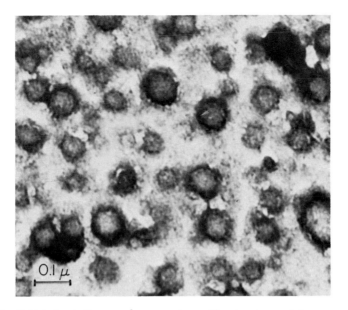

FIG. 5. Isolated rat liver nuclear envelope. Fixation in acrolein vapor. Positive contrast with uranyl acetate. Nuclear pores are clearly seen.

sedimented from the suspension by centrifugation for 3 minutes at 200g. The pellet is discarded, and the supernatant is centrifuged at 1500g for 8 minutes. The pellet containing membraneous material is washed two more times by repeated suspension and centrifugation at 1500g for 8 minutes (Franke, 1966).

Nuclear envelopes of higher purity for biochemical studies or electron

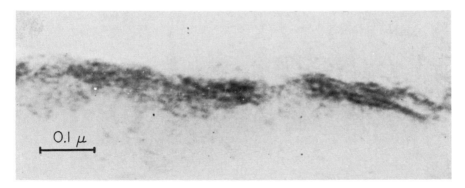

FIG. 6. Ultrathin section of a piece of isolated rat liver nuclear envelope. Both nuclear membranes and a pore are seen.

microscopy can be obtained by centrifugation of a freshly sonicated suspension without fixation, layered over a 62% (w/v) sucrose solution for 50 minutes at 3000g. The nuclear membrane layer over the sucrose solution is collected and washed in distilled water. The procedure of centrifugation over 62% sucrose and washing may be repeated (Franke, 1967).

C.   Isolation of Nuclear Membranes

An alternate procedure for the method described by us (Zbarsky et al., 1967, 1969; Perevoshchikova et al., 1968) allows the isolation of two nuclear membrane fractions, a heavier and a lighter one. Both of these fractions contain only a single membrane and are believed to represent the inner and outer nuclear membranes, respectively (see Sections VII and VIII).

The purified nuclear pellet is washed by suspension in 4 volumes of 0.25 M sucrose solution without calcium and centrifuged for 5 minutes at 400g. The pellet is gently homogenized in a loosely fitted glass Potter–Elvejhem homogenizer with 5 volumes of 0.25 M sucrose solution without calcium and allowed to stand in a cold room at 4°C for 10–15 minutes. The nuclei swell in this solution and the nuclear envelope becomes partly detached. The nuclear suspension is now sonicated at 20 kc for 15 to 20 seconds (MSE sonifier type L667, amplitude 6.5). The degree of disintegration of nuclei is monitored with light microscopy of smears stained with methylene blue. The sonication is repeated if more than 30% of nuclei remain undamaged.

After the sonication the suspension is centrifuged for 7 minutes at 400g. The pellet of nuclei and nuclear debris is discarded, and the supernatant is thoroughly mixed with 2 volumes of 69% sucrose solution. Aliquots (7.5 ml) of this suspension are transferred to each centrifuge tube of a Spinco No. 40 angle rotor, 1 ml of sucrose solution $d = 1.20$ is overlayered followed by the above-mentioned (see Section IV,B) discontinuous sucrose gradient, and the procedure described under Section IV,B for the isolation of nuclear envelope is followed. This procedure (beginning with sonication) results in the isolation of two membrane bands which are found between the sucrose layers $d = 1.19$ and $d = 1.20$ and $d = 1.16$ and $d = 1.14$ (membranes "ρ 1.19" and "ρ 1.16"). Each of these layers is collected separately, washed with distilled water by centrifugation for 30 minutes at 23,000g, purified in a discontinuous sucrose gradient in a SW-39 rotor, and washed again as was described for isolation of the nuclear envelope in Section IV,B (Fig. 7). Similar nuclear membrane fractions can be isolated from Ehrlich ascites carcinoma cells

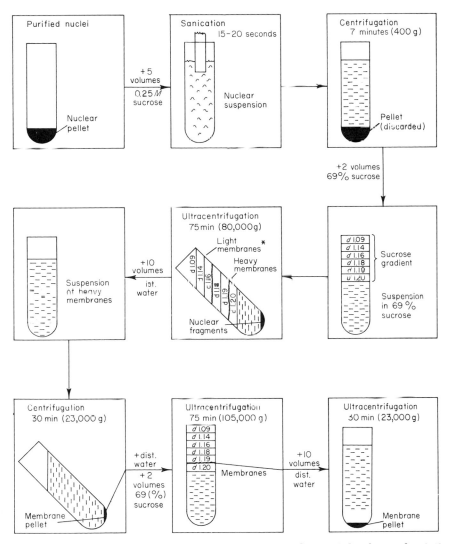

FIG. 7. A scheme for rat liver nuclear membrane isolation. (Zbarsky *et al.*, 1967, 1969; Perevoshchikova *et al.*, 1968.) *, Light nuclear membranes are treated separately in the same way.

and Zajdela ascites hepatoma and probably also from cell nuclei of different other tissues.

The procedure of membrane isolation from ascites tumor nuclei is essentially the same as for rat liver, but it was found empirically that sonication of nuclei proceeded better if the nuclear pellet was mixed with an equal volume of 0.25 M sucrose. After sonication the nuclear sus-

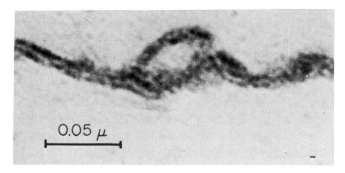

FIG. 8. Ultrathin section of an Ehrlich ascites cell light nuclear membrane. High magnification. A three-layered structure is clearly seen.

pension is directly (without centrifugation) mixed with 2 volumes of 69% sucrose solution and centrifuged with the overlayered discontinuous sucrose gradient. The isolation procedure otherwise is identical to that for rat liver.

Sometimes after the gradient centrifugation, both nuclear membrane layers of ascites tumors reveal a somewhat lower buoyant density, the light fraction bands at the interface of $d = 1.16$ and $d = 1.14$ sucrose layers and the heavy one inside the sucrose layer $d = 1.19$.

Both nuclear membrane fractions consist of single membranes. The yield of heavier membranes is somewhat higher (about 1.5 times by weight) than that of lighter membranes. Electron microscopy of ultrathin nuclear membrane sections reveals a characteristic three-layer structure: two electron-dense sheaths about 30 Å thick each and a less dense layer between them, some 40 Å in diameter (Fig. 8).

Both fractions differ biochemically (see Section VII) as well as mor-

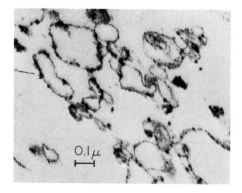

FIG. 9. Ultrathin section of a rat liver heavy nuclear membrane. Small granules, approximately 100 Å in diameter, are seen.

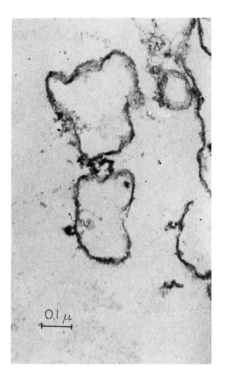

FIG. 10. Ultrathin section of a rat Zajdela hepatoma light nuclear membrane. Ringlike formations, apparently the remnants of pore complexes, are seen.

phologically. In ultrathin sections of the layers ρ 1.19, electron-dense granules about 100 Å in diameter can be seen (Fig. 9), which are absent from the fraction ρ 1.16. Moreover, dispersed thick areas with ringlike structures that probably represent the remnants of the pore complexes (Fig. 10) are observed in nuclear membrane sections, predominantly in the layer ρ 1.16. The nuclear membrane fractions from ascites tumors have a similar appearance but the two fractions are morphologically less different.

V. Isolation of the Nuclear Envelope by Salt Extraction at High Ionic Strength

These methods are based on the ability of salts, at high ionic strength, to solubilize deoxyribonucleoproteins of chromatin and nucleoplasm, leaving the nuclear envelopes as an insoluble residue (see Section III,A). Several methods based on this principle differ in the salt used, application of DNase or sonication, etc.

A.   Methods with Deoxyribonuclease

DNase treatment for isolation of nuclear membranes was first used by Widnell and Siekevitz (1967) for a comparative study of protein and lipid turnover in different cell membranes, but no details of the method were published.

A large-scale preparation of nuclear membranes from bovine liver is reported by Berezney et al. (1970). For isolation of the nuclei these authors homogenize large quantities (approximately 10 pounds) of bovine liver cut into pieces of about 1 cm^3 in 3 volumes of "H" buffer (0.05 M tris-HCl buffer, pH 7.5, 0.25 M sucrose, 0.025 M KCl, and 0.005 M MgCl$_2$) in a Waring blendor at 19,000 rpm for 30 seconds. Then, the homogenate is filtered sequentially through 2, 4, and 8 layers of cheesecloth and centrifuged at 1800g for 20 minutes. The crude nuclear pellet is then washed by suspension in H buffer and resedimented at 1800g for 20 minutes, resuspended in the H buffer containing 2.3 M sucrose, and filtered through 12 layers of cheesecloth. This filtrate is adjusted with 2.3 M sucrose–H buffer to 0.3 volume of the original filtered homogenate, and centrifuged for 30 minutes at 78,480g in a Spinco No. 30 rotor. The nuclear pellets are washed a few times with H buffer by resuspending and centrifuging for 5 minutes at 3600g.

To isolate the nuclear membranes the purified nuclei are suspended in H buffer to a concentration of 20 mg of nuclear protein per milliliter and are digested with DNase (50 μg/ml) at 2°C for 14 hours. The digested nuclei are washed twice with H buffer, resuspended in H buffer, and brought to 0.5 M in MgCl$_2$. This medium of high ionic strength solubilizes most of nuclear material. The nuclear membrane pellet is sedimented at 27,000g for 15 minutes. The process of extraction with 0.5 M MgCl$_2$–H buffer is repeated, and the nuclear membrane fractions recovered by similar centrifugation are washed twice in H buffer.

According to the authors, recovery of nuclear protein in the nuclear membrane preparation amounted to 10%. The membrane material contained about 75% protein, 9% RNA, and only approximately 14% phospholipid. Electron micrographs presented by the authors show the presence of both the inner and outer nuclear membranes and of the nuclear pores. Nevertheless, significant contamination of membranes with ribosomes, nuclear and nucleolar fragments, and other undefined material is seen.

A slightly different procedure was used by Ueda et al. (1969). These authors employed calf thymus nuclei obtained by Triton purification of a nuclear fraction isolated in 0.25 M sucrose and 0.003 M CaCl$_2$ according to Allfrey et al. (1957). The pellet is mixed with 2 volumes of a 2.3 M sucrose–0.003 M CaCl$_2$ solution, layered over a 1.95 M sucrose–0.003 M CaCl$_2$ solution, and centrifuged at 7500g for 30 minutes. The nuclear pellet is resuspended in 0.25 M sucrose–0.003 M CaCl$_2$ containing 0.25% Triton X-100 and recentrifuged immediately. The resulting "Triton nuclei" presumably contain only the inner nuclear membrane.

The Triton nuclei are digested at 0°C for 16 hours with DNase (15 μg/gm protein) in tris-HCl buffer, pH 7.4, containing 5 mM MgSO$_4$. The digest is mixed with an equal volume of 2 M NaCl and centrifuged for 10 minutes at 7500g. The pellet resuspended in 0.05 M tris buffer, pH 7.4, is called the "heavy nuclear membraneous fraction." The supernatant recentrifuged at 78,000g for 90 minutes gave another pellet—the "light nuclear membraneous fraction," which was resuspended in the same buffer. Both fractions were shown to have a high content of cytochromes comparable with mitochondria. No other chemical or morphological characterization was given. Both fractions represent probably the larger and smaller pieces of the inner nuclear membranes.

B. Methods without Deoxyribonuclease

A description of such a method for rat liver nuclear membranes has been published recently (Kashnig and Kasper, 1969). The method of Blobel and Potter (1966), adapted for large scale preparation, was used for nuclei isolation.

To a nuclear pellet containing approximately 16 mg protein, 10 ml of TKM buffer (0.05 M tris-HCl, pH 7.5, containing 0.025 M KCl and 0.005 M MgCl$_2$) are added. The nuclei are disrupted by sonication of the suspension for 10 to 15 seconds (setting 6.5, Sonifer cell disrupter W 140-D, Heat Systems, Melville, New York). The sonic treatment is repeated if more than 30% of nuclei remain undamaged (examination with a phase-contrast microscope). These conditions are crucial, since higher doses of sonication can destroy the nuclear membranes.

Solid potassium citrate is then added with stirring to the final concentration of 10% (w/v) to dissolve the bulk of nucleoplasm. Then the suspension is centrifuged for 45 minutes at 39,000g at 3°C and the supernatant discarded. The pellet is suspended in 3 ml of sucrose–TKM buffer–10% citrate solution with the final density 1.22 and transferred to a cellulose nitrate tube of a Spinco SW-25 rotor. The rest of the pellet is rinsed with 7 ml of sucrose solution of $d = 1.22$ and added to the tube. Over the membrane suspension a discontinuous gradient consisting of sucrose–TKM buffer–10% citrate solution is layered with densities of 1.20, 1.18, and 1.16, respectively. The volume of each layer is 7 ml. The gradient is centrifuged at 100,000g for 10 to 14 hours at 3°C. Nuclear membranes are collected at interfaces between the layers with $d = 1.18$ and 1.20 and $d = 1.16$ to 1.18. Both layers of membranes are harvested separately and purified by suspension in 2 volumes of TKM buffer–10% potassium citrate and centrifugation at 100,000g for 60 minutes.

Both membrane bands under the electron microscope appear as pieces of the nuclear envelope and exhibit almost identical morphology. Their chemical composition and enzymatic activity were also essentially similar. The heavier and the lighter fractions of the nuclear envelope contain no DNA and 6.1 and 3.1% of RNA, respectively.

Another method for isolation of nuclear membranes from rat and pig liver was quite recently described by Franke $et\ al.$ (1970).

Immediately after sacrifice of the animals, the livers are removed, cut into slices, and incubated for 2 to 8 hours in a medium containing 0.4 M sucrose, 0.07 M KCl, 2% gum arabic, 0.004 M n-octanol, and 0.01 M tris-acetate buffer, pH 7.2 (medium A). The material is then homogenized with a rotating knife homogenizer (Fa. E. Bühler, Tübingen, FRG) four times, 2–3 seconds each at 40,000 rpm, and centrifuged at 1500g for 20 minutes. The nuclear pellet is resuspended in medium B (the same as "A" but with 1.0 M sucrose and 1% gum arabic), slightly homogenized in a Potter-Elvejhem homogenizer, and centrifuged at 2000g for 30 minutes. This pellet is mixed with 1.4 volumes of medium C (the same as "A", but without gum arabic and with 3.0 M sucrose) and spun at 110,000g for 40 minutes. The pellet is resuspended in medium A and centrifuged for 15 minutes at 2000g. This pellet is resuspended in a small volume of medium A, thoroughly mixed with 4 volumes of medium D (same as C, but with 2.2 M sucrose), and layered over an equal volume of nondiluted medium D. The tubes containing these two layers are centrifuged for 90 minutes at 100,000g. The sediment of purified nuclei is collected, diluted with 10 (or more) volumes of medium A, and sedimented at 1500g for 20 minutes. This nuclear preparation is checked by phase-contrast and electron microscopy and is used for nuclear membrane isolation.

The nuclear preparation is suspended in 0.3 M sucrose containing 0.135 M KCl and 0.01 M McIlvaine's citrate–phosphate buffer, pH 7.4 (medium E), and sonicated ten times for 2 seconds each (with 15-second cooling intervals) using a Branson sonifier S125, position 5. After sonication the suspension is mixed with 20 volumes of medium E containing 1.5 M KCl and stirred in a cold room at 4°C for 5 to 6 hours. The nuclei and nuclear debris are eliminated by centrifugation at 1000g

for 20 minutes and discarded. The supernatant is spun at 110,000g for 1 hour. The supernatant is discarded and the pellet resuspended in a small amount of medium E, slightly sonicated, and layered on the top of a linear continuous sucrose gradient (30–70% w/v) containing 0.07 M KCl and 0.01 M McIlvaine's buffer, pH 7.4. The gradient is spun at 80,000g for 5 to 6 hours, and the sharp bands of nuclear membrane material with buoyant densities (ρ_4^{22}) 1.215 (for rat) and 1.203 (pig) are collected. Electron microscopy of membranes shows that they consist of nuclear envelope fragments of varying sizes. Chemical and enzymatic analyses were also made.

VI.   Partial Isolation of the Nuclear Envelope by Citric Acid or Detergents

It was shown that nuclei isolated in citric acid media did not retain the outer nuclear membrane (Fig. 11). Only a dense layer around the

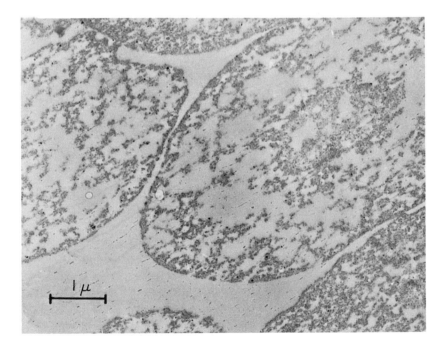

FIG. 11. Ultrathin section of rat liver nuclei isolated in dilute citric acid medium. The outer nuclear membrane is absent. Only the inner nuclear membrane is preserved.

nuclei thought to represent the inner nuclear membrane, remained (Gurr *et al.,* 1963; Kuzmina *et al.,* 1969a). Some detergents, e.g., Triton X-100 (Blobel and Potter, 1966), exert similar action on isolated nuclei. This property has been used by several authors to isolate nuclear membranes by the action of these agents upon nuclei isolated in sucrose media.

A. Use of Citric Acid

The method of this type is reported by Bornens (1968). Rat liver nuclei are isolated in concentrated sucrose by the Chauveau *et al.* (1956) procedure. These nuclei, retaining both nuclear membranes, are extracted at first with 0.1% and then several times with 0.02% citric acid in 0.25 M sucrose, pH 3.6 to 3.8. The material not solubilized by this treatment is sedimented by centrifugation at 2000g for 20 minutes. The membraneous fraction is isolated from the supernatant by centrifugation at 150,000g for 3 hours. This material exhibits a morphology and chemical composition characteristic for the membranes, but the very high RNA content (10%) throws some doubt on its purity.

Bornens (1968) believes that citric acid nuclei do not contain membranes at all and that the membranes isolated by this method represent the whole nuclear envelope. This latter point is doubtful, since there is no proof that the remaining nuclei were free from the inner nuclear membrane.

A similar method using more concentrated (2.5%) citric acid for treatment of nuclei isolated by the Chauveau procedure is described by Smith *et al.* (1969). The isolated membranes are thought to represent the outer layer of the nuclear envelope with attached ribosomes.

The preparation of membranes from rat liver nuclei isolated in 1% citric acid medium by the method described under Section IV,C, in our laboratory, contained only the heavier (inner) nuclear membrane (Zbarsky, 1972a).

B. Use of Detergents

Isolation of the outer nuclear membrane from rat liver nuclei was reported by Whittle *et al.* (1968). The nuclei obtained according to Blobel and Potter (1966) were washed with 1% Triton X-100 in the presence of a cytoplasmic ribonuclease inhibitor. After removing nuclei by sedimentation at 600g for 15 minutes, the outer nuclear membrane was pelleted from the supernatant by centrifugation at 105,000g for 4 hours. No morphological or chemical characteristics of isolated membranes were given except the RNA content, which equaled only 0.2% of the total cellular RNA.

Triton X-100 was also used by other authors for the isolation of nuclear membrane fractions (Sadowski and Howden, 1968).

VII.   Some Properties of Isolated Nuclear Membranes

Isolated nuclear membranes have a composition typical of membrane structures and are rich in various enzymes. Their most remarkable feature is the high activity of some oxidative enzymes, particularly cytochrome oxidase (Zbarsky et al., 1968, 1969) and monoamine oxidase (Gorkin et al., 1970), as well as ATPase (Delektorskaya and Perevoshchikova, 1969). Two nuclear membrane fractions, presumably the inner and the outer nuclear membranes, differ in their chemical composition and enzymatic profile (Zbarsky et al., 1969; Zbarsky, 1972a; Pokrovsky et al., 1970).

The isolated nuclear envelopes exhibit a high level of respiration (Kuzmina et al., 1969b) and oxidative phosphorylation (Zbarsky, 1972a).

A.   Composition of Nuclear Membranes

The gross chemical composition of nuclear membranes isolated by different methods is characteristic for membraneous structures (Table I). Nevertheless, the values reported by different authors are not in good agreement. This may stem from differences in methods of membrane isolation and determination of their chemical components.

Nuclear envelopes isolated by different authors, as well as in our laboratory, using salt extraction, have a higher percentage of protein and RNA and lower content of phospholipids than the nuclear envelopes isolated without treatment with salt solutions at high ionic strength.

Heavy nuclear membranes (ρ 1.19) isolated in our laboratory are richer in protein and DNA, but contain less carbohydrate than the light fraction (ρ 1.16). Nuclear membranes isolated by identical procedures from different cells are similar in their composition (Table I).

Phospholipid compositions studied in three laboratories agree sufficiently well (Table II). Although the nuclear envelope contains about one-half of the total nuclear lipids (Kuzmina et al., 1969a, 1970), its phospholipid composition is essentially the same as that for the whole nuclei. When compared with other membraneous structures it resembles

TABLE I

CHEMICAL COMPOSITION OF NUCLEAR ENVELOPE AND NUCLEAR MEMBRANES[a,b]

Chemical components	Nuclear envelope — Osmotic shock rat liver (8)	Nuclear envelope — Salt extraction, Rat liver, ρ 1.18–1.20 (4)	Nuclear envelope — Salt extraction, Rat liver, ρ 1.16–1.18 (4)	Nuclear envelope — Salt extraction, Rat liver, (8)	Nuclear envelope — Salt extraction, Pig liver (3)	Nuclear envelope — DNase and salt extraction bovine liver (1, 5)	Inner nuclear membrane — Sonification and gradient centrifugation, fraction ρ 1.19 (6, 7, 8) — Rat liver	Inner — Ehrlich ascites carcinoma	Inner — Zajdela hepatoma	Extract, with citric acid (both membranes ?) rat liver (2)	Outer nuclear membrane — Sonification and gradient centrifugation, fraction ρ 1.16 (6, 7, 8) — Rat liver	Outer — Ehrlich ascites carcinoma	Outer — Zajdela hepatoma
Protein	33	62	59	61	75	75	48	47	47	52	27	24	29
Total lipids	37	28	35	41	18	23	44	36	45	37.5	34	39	35
Phospholipids	26	24	29			14							
Glycerides	6	3.6	6	2.8 }	3.0 }	1.8							
Cholesterol	1.3	3.9	2.9			3.7							
Total carbohydrate							4.6	4.5			9.3	10.3	
RNA	2.7	6.1	3.1	1.0	2.8	8.9	3.3	1.4	2.3	10	2.1	1.1	2.3
DNA	1.0	0.0	0.0	2.1	1.2	0.9	1.3	0.9	1.5		0.2	0.2	0.5
Sialic acid		0.09	0.06				0.09	0.08			1.0	1.0	

(In the original, the "2.8" and "3.0" glyceride/cholesterol values are shown with braces indicating combined glycerides + cholesterol.)

[a] In percentage of dry weight.

[b] Numbers in parenthesis refer to the following references: (1) Berezney et al. (1970); (2) Bornens (1968); (3) Franke et al. (1970); (4) Kashnig and Kasper (1969); (5) Keenan et al. (1970); (6) Perevoshchikova et al. (1968); (7) Zbarsky (1969); (8) Zbarsky (1972a).

TABLE II

Phospholipid Composition of Liver Nuclear Envelope and Nuclei[a]

Preparation	Total phospholipid (% of total lipids)	Cholesterol (% of total lipids)	Cardiolipin	Phosphatidyl-ethanolamine	Phosphatidyl-choline	Sphyngomyelin	Phosphatidyl-serine	Phosphatidyl-inositol	Lyso-lecithin
Rat liver nuclei (Kuzmina et al., 1969a)	47.0		3.1	24.7	53.7	5.4	3.4	7.4	2.3
Rat liver citric acid nuclei (Kuzmina et al., 1969a)			0	27.6	52.5	4.9	7.3	4.1	3.6
Rat liver citric acid nuclei (Gurr et al., 1963)				25.1	52.2	6.3	5.6	4.1	
Rat liver nuclear envelope (Zbarsky, 1972a)	60.5	3.5	2.9	23.3	54.4	9.8	4.3	6.6	
Rat liver nuclear envelope (Keenan et al., 1970)	62.2	16.0	2.1	22.5	54.4	5.8	6.0	7.6	1.6
Rat liver nuclei (Kleinig, 1970)			0	22.0	61.1	2.1	4.0	8.4	1.2
Rat liver nuclear envelope (Kleinig, 1970)			0	22.7	61.4	3.2	3.6	8.6	1.5
Pig liver nuclei (Kleinig, 1970)		2.5[b]	0	25.0	58.2	2.0	4.8	8.8	1.0
Pig liver nuclear envelope (Kleinig, 1970)		3.2[b]	0	25.9	58.2	2.4	4.4	8.9	<1.0

[a] Percent of phosphorus in each fraction to the total phospholipid phosphorus.
[b] Approximate value calculated by us.

the membranes of microsomes but markedly differs from those of mitochondria and plasma membranes (Kuzmina *et al.*, 1969a; Kleinig, 1970).

B. Enzymatic Equipment of the Nuclear Membranes

Analysis of the enzymatic activities of the nuclear envelope (Table
III) demonstrates the presence and high concentration of a number of
enzymes; this might be indicative of its diverse functions.

Specific adenosine triphosphatase and monoamine oxidase activities of
the nuclear membranes appear to be higher than in any other subcellular
structure, while cytochrome oxidase activity and the concentration of
some other oxidative enzymes, as well as respiration and oxidation phosphorylation rates, approach those of mitochondria. The contents of cytochromes a, a_3, b, and c in the nuclear envelopes (Ueda *et al.*, 1969;
Conover, 1970b; Betel, 1972; Zbarsky, 1972a) also approaches those of
mitochondria.

The nuclear membranes have a characteristic enzymatic pattern different from that of any other membraneous or nonmembraneous cell
structural component. Two nuclear membrane fractions, the heavy
("ρ 1.19") and the light ("ρ 1.16") one, isolated by sonication and
gradient centrifugation, and probably corresponding to the inner and
the outer nuclear membranes, respectively, are fundamentally different
in their enzymatic profiles.

Differences in enzymatic activity of nuclear membranes reported by
different authors apparently depend first of all on the isolation procedures. The treatment with salts at high ionic strength results in marked
decrease of ATPase activity, while cytochrome oxidase remains apparently undamaged.

VIII. Conclusion

Isolation of the nuclear envelope and nuclear membranes has been
accomplished only during recent years, and the elaboration of relevant
methods is being continued. Accordingly, the biochemical characterization of nuclear membranes is as yet poor and demands more detailed
study. Nevertheless, the present data indicate diverse functions and high
metabolic activity of the nuclear envelope. It also becomes clear that the
nuclear envelope, exhibiting high respiration and oxidation phosphorylation, does supply energy for intranuclear synthetic and metabolic processes, as well as for the processes of nucleocytoplasmic interchange.

TABLE III

ENZYMATIC ACTIVITIES OF RAT LIVER NUCLEAR ENVELOPE, NUCLEI, AND SOME OTHER SUBCELLULAR FRACTIONS[a,b]

Enzymatic activity	Nuclear envelope			Nuclear membranes		Nuclei			Mitochondria + lysosomes	Microsomes	
				ρ 1.19	ρ 1.16						
Adenosine triphosphatase	29 (9)[c]	39 (9)[d]	92 (10)	233 (1)	393 (1)	36 (1)	22 (10)	18 (9)	163 (1)	73 (1)	122 (10)
Glucose-6-phosphatase	354 (9)[c]	254 (9)[d]	1.3 (10)	2.8 (2)	0 (2)	2 (2)	32 (10)	35 (9)	125 (2)	199 (2)	122 (10)
Arylsulfatase A and B				9.3 (2)	1.6 (2)	2.2 (2)			21 (2)	1.6 (2)	2.1 (2)
Proteinase				0 (2)	0 (2)	2.1 (2)			10 (2)		
Acetyl esterase				113 (3)	19 (3)	58 (3)			629 (3)		
Carboxyesterase		82 (8)		115 (3)	0 (3)	64 (3)			481 (3)		
Cytochrome c oxidase				38 (4)	31 (4)	6 (4)			59 (4)	6 (4)	
Glutamate dehydrogenase			32 (10)[e]	48 (4)	27 (4)	15 (4)	28 (10)[e]		118 (4)	4 (4)	40 (10)[e]
NADH-cytochrome c reductase	381 (9)[c]	379 (9)[d]	10 (10)	60 (4)	16 (4)	21 (4)			62 (4)	283 (4)	350 (10)
NADPH-cytochrome c reductase			18 (10)	7 (4)	8 (4)	2 (4)			11 (4)	18 (4)	49 (10)
Monoamine oxidase			16 (5)			1.6 (5)			4 (5)		
Respiration (endogenous)			16 (6)			3 (6)					
Respiration (in the presence of NADH and cytochrome c)			54 (6)			9 (6)					
Phosphorylation (in the presence of NADH and cytochrome c)			11 (7)			.4 (7)					
Phosphorylation (in the presence of NADH and cytochrome c and optimal Mg^{2+} ion concentration)			40 (7)								

[a] μmoles of substrate or reaction product per 1 gm/1 minute.

[b] Numbers in parentheses correspond to the following references: (1) Delektorskaya and Perevoshchikova (1969); (2) Pokrovsky et al. (1968); (3) Pokrovsky et al. (1970); (4) Zbarsky et al. (1968); (5) Gorkin et al. (1970); (6) Kuzmina et al. (1969b); (7) Zbarsky (1972a); (8) Conover (1970b); Nuclear fraction enriched with membranes. (9) Kashnig and Kasper (1969); (10) Franke et al. (1970).

[c] Fraction ρ 1.16–1.18.

[d] Fraction ρ 1.13–1.20.

[e] Pig liver.

Treatment of nuclei with hypotonic solutions or with very low doses of ultrasound, with subsequent gradient centrifugation (Section IV,C), apparently results in isolation of nuclear envelope fragments retaining both nuclear membranes and nuclear pores.

However, higher doses of sonication probably lead to detachment of the outer nuclear membrane from the inner one and result in the isolation of two separated membrane fractions: the heavy (ρ 1.19) and the light (ρ 1.16) nuclear membrane fractions (see Section IV,C). These two fractions are different in composition and enzymatic activity. The yield of the heavy fraction is somewhat higher, and it is richer in protein and DNA, as well as in oxidative and some hydrolytic enzymes, but less rich in ATPase. The light fraction resembles the endoplasmic reticulum in its composition and buoyant density. Electron microscopy shows that the pore complexes are apparently better preserved in the light fraction than in the heavy one. These characteristics, when considered with the fact that only the heavy fraction can be obtained from the nuclei isolated in citric acid media (retaining only the inner nuclear membrane), allow one to suppose that the heavy fraction represents the inner nuclear membrane and the light fraction the outer nuclear membrane.

Treatment of isolated nuclei with salt solutions of high ionic strength, and later in combination with deoxyribonuclease treatment or sonication (see Section V), apparently results in isolation of the total nuclear envelope fragments. Such procedures lead to a markedly higher yield of membraneous material. Nevertheless, these preparations have a comparatively high content of protein and are poor in phospholipids, while studies by electron microscopy reveal significant contamination with nonmembraneous admixtures. This may be considered as evidence of their impurity. The low activity of several enzymes (e.g., ATPase) might indicate that extensive denaturation or loss of some essential components from membranes may take place in the course of their isolation.

An approximate evaluation of the percentage of total nuclear mass in the nuclear envelope, calculated by three independent methods (Zbarsky, 1972b), gives a value of about 7%. In this connection the very high yield of nuclear membranes exceeding 10% of the total nuclear protein reported by some authors (Kashnig and Kasper, 1969; Berezney et al., 1970) is evidence for their low purity. It is worth mentioning, however, that examination of these methods in our laboratory resulted in yields of membraneous material amounting to 1–2% of dry nuclear matter.

However, the methods developed in our laboratory (Section IV,B and C) result in a low yield of the nuclear envelope or nuclear membranes (a few tenths of a percent of dry nuclear mass). This might indicate a significant loss of the material during the isolation procedure. These preparations, however, are considerably more pure and are devoid of contamination as shown by electron microscopy. They reveal also a higher enzymatic, particularly ATPase, activity (Delektorskaya and Perevoshchikova, 1969).

Contamination of nuclear membrane preparations follows also from significant glucose-6-phosphatase activity (Kashnig and Kasper, 1969). Low ATPase and NADPH-cytochrome c reductase activities (Kashnig and Kasper, 1969; Berezney et al., 1970) probably suggest extraction or denaturation of a part of the membraneous material. Among the three methods for the isolation of nuclear membranes (nuclear envelope) by salt solutions at high ionic strength reported in the literature, the method of Franke et al. (1970) appears to render the most truly "native" preparations.

The purity of the initial isolated nuclei is a prime prerequisite for the isolation of clean nuclear envelope and nuclear membranes. The following criteria may be used to characterize the purity of isolated nuclear envelope and nuclear membranes.

1. The absence of visible contamination as judged by electron microscopy of ultrathin sections

2. Characteristic chemical and phospholipid composition along with a low nucleic acid content (see Section VII,A).

3. Characteristic enzymatic profile, especially the absence of glucose-6-phosphatase and succinate dehydrogenase; high cytochrome oxidase and monoamine oxidase activity (see Section VII,B).

It may be concluded that pure and native preparations are necessary for the study of fine structure and molecular organization, as well as of functions and enzymatic activities of the nuclear envelope and nuclear membranes. Therefore, the use of the methods described under Section IV,B and C is to be preferred. In some cases, however, the mildest methods using salts at high ionic strength (Franke et al., 1970) may be used.

Nevertheless, if the components to be studied are firmly bound to nuclear membranes and not damaged severely in the course of isolation (e.g., lipid or phospholipid composition, cytochrome oxidase activity) or if contaminants of nuclear origin are tolerable, particularly in comparative studies, the methods involving the use of salts at high ionic strength, giving higher yields (see Section V,A and B) and requiring less time, may be used. Sometimes, the use of nuclear fractions enriched with nuclear membranes may also be recommended (Section III,B and C).

ACKNOWLEDGMENTS

The author is greatly indebted to Dr. K. A. Perevoshchikova and Dr. S. N. Kuzmina for their help in the preparation of the manuscript, to Dr. G. G. Gause,

Jr., for correction of the English text, and to Dr. V. V. Delektorsky for the electron micrographs.

REFERENCES

Allfrey, V. G., Osawa, S., and Mirsky, A. E. (1957). *J. Gen. Physiol.* **40,** 451.

Berezney, R., Funk, L. K., and Crane, F. L. (1970). *Biochim. Biophys. Acta* **203,** 531.

Betel, I. (1972). *In* "The Cell Nucleus. Structure and Function" (I. B. Zbarsky and G. P. Georgiev, eds.), in press. Nauka, Moscow.

Birnstiel, M. L., Rho, J. H., and Chipchase, M. (1962). *Biochim. Biophys. Acta* **55,** 734.

Blobel, G., and Potter, V. R. (1966). *Science* **154,** 1662.

Bornens, M. (1968). *C. R. Acad. Sci. (Paris)* **D266,** 596.

Busch, H. (1967). *In* "Methods in Enzymology" (S. P. Colowick and N. O. Kaplan, eds.), Vol. 12, pp. 421–464. Academic Press, New York.

Busch, H., Lane, M., Adams, H. R., De Bakay, M. E., and Muramatsu, M. (1965). *Cancer Res.* **25,** 225.

Callan, H. G., and Tomlin, S. G. (1950). *Proc. Roy. Soc. (London)* **B137,** 367.

Chauveau, J., Moulé, Y., and Rouiller, C. (1956). *Exp. Cell Res.* **11,** 317.

Conover, T. E. (1970a). *Arch. Biochem. Biophys.* **136,** 541.

Conover, T. E. (1970b). *Arch. Biochem. Biophys.* **136,** 551.

Delektorskaya, L. N., and Perevoshchikova, K. A. (1969). *Biokhimiya* **34,** 199.

Di Girolamo, A., Henshaw, E. C., and Hiatt, H. (1964). *J. Mol. Biol.* **8,** 479.

Fisher, H. W., and Harris, H. (1962). *Proc. Roy. Soc. (London)* **B156,** 521.

Franke, W. W. (1966). *J. Cell. Biol.* **31,** 619.

Franke, W. W. (1967). *Z. Zellforsch. Mikrosk. Anat.* **80,** 585.

Franke, W. W. (1970). *Z. Zellforsch. Mikrosk. Anat.* **105,** 405.

Franke, W. W., and Scheer, U. (1970). *J. Ultrastr. Res.* **30,** 288.

Franke, W. W., Deumling, B., Ermen, B., Jarasch, E. D., and Kleinig, H. (1970). *J. Cell Biol.* **46,** 379.

Frenster, J. H., Allfrey, V. G., and Mirsky, A. E. (1963). *Proc. Nat. Acad. Sci. U. S.* **50,** 1026.

Gall, J. G. (1954). *Exp. Cell Res.* **7,** 197.

Gall, J. G. (1964). *In* "Protoplasmatologia" (A. E. Mirsky, ed.), Vol. V, No. 2, pp. 4–25. Springer Verlag, Vienna.

Gall, J. G. (1967). *J. Cell Biol.* **32,** 391.

Georgiev, G. P. (1958). *Biokhimiya* **23,** 700.

Georgiev, G. P., Yermolaeva, L. P., and Zbarsky, I. B. (1960). *Biokhimiya* **25,** 318.

Gorkin, V. Z., Kuzmina, S. N., and Zbarsky, I. B. (1970). *Dok. Akad. Nauk SSSR (Biochemistry)* **191,** 472.

Gurr, M. I., Finean, J. B., and Hawthorne, J. N. (1963). *Biochim. Biophys. Acta* **70,** 406.

Kashnig, D. M., and Kasper, C. B. (1969). *J. Biol. Chem.* **244,** 3786.

Keenan, T. W., Berezney, R., Funk, L. K., and Crane, F. L. (1970). *Biochim. Biophys. Acta* **203,** 547.

Kleinig, H. (1970). *J. Cell Biol.* **46,** 396.

Kuehl, L. (1964). *Z. Naturforsch.* **19b,** 525.

Kuzmina, S. N., Troitzkaya, L. P., Zbarsky, I. B., Diatlovitzkaya, E. V., Torkhovskaya, T. I., and Bergelson, L. D. (1969a). *Biokhimiya* **34,** 763.

Kuzmina, S. N., Zbarsky, I. B., Monakhov, N. K., Gaitzkhoki, V. S., and Neifakh, S. A. (1969b). *FEBS Lett.* **5**, 34

Kuzmina, S. N., Zbarsky, I. B., Diatlovitzkaya, E. V., Torkhovskaya, T. I., and Bergelson, L. D. (1970). *In* "The Cell Nucleus and its Ultrastructures" (I. B. Zbarsky, ed.), pp. 220–225. Nauka, Moscow.

Mirsky, A. E., and Pollister, A. W. (1946). *J. Gen. Physiol.* **30**, 117.

Perevoshchikova, K. A., Delektorskaya, L. N., Delektorsky, V. V., and Zbarsky, I. B. (1968). *Tsitologiya* **10**, 573.

Pokrovsky, A. A., Zbarsky, I. B., Tutelyan, V. A., Perevoshchikova, K. A., Lashneva, N. V., and Delektorskaya, L. N. (1968). *Dokl. Akad. Nauk SSSR* (*Biochemistry*) **181**, 1280.

Pokrovsky, A. A., Zbarsky, I. B., Ponomareva, I. G., Gapparov, M. M., Tutelyan, V. A., Lashneva, N. V., Perevoshchikova, K. A., and Delektorskaya, L. N. (1970). *Biokhimiya* **35**, 343.

Roodyn, D. B. (1969). *In* "Subcellular Components" (G. D. Birnie and S. M. Fox, eds.), pp. 15–42. Butterworth, London and Washington, D. C.

Sadowski, P. D., and Howden, J. A. (1968). *J. Cell. Biol.* **37**, 163.

Smith, S. J., Adams, H. R., Smetana, K., and Busch, H. (1969). *Exp. Cell Res.* **55**, 185.

Soudek, D., and Benců, L. (1955). *Folia Biol.* (*Prague*) 1, 261

Soudek, D., and Nečas, O. (1963). *Folia Biol.* (*Prague*) **9**, 447.

Stedman, E., and Stedman, E. (1943). *Nature* (*London*) 152, 267.

Steele, W. J., and Busch, H. (1964). *Exp. Cell Res.* **33**, 68.

Ueda, K., Matsuura, T., Dale, N., and Kawai, K. (1969). *Biochem. Biophys. Res. Commun.* **34**, 322.

Whitfield, J. F., and Perris, A. D. (1968). *Exp. Cell Res.* **49**, 359.

Whittle, E. D., Bushnell, D. E., and Potter, V. R. (1968). *Biochim. Biophys. Acta* **161**, 41.

Widnell, C. C., and Siekevitz, Ph. (1967). *J. Cell. Biol.* **45**, 142A.

Yoo, B. Y., and Bayley, S. T. (1967). *J. Ultrastr. Res.* **18**, 651.

Zbarsky, I. B. (1950). *In* "Uspekhi Biologicheskoy Khimii" (I. B. Zbarsky, ed.), Vol. 1, pp. 91–115. Acad. Med. Sci., Moscow.

Zbarsky, I. B. (1963). *In* "Functional Biochemistry of Cell Structures" (O. Lindberg, ed.), Vol. 22, pp. 116–126. Pergamon, New York.

Zbarsky, I. B. (1969). *Usp. Sovrem. Biol.* **67**, 323.

Zbarsky, I. B. (1972a). *In* "The Cell Nucleus. Structure and Functions" (I. B. Zbarsky and G. P. Georgiev, eds.), in press. Nauka, Moscow.

Zbarsky, I. B. (1972b). *In* "Biological Membranes" (L. F. Panchenko, ed.), in press. Medicina, Moscow.

Zbarsky, I. B., and Debov, S. S. (1948). *Dokl. Akad. Nauk SSSR* (*Biochemistry*) **72**, 795.

Zbarsky, I. B., and Debov, S. S. (1951). *Biokhimiya* **16**, 390.

Zbarsky, I. B., and Georgiev, G. P. (1959a). *Biokhimiya* **24**, 192.

Zbarsky, I. B., and Georgiev, G. P. (1959b). *Biochim. Biophys. Acta* **32**, 306.

Zbarsky, I. B., and Perevoshchikova, K. A. (1948). *Dokl. Akad. Nauk SSSR* (*Biochemistry*) **60**, 77.

Zbarsky, I. B., and Perevoshchikova, K. A. (1951). *Biokhimiya* **16**, 112.

Zbarsky, I. B., and Yermolaeva, L. P. (1968). *In* "Biochemistry of Ribosomes and Messenger-RNA" (R. Lindigkeit, P. Langen, and J. Richter, eds.), pp. 337–341. Akademie Verlag, Berlin.

198 I. B. ZBARSKY

Zbarsky, I. B., Dmitrieva, N. P., and Yermolaeva, L. P. (1962a). *Exp. Cell Res.* **27**, 573.
Zbarsky, I. B., Yermolaeva, L. P., and Dmitrieva, N. P. (1962b). *Vopr. Med. Khim.* **8**, 218.
Zbarsky, I. B., Perevoshchikova, K. A., and Delektorskaya, L. N. (1967). *Dokl. Akad. Nauk SSSR (Cytology)* **177**, 445.
Zbarsky, I. B., Pokrovsky, A. A., Perevoshchikova, K. A., Gapparov, M. M., Lashneva, N. V., and Delektorskaya, L. N. (1968). *Dokl. Akad. Nauk SSSR (Biochemistry)* **181**, 993.
Zbarsky, I. B., Perevoshchikova, K. A., Delektorskaya, L. N., and Delektorsky, V. V. (1969). *Nature (London)* **221**, 257.

Addendum

During the last year a series of new publications appeared involving the isolation and properties of the nuclear envelope. The method of Franke was successfully applied to nuclear envelope isolation from nucleated hen erythrocytes (Zentgraf *et al.*, 1971). New data have been published on phospholipids (Lemarchal and Bornens, 1969; Kleinig *et al.*, 1971) and fatty acids (Stadler and Kleinig, 1971) of the nuclear envelope. There are also data on the initiation of DNA synthesis in association with the nuclear membrane (Hanaoka and Yamada, 1971; Yoshida *et al.*, 1971; Mizuno *et al.*, 1971a; 1971b). The last two papers also suggest that the heavier nuclear membrane fraction corresponds to the inner membrane and the lighter to the outer membrane of the nucleus.

REFERENCES

Hanaoka, F., and Yamada, M. (1971). *Biochem. Biophys. Res. Commun.* **42**, 647.
Kleinig, H., Zentgraf, H., Comes, P., and Stadler, J. (1971). *J. Biol. Chem.* **246**, 2996.
Lemarchal, P., and Bornens, M. (1969). *Bull. Soc. Chim. Biol.* **51**, 1021.
Mizuno, N. S., Stoops, C. E., and Sinha, A. A. (1971a). *Nature (London), New Biol.* **229**, 22.
Mizuno, N. S., Stoops, C. E., and Pfeiffer, R. L., Jr. (1971b). *J. Mol. Biol.* **59**, 517.
Stadler, J., and Kleinig, H. (1971). *Biochim. Biophys. Acta* **233**, 315.
Yoshida, S., Modak, M. J., and Yagi, K. (1971). *Biochem. Biophys. Res. Commun.* **43**, 1408.
Zentgraf, H., Deumling, B., Jarasch, E.-D., and Franke, W. W. (1971). *J. Biol. Chem.* **246**, 2986.

Chapter 6

Macro- and Micro-Oxygen Electrode Techniques for Cell Measurement[1]

MILTON A. LESSLER

*Department of Physiology, Ohio State University,
College of Medicine, Columbus, Ohio*

I. Introduction

Until the development of reliable oxygen electrodes, manometry was the method of choice for the study of oxygen uptake or evolution in biological systems. Some of the difficulties inherent in manometric techniques, such as the inability to follow rapid changes, the time necessary to prepare and equilibrate samples, and the problems in measuring differences in oxygen utilization in different parts of biological systems, were eliminated with the advent of rapidly responding oxygen electrodes. This is not to say that manometric techniques are not useful for special

[1] Supported in part by the International Lead Zinc Research Organization and a USPHS General Research Support Grant No. 5409 (NIH).

types of measurements, but that the oxygen electrode has come to replace manometry because of its ability to give continuous and reliable information with both small and large samples. Recent advances in the construction of oxygen electrodes of increased ruggedness and dependability coupled with sophisticated systems for recording the data have led to their general acceptance for the measurement of oxygen exchange and oxygen content of solutions. Excellent reviews on the history and development of the oxygen cathode can be found in Davies (1962), Payne and Hill (1966), Silver (1967), and Lessler and Brierley (1969).

Investigators often refer to the oxygen electrode as the "polarographic method for oxygen determination" although the original polarographic design is seldom used. More recently the terms oxygen cathode, oxygen electrode, or Clark electrode have been used interchangeably to refer to electrodes of widely differing design and response. In fact, the original Clark electrode design (Clark, 1956) is rarely used today, but electrodes, employing the Clark principle of placing the cathode and anode under a thin, oxygen-permeable membrane, are used in many applications for intermittent or continuous measurement of the oxygen content of solutions. Oxygen electrodes may be naked, film-covered, or of the Clark-

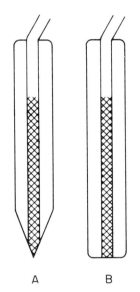

A   B

FIG. 1. Naked cathode oxygen electrodes. (A) Needle type for insertion into cells or tissues. (B) Flush type used as vibrating electrode, or on the surface of cells or tissues. The cross-hatched area indicates a thin Pt or Au wire which is usually encased in glass or epoxy. Electrodes of this type are used with a remote Ag–AgCl anode.

type, which may be covered with a wide variety of plastic membranes such as Saran, nylon, Teflon, cellophane, polyethylene, mylar, and silicone rubber or any other material in which oxygen is reasonably soluble. Oxygen cathodes are usually made of platinum or gold and are encased in glass, plastics, or epoxy to insulate them from the anode which is usually Ag–AgCl (Figs. 1 and 2).

Oxygen electrode measurements depend on the electrolysis of oxygen at a weakly negative cathode (0.4–0.8 V). The electrode is polarized by a constant voltage supply, and it produces a current proportional to the number of oxygen atoms reduced at the cathode. This current (usually in microamperes) is fed to a dc amplifier and may be read on a micro-ammeter or recorded by a potentiometric recorder to obtain continuous information of the oxygen tension (P_{O_2}). The electrode current depends on the solubility and rate of diffusion of oxygen, and, hence, is sensitive to temperature and pressure. Care must be exercised to maintain a constant temperature and pressure, or to correct for variations that may occur during calibration or measurement (see Lessler and Brierley, 1969). The readings from an oxygen electrode can be related directly to the amount of oxygen in solution only after careful calibration in air, 100% oxygen, or gas mixtures of known oxygen content (see Section III). The linear response of a calibrated oxygen electrode at constant temperature permits direct reading of P_{O_2} in solution and the calculation of the oxygen content. It provides the investigator with a means of obtaining a continuous record of changes (limited only by the response time of the electrode) due to the activity of cells and cell fractions in media of variable composition, gas content, and pH.

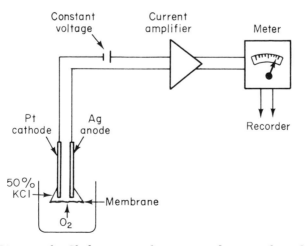

FIG. 2.   Diagram of a Clark-type, membrane-covered oxygen electrode system.

II. Types of Electrode Systems

Electrodes may be classified as separated cathode and anode or combined cathode and anode and may be covered or uncovered. They can also be classified as microcathodes (diameter of cathode less than $50\,\mu$) or macrocathodes. Because bare metal electrodes may be rapidly poisoned, they require special techniques for the maintenance of their sensitivity and linearity. Most investigators protect bare electrodes with collodion, Formvar, or some other thin membrane-forming material such as Rhoplex (Bicher and Knisely, 1970). The current available from an electrode is greatly increased in a moving medium because oxygen distribution by convection is much more rapid than by simple diffusion. A number of rotating and vibrating electrode systems have been designed (Chance and Williams, 1955a,b; Davies, 1962; Hagihara, 1961) to replace the solution in contact with the electrode, which is rapidly depleted in oxygen content. Movement or stirring of the medium also keeps the biological material in a more uniform suspension resulting in enhanced measurement of oxygen utilization. The stirring must be controlled within small limits, in order to obtain reproducible results. There is no doubt that either moving the electrode or flowing the medium past the electrode surface results in larger oxygen currents. Under these conditions, there is no delay in the establishment of a steady state current at the cathode surface and diffusion transients are minimized (Grunewald, 1970). If the movement is too rapid, however, movement artifact occurs, which interferes with the ability of the cathode to accurately record the oxygen current. Chappell (1964) and a number of other investigators have found that sulfides, phenylenediamine, and indophenol interfered with oxygen measurements with the bare vibrating Pt electrode. This interference was minimal with the Clark-type electrode, because the membrane does not allow most substances to reach the surface of the cathode or anode. There are, however, a few substances which do cause difficulty with membrane-covered electrodes by depositing on the surface of the membrane or slowly diffusing into the electrolyte chamber.

Major advantages of the Clark-type, membrane-covered electrodes (Fig. 2) are their ability to operate in solutions of variable temperature, composition, pH, gas content and to be relatively insensitive to movement artifact (Fatt, 1968). Small amounts of metabolic intermediates may be added during the course of a measurement provided that the material is temperature equilibrated and represents only a very small increase in the volume of the medium. Although Teflon or polyethylene membranes do reduce electrode poisoning and aging, they cannot completely protect the electrode. Sulfides may slowly diffuse through the membrane and deposit on the anode, and lipids can block the surface of the membrane (Hendler *et al.*, 1970). Deposits on a Pt cathode or a Ag anode reduce the effective area for current flow and change the electrode's characteristics. Thus, bare or membrane-covered electrodes must be looked at, checked, calibrated and reconditioned, or replaced from time to time, to ensure optimal response time and linearity.

The presence of a membrane reduces the response time of an electrode depending on O_2 solubility in the membrane (Fig. 3). The speed of response of a membrane-covered electrode is slower than that of a naked electrode; an important point to consider when measurements must be made at relatively low temperatures where the electrode response is slowed. However, an adequately rapid response, down to approximately 0°C, can be obtained with some types of Clark electrodes by using a very thin membrane. Standard plastic films are about 5–25 μ thick, but it is possible to obtain very rapid response times by stretching the film very thin during application (Friesen and McIlroy, 1970). Such membranes

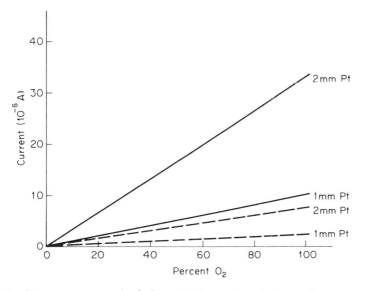

FIG. 3.   Linear response of Clark electrodes of two different diameters covered with films of the same thickness. Note the difference in oxygen current of a 1 mm as compared to a 2 mm diameter platinum electrode. The dashed lines represent the response of electrodes covered with 25 μ polyethylene and the solid lines electrodes covered with 25 μ Teflon. (Redrawn from the Severinghaus and Bradley USPHS exhibit.)

are relatively fragile and need special care to prevent them from leaking.

Recent modifications of the Clark-type electrode, in which the Pt or Au cathode and Ag anode are cast in epoxy resin, have proved to be more durable and stable than the earlier glass electrodes (Lessler *et al.*, 1965). These electrodes have flush metal surfaces exposed under a tightly stretched membrane, and the electrical circuit is completed by a drop or two of electrolyte placed on the electrode surface before the membrane is stretched (Fig. 2). A variety of electrolytes, such as saturated KCl, 50% KCl, 0.1 M KCl in pH 9.5 borate buffer (Hill and Fatt, 1963), 1 M KOH (Carey and Teal, 1965) have been recommended for different electrodes. Although the electrolyte concentration is not critical to the functioning of the oxygen electrode, following the instructions of the manufacturer on the type of electrolyte and its pH will ensure an optimal response.

III.   Evaluation and Calibration

The partial pressure at the cathodal surface of an oxygen electrode depends on the mean oxygen tension in the reaction chamber, provided that it is well mixed. When no oxygen diffuses into the sample chamber during a measurement, the velocity of reduction is limited mainly by the maximum rate of movement of oxygen to the cathode, because the oxygen concentration at the surface of the cathode is maintained essentially at zero. Minute amounts of H_2O_2 appear in the electrolyte indicating that at least two steps occur at the cathode during reduction of oxygen (Davies, 1962):

$$O_2 + 2H_2 + 2e^- \rightarrow H_2O_2 + 2OH^-$$
$$H_2O_2 + 2e^- \rightarrow 2OH^-$$

The maximum rate at which an oxygen electrode can follow changes of oxygen tension in solution depends on the size of the cathode, the solubility of oxygen in the membrane, the time constant of the amplifier–recorder system, and other factors. The response time of early Clark electrodes was about 30 seconds for a 90% response, but improvements in design have led to the development of commercially available oxygen electrodes with 90% response times of 10 seconds or less. For example, a YSI Model 53 electrode with a 600 μ diameter Pt cathode, covered by a 25 μ Teflon membrane at 37°C, has a 90% response time of 10 seconds. Other manufacturers, such as Beckman, IL, and Radiometer, have commercially available electrodes with similar characteristics. Hill and Fatt

(1963) have used a microelectrode with a 25 μ diameter Pt cathode covered by a 12 μ thick polyethylene membrane that has a 90% response time of 2 seconds. Ultramicroelectrodes with cathodes from 25 μ down to 1 μ diameter (Fatt, 1964; Fatt and St. Helen, 1969) have been designed which give very rapid response times. If very fast reactions involving oxygen must be studied, the instrument of choice is usually the recording spectrophotometer. Reactions that are completed in several seconds can be monitored with an uncovered vibrating Pt electrode (Chance and Williams, 1955a,b) or a miniaturized Clark-type oxygen electrode with a microcathode of less than 50 μ in diameter (Fatt, 1968).

The speed of response of an oxygen electrode system is inversely proportional to the diameter of the electrode and directly proportional to the solubility of oxygen in the membrane material and its thinness (Fig. 3). Response time is also influenced by the time constant of the amplifier–recorder system used. It is beyond the scope of this chapter to discuss the details of response time, but the investigator should have a good idea of the response time of the biological system being studied, so that it can be matched to the electrode–amplifier–recorder response time. When micro-oxygen cathodes are used (less than 50 μ diameter), the current produced is small, and sophisticated amplifiers and recorders are required, thereby increasing the cost of the system. These factors are discussed in the review articles, cited in the introduction, and by Fatt and St. Helen (1969).

A Pt cathode maintained at 0.4 to 0.8 V negative with respect to a reference electrode (Ag–AgCl) consumes oxygen at a rate dependent on its design, diameter, and the oxygen concentration of the solution (Fig. 3). An electrode current of 1 μA, in air equilibrated solution at 37°C, 760 mm Hg, represents an oxygen reduction of 0.3×10^{-6} gm of oxygen or approximately 0.24 μl/hour. If the solution were equilibrated with 100% oxygen, the oxygen consumption of the electrode would be approximately five times greater or 1.2 μl/hour. This tendency of the oxygen electrode itself to remove oxygen from a solution must be considered as a necessary correction factor when measurements are made with cells that metabolize slowly. Note that the electrode error is greater at higher oxygen tensions, and a significant improvement in accuracy can be obtained by equilibrating the medium with air rather than 100% oxygen.

In practice, the current produced by an electrode is translated into concentration of oxygen by calibration in a medium equilibrated at constant temperature, with known amounts of oxygen (Fig. 4). Often this is done with air, which has a constant composition (Carpenter, 1937; Machta and Hughes, 1970), 100% oxygen, or analyzed gas mixtures (LeFevre, 1969). Because good oxygen electrodes are linear with the

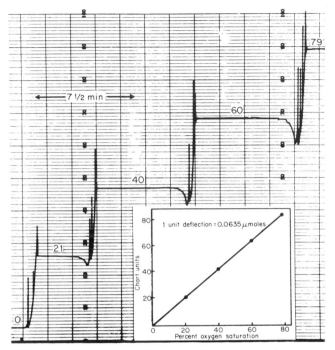

FIG. 4. Calibration of an oxygen electrode. Recorder record shows the response of the electrode in water equilibrated with known oxygen concentrations. Insert shows a plot of percentage oxygen saturation against chart units. (LeFevre *et al.*, 1970a, used by permission of the author.)

partial pressure of oxygen, only one or two concentrations of oxygen are required to establish the calibration curve (Figs. 3 and 4). It should be remembered, however, that some solutions contain appreciable quantities of salts, protein, or hemoglobin which alter the solubility of oxygen in the medium. This makes it desirable to experimentally determine the amount of oxygen in a particular solution, at given partial pressures of oxygen, both to establish the calibration curve of the electrode and to test for linearity of response.

Chappell (1964) introduced a method, using particles from disintegrated mitochondria, which has been widely used for electrode calibration. Recently a number of investigators have described more convenient methods for calibration of electrodes which do not require mitochondrial isolation. These methods involve the use of xanthine oxidase (Billiar *et al.*, 1970), glucose oxidase (Ghosh *et al.*, 1970), or *N*-methylphenazonium methosulfate-NAD-catalase (Robinson and Cooper, 1970). An ingenious method using the electrolysis of water to gradually increase the oxygen content of the calibrating solution was reported by Longmuir and Chow (1970), and Lynn (1970) used anaerobic chloroplasts with known amounts of ferricyanide added to calibrate his electrodes. There are also methods of calibration based on calculation of the oxygen content by means of the Bunsen coefficient (α) for the solubility of oxygen at a given temperature and pressure with corrections for the diminished oxygen solubility due to the presence of salts in solution. Solubility constants (α) for O_2 can be found in most handbooks of physics and chemistry, but many investigators refer to Umbreit et al. (1964) for tables on the solubility of oxygen in water or Ringer's solution (see Table I).

Recently the topic of oxygen solubility has been reopened, and determinations have been made in a number of usual and unusual solutions to provide a broad base for recalculation of the amount of oxygen dissolved in a given volume of medium. Although most cellular studies are done in aqueous saline solutions, where standard calibration techniques are adequate, the recent interest in fluorocarbons for oxygen transport resulted in the determination of oxygen solubility in a number of partly aqueous and nonpolar solutions (Battino and Clever, 1966). In a recent symposium entitled "Inert Organic Liquids for Biological Oxygen Transport," Osburn (1970) explores the relationship between surface tension and oxygen solubility and indicates that standard calibration methods for oxygen determinations in fluorocarbon emulsions may not be adequate. There are also a number of commonly used solutions in which special

TABLE I
VOLUME OF OXYGEN DISSOLVED IN AQUEOUS MEDIUM[a]

Temp. (°C)	Equilibrated with 100% O_2		Equilibrated with air (21% O_2)	
	H_2O[b]	Ringer soln.[c]	H_2O	Ringer soln.[c]
15	34.2	34.0	7.18	7.14
20	31.0	31.0	6.51	6.51
25	28.5	28.2	5.98	5.92
28	26.9	26.5	5.65	5.56
30	26.1	26.0	5.48	5.46
35	24.5	24.5	5.14	5.14
37	23.9	23.9	5.02	5.02
40	23.1	23.0	4.85	4.83

[a] Microliters of oxygen per milliliter at 1 atm.

[b] From Handbook of Chemistry and Physics, 40th Ed., Chemical Rubber Pub. Co., Cleveland, 1958–1959.

[c] Recalculated from Umbreit et al. (1964).

calibration techniques are required because salts, hemoglobin, and other proteins have the ability to modify the solubility and diffusibility of oxygen (Goldstick and Fatt, 1970). Special methods have been developed for the determination of oxygen in blood (Severinghaus, 1968; Hulands *et al.*, 1970; Holmes *et al.*, 1970), and Christoforides *et al.* (1969; Christoforides and Hedley-Whyte, 1969) have reinvestigated the solubility of oxygen in solutions containing plasma and hemoglobin. The reader is referred to LeFevre *et al.* (1970b) and Lessler and Brierley (1969) for standard methods of calibration, and to the above-mentioned studies for special methods.

IV. Calculation of Oxygen Content

The response of a good oxygen electrode is linear with the change in partial pressure of oxygen where the measurement is not limited by diffusion factors (Fig. 4). The initial step in calculating the oxygen uptake from a record is to know the time base and the calibration of the electrode, so that each division on the abscissa of the record can be expressed in time units and each division on the ordinate in oxygen units. Although the oxygen electrode recording is linear for P_{O_2} in mm Hg, we usually wish to express oxygen uptake as an amount rather than as a partial pressure differential. This can be done by calculating the results in microliters of oxygen (μl O_2), microgram atoms of oxygen (μg atoms O_2), or millimoles of oxygen. To do this, it is necessary to know the amount of oxygen dissolved per milliliter of medium, the number of milliliters of medium in the reaction vessel, and the temperature at which the

TABLE II

SOLUBILITY OF O_2 IN BUFFERED MITOCHONDRIAL MEDIUM
EQUILIBRATED WITH AIR (20.9% O_2)

Temp. (°C)	μg Atoms O_2/ml[a]	$\mu moles/ml$ (mM)
15	0.575	0.288
20	0.510	0.255
25	0.474	0.237
30	0.445	0.223
35	0.410	0.205
37	0.398	0.199
40	0.380	0.190

[a] Solubility of O_2 experimentally determined by Chappell (1964) in a buffered mitochondrial medium containing $NADH_2$, inorganic phosphate, and isolated mitochondria.

medium was equilibrated. Tables I and II give the oxygen content of media equilibrated with air or 100% oxygen. It is possible to verify these values and determine the oxygen content of other media by the calibration techniques referred to in Section III, but this is usually not necessary unless they vary considerably in salt or protein content. A typical experimental record is shown in Fig. 5, along with a method for the calculation of oxygen uptake of reticulocytes. Note that the oxidative activity of the biological sample requires a specific parameter of expression, such as number of cells, milligrams dry weight, grams wet weight, protein nitrogen, or other measurements.

The recording from a carefully calibrated oxygen electrode may be either straight or curved. Calculation of oxygen uptake from a linear record presents no difficulty (see Fig. 5), but if the recording becomes curvilinear, special methods may be needed to obtain an accurate oxygen uptake value. First a straight line approximation should be attempted—if the unenclosed area above the straight line is less than 5% of the area under the curve, it is possible to get reasonably accurate results using a part of or the entire straight line for calculation. Another method is to use only the linear record obtained during the first 5–10 minutes and

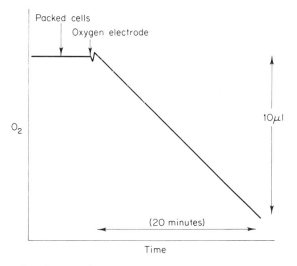

FIG. 5. Record and method for calculation of oxygen consumption of mammalian reticulocytes at 37°C. A 0.2 ml sample of packed red cells containing 50% reticulocytes (the actively respiring cells) was suspended in 3.8 ml of incubation medium. A Coulter counter determination of the number of cells in the packed cell sample shows 10×10^9 cells/ml.

Therefore: $(0.50) (0.2) (10 \times 10^9) = 1 \times 10^9$ reticulocytes.

Oxygen utilization (from record) = 10 μl/20 minutes or 30 μl/hour.

Reticulocyte O_2 consumption = 30 μl O_2/10^9 reticulocytes/hour.

calculate the immediate or (maximal) oxygen uptake of the sample (Kollarits *et al.*, 1970). This cannot be done when the oxygen uptake period record is curvilinear, or when the total period of oxygen uptake is important. Under these conditions the oxygen uptake may be determined by measuring the area under the curve with a polar planimeter and calculating the uptake from this value. The polar planimeter calculation is time-consuming and, in many cases, can be avoided by readjusting the speed of the recorder and/or the concentration of the reactants to obtain a more linear record. In most cases, these readjustments speed up the experimental work, and do not prejudice the course of the experiments.

V.   Measurements of Cells and Cell Fractions

A.   Preparation

Tissue slice techniques, developed by Warburg, were `used to study the physiological processes of cells until increased interest in the individual cell types led to the development of methods for cell separation and cell fractionation (Umbreit *et al.*, 1964). A number of techniques were developed to weaken the intracellular material with chelating agents or enzymes so that the cells could be separated and isolated by differential or gradient centrifugation. Tissues also can be prepared by mincing or gentle homogenization to produce a brei, which provides for a more intimate contact between the cells and the medium. Cells are usually studied in some variation of the buffered physiological saline solutions developed by Krebs (1950), either with or without added metabolic intermediates. Mitochondria and other cell fractions require special media which contain either sucrose or bovine serum albumin or both for optimal activity (Chance and Williams, 1955a; Chappell, 1964). No matter what type of preparation is used, it cannot be assumed that the substrate is freely diffusable into the cells or cell fractions unless tests of endogenous and substrate-stimulated metabolism are made. It is beyond the scope of this chapter to discuss cell fractionation procedures, but it should be recognized that the activity of isolated cells and cell fractions are measured under different physiological conditions than exist *in vivo*.

B.   General Methods

Many methods have been described for the study of cells and cell fractions with oxygen electrodes (Clark and Sachs, 1968). The literature in this area is so extensive that I shall attempt to describe only a few of the more important developments that have occurred since Blinks and Skow (1938) replaced the dropping Hg electrode with a Pt cathode to study the oxygen exchange of green plant cells. The essentials of a typical system are shown in Fig. 2, and with only slight modification, similar systems have been applied to a wide variety of biological and biochemical measurements (Lessler and Brierley, 1969). Many different oxygen electrode and reaction vessel designs are described in the literature, and some of these are available commercially. They are usually made of metal, glass, and plastic with some facility for temperature regulation and stirring that permits little or no back diffusion of oxygen from the atmosphere into the reaction vessel. An important factor in accurate measurements of slow reactions, or oxygen evolution studies where the reactants are initially anaerobic, is the permeability of almost all plastics to oxygen. Anaerobic or slow reactions should be carried out in glass reaction chambers because plastics usually contain substantial amounts of dissolved oxygen (Chapman et al., 1970; Davies and Baker, 1970).

A number of commercially available macro- and microelectrode systems designed to measure the pH, P_{O_2}, and P_{CO_2} of blood have been used for a variety of metabolic studies (Severinghaus, 1968). Blood oxygen electrode systems are ideal for measuring the oxygen content of fluids that perfuse a tissue or reaction chamber, but are not convenient for the direct measurement of cell metabolism. The electrodes used in these systems, however, are reliable and have been adapted to many types of cell and biochemical measurements. Recently, Fleischman and Hershey (1970) and Koch and Hershey (1970) described a two-oxygen electrode system for simultaneous measurement of the oxygen content in the gas and liquid phases of a closed vessel. They studied the oxygen transport of red cells, but a similar apparatus could be adapted to the study of oxygen transport in other animal or plant cells. It is possible to make oxygen measurements of very small samples or samples with very low respiration rates with oxygen electrode chambers of small volume such as those described by Foster (1966), Clark and Sachs (1968), Novák et al. (1969; Novák and Monkus, 1970), and others. Because reliable oxygen electrodes are readily available, adaptations of this type depend only on the ingenuity of the investigator.

One of the difficulties inherent in closed-chamber oxygen electrode measurements is that, in the absence of a gas phase to replenish the oxygen in solution, cells or cell fractions are subjected to a constantly decreasing oxygen tension during study. Garrison and Ford (1970) described a two-part Ussing-type transport chamber that uses a feedback loop, activated by an oxygen electrode, to maintain a constant oxygen tension in the medium. A decrease in P_{O_2} activates a step-motor which injects small volumes of oxygen-saturated solution into the measuring chamber to record the oxygen uptake and replace the oxygen used by the cells. Clark and Sachs (1968) call this an oxystat procedure and point out that it eliminates metabolic shifts encountered when cells are sensitive to changes in oxygen tension during study. Hodson (1970) described a system for the continuous readout of both gas and liquid phases of shaken fermentation flasks, using the steam sterilizable oxygen electrodes of Johnson *et al.* (1967), which could be applied to cell synchronization and cell culture studies. It should be remembered that the metabolism of some cells is P_{O_2} dependent (Pietra and Cappelli, 1970) and accurate metabolic measurements may require a system to maintain the oxygen in solution within small limits.

C. Photosynthetic Measurements

Plant cell respiration and oxygen evolution during photosynthesis have been measured with the oxygen electrode systems described above. The only modification necessary for the study of chloroplasts or green plant cells is some facility to illuminate and darken the reaction vessel. A recent innovation in this area is the Joliot electrode (Joliot and Joliot, 1966, 1968) on which a single layer of plant cells or chloroplasts can be deposited and illuminated with an intensity-modulated light beam of a

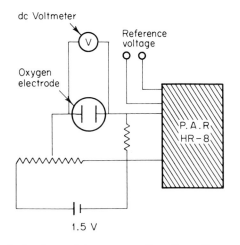

Fig. 6. Electrical circuit used to measure the alternating current from a Joliot (Joliot and Joliot, 1968) modulated oxygen electrode. The HR-8 lock-in amplifier was used with a type A high impedance preamplifier. (Used by permission of the author, Sinclair, 1969.)

given wavelength (Fig. 6). The readout from this system is an ac current produced by the photosynthetic activity of the sample, and it is possible to record the oxygen produced by photosynthesis independently of other types of oxygen production and consumption. Using the Joliot electrode, Bonaventura and Myers (1969) made simultaneous measurements of photosynthetic oxygen production and fluorescence, and there is reason to believe that further developments may lead to other sophisticated simultaneous measurements of cell metabolism during photosynthetic activity.

D. Multiparameter Measurements

The membrane-covered, Clark-type oxygen electrode is isolated both electrically and chemically from the biological material being studied. This permits the use of other measuring electrodes in the medium provided that they do not interact with each other and do not interfere with the oxygen electrode measurements. By careful choice of electrodes, amplifiers, and recorders, a number of investigators have been able to

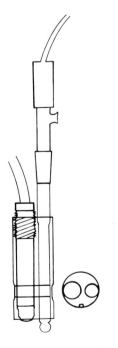

FIG. 7. An oxygen electrode and a combined pH electrode set in the same Lucite plunger for simultaneous measurement of P_{O_2} and pH.

make direct and continuous analyses of several biochemical phenomena that were not readily analyzable until simultaneous multiparameter measurements became possible.

One of the first multiparameter adaptations was to place a pH electrode in the medium to monitor H^+ ion changes during oxygen uptake or oxygen evolution studies. A convenient development, shown in Fig. 7, was to mount a small diameter combined pH electrode directly in the oxygen electrode holder. Although some care is required in the choice of pH electrode, several designs are available that have a more or less linear configuration and a long enough stem (such as the Thomas 4858) so that they can fit into a hole machined in the oxygen electrode carrier. Brierley (1969) used a reaction vessel which fits into an Eppendorf photometer that can monitor pH, oxygen uptake, and swelling of mitochondria (Fig. 8). The addition of an ion-specific electrode to the chamber makes it possible to measure simultaneous changes in four different characteristics of the system. Although this chamber was designed for mitochondrial work, it can be adapted to the measurement of either plant or animal cells, and other cell fractions.

An interesting development is the construction of an oxygen electrode reaction vessel which can be used on a microscope stage. Rikmenspoel *et al.* (1969) designed a chamber in which simultaneous measurements of pH, oxygen uptake, sperm flagella motility, and fructolysis could be

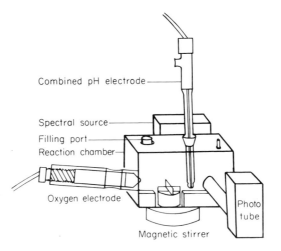

FIG. 8. Diagram of a multiparameter reaction chamber which permits P_{O_2}, pH, and photometric changes to be studied simultaneously in a stirred medium. The addition of an ion-specific electrode (e.g., Ca^{2+}) to the filling port enables four types of measurements to be made.

made. While quantitating flagella motility with an integrating photo-electronic device that scanned a small area of the microscope chamber, pH and oxygen tension were monitored with microelectrodes, and fructolysis was followed by removing sequential samples with a microsyringe. The adaptation of oxygen electrodes to multiparameter measurements has opened up a new area of dynamic studies which can lead to the resolution of many previously unsolved problems in cell biology.

VI. Microelectrodes

It has been pointed out that the response time of an oxygen electrode becomes more rapid as the size of the cathode is reduced along with a reduction in oxygen current (Fatt, 1964, 1968). This occurs because microcathodes remove oxygen from a hemisphere of medium only five to ten times the diameter of the cathode. Thus, without stirring, the oxygen current of a microelectrode is greatly influenced by the exact location of the active cathode in the medium (Clark and Sachs, 1968). With the advent of stable and reliable electronic devices for linear dc amplification, the unique characteristics of microelectrodes have been exploited for both extra- and intracellular oxygen tension measurements. Microelectrodes often are used in a manner that prevents direct calibration at the site being measured; therefore, a good deal of care is required to check the temperature response, calibration, and linearity of these electrodes (see Section III), before and after use, and to apply corrections for variations that cannot be directly controlled.

Oxygen tension changes in brain, kidney, liver, spleen, eye, and other tissues have been studied with oxygen microelectrodes (Clark and Sachs, 1968). Most of these measurements were made by applying the microelectrode to the surface of the tissue, but recently ultramicro-oxygen electrodes have been devised which are small enough to be placed within muscle or nerve cells for oxygen tension determinations. Lübbers (1966) pioneered in the design and use of microelectrodes for oxygen measurements in intact cells and tissues, and the reader is referred to his excellent review of this area. Most investigators of intracellular oxygen tension make their own electrodes using a bare noble metal cathode with a remote Ag–AgCl anode. Examples of these are the 200 μ araldite-insulated Au cathode of Baker and Lindop (1970), the 125 μ enamel-insulated Pt cathode of Moss et al. (1969), and the 25 μ glass-insulated Pt cathode of Naylor and Evans (1970). It is difficult to make very small electrodes, but Whalen and Nair (1967, 1970) have developed a method for pulling a glass capillary filled with Wood's metal into tips as small as 1 μ which can be gold plated to make good intracellular electrodes. They have been successful in measuring the P_{O_2} changes in skeletal muscle cells with these electrodes. Because of the difficulties inherent in measuring intracellular oxygen tension with a remote Ag–AgCl anode, Speckmann and Caspers (1970) designed a glass-enclosed Pt cathode and an Ag–AgCl anode which were fused together with plastic to provide a rigid electrode. The fused cathode and anode are less than 5 μ in diameter and can be used to penetrate cells or tissues for study of intracellular oxygen tension. Recently, prefabricated ultramicroelectrodes for intracellular work became available (Transidyne General, Ann Arbor, Michigan), and early reports indicate that these electrodes are stable and useful (Bicher and Knisely, 1970).

A number of investigators have worked on designs for miniature oxygen electrodes small enough to place into catheters and needles that could be used in the vascular system or introduced into tissues. This resulted in the development of commercially available membrane-covered electrodes (Beckman-Spinco, Palo Alto, California) which could be fitted into an 18-gauge needle. Electrodes of this type are also available from other manufacturers (Radiometer and IL) and have been used for intratissue (Jones et al., 1969) and intravascular measurements. There are apparently some difficulties with this type of microelectrode that require careful attention in calibration and measurement (Gruber and Amato, 1970). The membrane-covered microelectrode designed by Fatt was applied by Hill and Fatt (1963) to studies of oxygen uptake from the corneal epithelium in situ. The rapid response time of this electrode made it possible to measure the minute changes in corneal oxygen tension that occur under a contact lens during blinking (Fatt and Hill, 1970). The development of modern and stable micro- and ultramicro-oxygen electrodes has opened up the possibility of many new kinds of cellular and subcellular measurements.

Acknowledgments

The author wishes to express his gratitude to Mrs. Karen Rheuban who prepared most of the illustrations and checked all the references generated by the MEDLARS search for publications on the O_2 electrode.

References

Baker, D. J., and Lindop, R. J. (1970). Phys. Med. Biol. 15, 263.
Battino, R., and Clever, H. L. (1966). Chem. Rev. 66, 395.
Bicher, H. I., and Knisely, M. H. (1970). J. Appl. Physiol. 28, 387.
Billiar, R. B., Knappenberger, M., and Little, B. (1970). Anal. Biochem. 36, 101.

Blinks, L. R., and Skow, R. K. (1938). *Proc. Nat. Acad. Sci. U. S.* **24**, 420.

Bonaventura, C., and Myers, J. (1969). *Biochim. Biophys. Acta* **189**, 366.

Brierley, G. P. (1969). *Biochem. Biophys. Res. Commun.* **35**, 396.

Carey, F. G., and Teal, J. M. (1965). *J. Appl. Physiol.* **20**, 1074.

Carpenter, T. M. (1937). *J. Amer. Chem. Soc.* **59**, 358.

Chance, B., and Williams, G. R. (1955a). *J. Biol. Chem.* **217**, 383.

Chance, B., and Williams, G. R. (1955b). *Nature (London)* **175**, 1120.

Chapman, J. D., Sturrock, J., Boag, J. W., and Crookall, J. O. (1970). *Int. J. Radiat. Biol.* **17**, 305.

Chappell, J. B. (1964). *Biochem. J.* **90**, 225.

Christoforides, C., and Hedley-Whyte, J. (1969). *J. Appl. Physiol.* **27**, 592.

Christoforides, C., Laasberg, L. H., and Hedley-Whyte, J. (1969). *J. Appl. Physiol.* **26**, 56.

Clark, L. C. (1956). *Trans. Amer. Soc. Artif. Intern. Organs* **2**, 41.

Clark, L. C., and Sachs, G. (1968). *Ann. N. Y. Acad. Sci.* **148**, 133.

Davies, P. W. (1962). *In* "Physical Techniques in Biological Research" (W. H. Nastuk, ed.), Vol. IV, p. 137. Academic Press, New York.

Davies, R. W., and Baker, D. J. (1970). *Brit. J. Radiol.* **43**, 496.

Fatt, I. (1964). *J. Appl. Physiol.* **19**, 326.

Fatt, I. (1968). *Ann. N. Y. Acad. Sci.* **148**, 81.

Fatt, I., and Hill, R. M. (1970). *J. Amer. Acad. Optometry* **47**, 50.

Fatt, I., and St. Helen, R. (1969). *J. Appl. Physiol.* **27**, 435.

Fleischman, M., and Hershey, D. (1970). *In* "Blood Oxygenation" (D. Hershey, ed.), p. 107. Plenum, New York.

Foster, J. M. (1966). *Anal. Biochem.* **14**, 22.

Friesen, W. O., and McIlroy, M. B. (1970). *J. Appl. Physiol.* **29**, 258.

Garrison, J. C., and Ford, G. D. (1970). *J. Appl. Physiol.* **28**, 685.

Ghosh, A., Janic, U., and Sloviter, H. (1970). *Anal. Biochem.* **38**, 270.

Goldstick, T. K., and Fatt, I. (1970). *Chem. Eng. Progr. Symp. Ser.* **66**, 101.

Gruber, R. B., and Amato, J. J. (1970). *Mil. Med.* **135**, 1036.

Grunewald, W. (1970). *Pflügers Arch.* **320**, 24.

Hagihara, B. (1961). *Biochim. Biophys. Acta* **46**, 134.

Hendler, R. W., Burgess, A. H., and Scharff, R. (1970). *J. Cell Biol.* **44**, 376.

Hill, R. M., and Fatt, I. (1963). *Nature (London)* **200**, 1011.

Hodson, P. H. (1970). *Appl. Microbiol.* **19**, 551.

Holmes, P. L., Green, H. E., and Lopez-Majano, V. (1970). *Amer. J. Clin. Pathol.* **54**, 566.

Hulands, G. H., Nunn, J. F., and Paterson, G. M. (1970). *Brit. J. Anaesth.* **42**, 9.

Johnson, M. J., Borkowski, J., and Engblom, C. (1967). *Biotechnol. Bioeng.* **9**, 635.

Joliot, P., and Joliot, A. (1966). *Brookhaven Symp. Biol.* **19**, 418.

Joliot, P., and Joliot, A. (1968). *Biochim. Biophys. Acta* **153**, 625.

Jones, C. E., Crowell, J. W., and Smith, E. E. (1969). *J. Appl. Physiol.* **26**, 630.

Koch, W., and Hershey, D. (1970). *In* "Blood Oxygenation" (D. Hershey, ed.), p. 222. Plenum, New York.

Kollarits, C. R., Lessler, M. A., Berggren, R. B., and Kollarits, F. J. (1970). *Cryobiology* **7**, 94.

Krebs, H. A. (1950). *In* "Metabolism and Function" (D. Nachmansohn, ed.), p. 249. Elsevier, Amsterdam.

LeFevre, M. E. (1969). *J. Appl. Physiol.* **26**, 844.

LeFevre, M. E., Gennaro, J. F., and Brodsky, W. A. (1970a). *Amer. J. Physiol.* **219**, 716.

LeFevre, M. E., Wyssbrod, H. R., and Brodsky, W. A. (1970b). *Bioscience* **20**, 761.

Lessler, M. A., and Brierley, G. P. (1969). *Methods Biochem. Anal.* **17**, 1.

Lessler, M. A., Molloy, E., and Schwab, C. M. (1965). *Fed. Proc.* **24**, 336.

Longmuir, I. S., and Chow, J. (1970). *J. Appl. Physiol.* **28**, 343.

Lübbers, D. W. (1966). *In* "Oxygen Measurements in Blood and Tissues and Their Significance" (J. P. Payne and D. W. Hill, eds.), Little Brown, Boston.

Lynn, W. S. (1970). *Arch. Biochem.* **136**, 268.

Machta, L., and Hughes, E. (1970). *Science* **168**, 1582.

Moss, A. J., Minken, S. L., Samuelson, P., and Angell, C. (1969). *J. Atheroscler. Res.* **10**, 11.

Naylor, P. F., and Evans, N. T. (1970). *Brit. J. Dermatol.* **82**, 600.

Novák, M., and Monkus, E. F. (1970). *Anal. Biochem.* **36**, 454.

Novák, M., Hahn, P., and Melichar, V. (1969). *Biol. Neonatorum* **14**, 203.

Osburn, J. O. (1970). *Fed. Proc.* **29**, 1704.

Payne, J. P., and Hill, D. W. (1966). "Oxygen Measurements in Blood and Tissues and Their Significance." Little Brown, Boston.

Pietra, P., and Cappelli, V. (1970). *Experientia* **26**, 514.

Rikmenspoel, R., Sinton, S., and Janick, J. (1969). *J. Gen. Physiol.* **54**, 782.

Robinson, J., and Cooper, J. M. (1970). *Anal. Biochem.* **33**, 390.

Severinghaus, J. W. (1968). *Ann. N. Y. Acad. Sci.* **148**, 115.

Silver, I. A. (1967). *Phys. Med. Biol.* **12**, 285.

Sinclair, J. (1969). *Biochim. Biophys. Acta* **189**, 60.

Speckmann, E. J., and Caspers, H. (1970). *Pflügers Arch.* **318**, 78.

Umbreit, W. W., Burris, R. H., and Stauffer, J. F. (1964). "Manometric Methods," 4th Ed. Burgess, Minneapolis, Minnesota.

Whalen, W. J., and Nair, P. (1967). *Circ. Res.* **21**, 251.

Whalen, W. J., and Nair, P. (1970). *Amer. J. Physiol.* **218**, 973.

Chapter 7

Methods with Tetrahymena[1]

L. P. EVERHART, JR.

Department of Molecular, Cellular and Developmental Biology,
University of Colorado, Boulder, Colorado

[1] Supported in part by NIH postdoctoral fellowship 1-F02-GM-50,354 and by Grant No. E-520-A from the American Cancer Society to D. M. Prescott.

I. Introduction

Tetrahymena pyriformis (Ehrenberg) has been widely used in diverse studies at the cellular level. In this review a number of methods of current interest to the cell biologist will be presented. Only *T. pyriformis* will be considered here, since the bulk of the literature is written on this species. For this reason the organism will often be referred to by its generic name only.

A. Suitability of *Tetrahymena* for Studies in Cell Biology

Tetrahymena is a ubiquitous, free living freshwater ciliate. The morphology and biology of this organism have been discussed by others

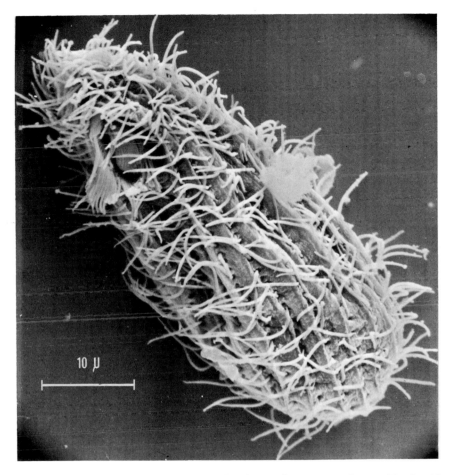

FIG. 1. Scanning electron micrograph of *Tetrahymena pyriformis*. Fixed with glutaraldehyde, postfixed with osmium, and gold shadowed after critical point drying. (Kindly provided by Dr. Kuruganti G. Murti.)

(see, for example, Elliott, 1959; Williams, 1964; Pitelka, 1968). A scanning electron micrograph of a *Tetrahymena* is shown in Fig. 1. This figure gives one some feeling for the complexity of the surface structures of the cell.

In nature *Tetrahymena* feed on bacteria, and they can be grown in this manner in the laboratory in a monoxenic culture (*mono* = one; *xenos* = stranger, i.e., a single species of bacteria). *Tetrahymena* can be grown in an axenic culture (*a* = without; *xenos* = stranger, i.e., in the absence of any other living organisms) on a variety of nutrient media and also in completely defined synthetic medium. Historically *Tetrahymena* had the distinction of being the first eukaryotic cell to be grown in a completely chemically defined medium. The organism has a short generation time comparing favorably with the faster growing bacteria and consequently large numbers of cells can be grown in short periods of time. This is of great value in the isolation of organelles or specific molecules.

Tetrahymena are large cells, about $20 \times 50 \mu$, which makes them favorable material for studies on individual cells. Individual cells can readily be selected with a micropipette and large numbers of hand-selected dividing cells can be used to start a synchronously dividing population. A large number of methods also exist for artificially synchronizing large populations of cells for biochemical studies of the cell cycle. In this respect it is noteworthy that *Tetrahymena* was the first cell to be artificially synchronized (Scherbaum and Zeuthen, 1954).

Tetrahymena cells have been used as a source for the isolation of a variety of cellular organelles including nuclei, mitochondria, and ribosomes. The *Tetrahymena* cortex is a particularly good source for the isolation of cilia. This cortex is a complex array of structures in which number and arrangement is precisely duplicated in each cell generation. Of particular interest is the relationship that exists between the cilium and the kinetosome, which is located at its base. The control of the assembly and duplication of the kinetosome and cilium is a fundamental problem in cell biology. The inheritance of patterns and arrangement of cilia and their associated structures is known as cortical inheritance and has been discussed for *Tetrahymena* by Nanney (1968).

Some aspects of the biology of ciliates make them unique among all other cells. For example, the structure of the *Tetrahymena* macronucleus and its manner of division are quite different from "normal" eukaryotic mitosis. On the one hand, this may be looked at as a disadvantage in that these cells are atypical and results obtained from studies on *Tetrahymena* may not be directly extrapolated to other eukaryotic cells. On the other hand, an understanding of these exceptions may lead to greater insight into normal cells and the processes that are common to all cells.

Some very elegant work on the genetics of *Tetrahymena* has been done. However, genetic analysis of *Tetrahymena* is complicated by the presence of two kinds of nuclei. These cells typically possess a macronucleus involved in vegetative growth and a micronucleus involved in sexual reproduction. The macronucleus is made up of many copies of the genetic material contained in the micronucleus. Many of the strains used in the laboratory have lost their micronucleus (amicronucleate). The phenotypic properties of these strains are very constant. This probably results from the high degree of endopolyploidy since a mutation in a given gene is swamped by the many normal copies of that same gene.

Even for the micronucleate strains, few mutants have been described (S. L. Allen, 1967).

Despite this lack of well-defined mutants there is a variety of phenotypic expression in the various strains of *Tetrahymena*. These include length of generation time, presence or absence of a micronucleus, length of various stages of the cell cycle, optimal temperatures of growth, amount of DNA, etc. One could no doubt take advantage of these differences in elucidating biochemical mechanisms.

Finally, and of great importance, there is an extremely large literature on this organism and a variety of techniques and methods are available for using *Tetrahymena* in studies related to cell physiology. It is the purpose of this paper to review and relate a number of these methods which may be of current interest.

B. Scope of This Review

It is impossible in the short space of this chapter to review in great detail the variety of experimental techniques that have been developed. Yet a cursory review of the field would be of little help to the investigator who is eager to try a new method.

Much has been written on the biology and methodology of *Tetrahymena*. What is the rationale for yet another chapter? In this present chapter I hope to bring together a number of methods which will be useful to the cell biologist trying to answer questions of current interest. The material included here is therefore based on my own concept of what questions are interesting. It is hoped that this chapter will stimulate the researcher unfamiliar with *Tetrahymena* to work with it; that it will guide him in selecting the method best suited for his needs; that it will get him started using that method, and lead him to the pertinent literature.

II. Media

Tetrahymena was the first protozoan to be grown in bacteria-free (axenic) culture (Lwoff, 1923). Similarly it was the first cell to be grown in chemically defined medium (Kidder and Dewey, 1951). Despite the rather complex nutritional requirements, which are very similar to those of many mammalian cells, *Tetrahymena* can be maintained in what seems, at first glance, to be a simple organic medium. That *Tetrahymena* can be grown in 2% proteose peptone alone, however, says more about the impurities found in the proteose peptone than it says about the nutritional requirements of *Tetrahymena*. Because of its fastidious nutritional requirements, *Tetrahymena* has been used as a tool in microbiological assays.

The simplest medium in which *Tetrahymena* can be grown, as suggested above, is a dilute solution of a protein hydrolyzate such as proteose peptone (Difco) or Bacto-tryptone (Difco). (Simple is used here in the sense that the medium contains only one component. From a nutritional standpoint these media are quite complex.) Both of these media provide the *Tetrahymena* with large amounts of amino acids, which serve both as carbon and nitrogen sources. Vitamins, minerals, and cofactors are present in these media coincidentally as contaminants. Growth rates in these media are slower than in enriched media (Prescott, 1958). Also, the maximal cell density is very dependent upon concentration of medium (Phelps, 1936), indicating that one growth-limiting factor is a nutrient substance in the medium. Increased growth rates as well as increased maximal cell densities can be obtained by supplementing these simple organic media with additives such as liver extract or yeast extract, substances that are rich in nucleic acids, vitamins, and cofactors.

Still further stimulation of growth can be achieved by the addition of various salts to these organic media. A widely used enriched medium containing proteose peptone, liver extract, and salts will be described below.

For washing cells free of nutrient media and for starvation studies it is often convenient to use a nutrient-free inorganic medium. Several of these will be discussed.

A. Organic Media

There are two widely used organic media for growth of *Tetrahymena*. The first of these uses 2% proteose peptone and is supplemented with 0.1% liver extract and salts. Its composition is given below in Table I. The procedure used in our laboratory for the preparation of this media is described: All ingredients are combined in a large beaker or flask. The solution is brought to 90°C with stirring to dissolve the solid components that would otherwise burn on the bottom. The solution is then allowed to cool to room temperature. It is next filtered through a coarse filter paper and then through a fine paper such as Whatman No. 1. Alternatively it can be filtered through Whatman No. 2 on a large Büchner funnel by suction filtration. Filtration is necessary to remove large particles from the medium that are ingested by the cell and are difficult to separate

TABLE I

COMPOSITION OF A PROTEOSE PEPTONE MEDIUM[a]

Component	Amount
Proteose peptone (Difco)	20.0 gm
Liver fraction L (Wilson Labs.)	1.0 gm
100X Chloride stock	10.0 ml
100X Sulfate stock	10.0 ml
Deionized glass-distilled water (GDW) to	1 liter
Chloride stock	
$CaCl_2 \cdot 2\ H_2O$	3.00 gm
$CuCl_2 \cdot 2\ H_2O$	0.30 gm
Deionized GDW to make	600 ml
Sulfate stock	
$MgSO_4 \cdot 7\ H_2O$	6.00 gm
$Fe(NH_4)2(SO_4)_2 \cdot 6\ H_2O$	1.50 gm
$MnCl_2 \cdot 4\ H_2O$	0.030 gm
$ZnCl_2$	0.003 gm
Deionized GDW to	600 ml

[a] This medium can also be prepared with other brands of peptone and liver extract.

from isolated cell fractions. Holm (1968) recommends allowing the filtered medium to sit for 14 days prior to use so that smaller particles will aggregate and settle out. It is interesting that cells grown to late log phase in unfiltered medium will be dark brown to black. As the population enters stationary phase the cells will lose this color and revert to the typical cream color. Cells grown in unfiltered 2% proteose peptone alone do not acquire these dark encrustations.

The medium is autoclaved under pressure at 115°C (10 psi) for 10 to 40 minutes depending upon the volume. This reduced pressure rather than the normal 15 psi and 121°C is recommended to prevent charring of the medium and the breakdown of complex molecules. Wille and Ehret (1968) have shown that *Tetrahymena*, grown in media autoclaved under different conditions, have different growth rates and different maximal densities. Stone (1968a) has demonstrated that longer autoclaving times (and, presumably, higher temperatures) destroy an unknown compound in the medium that inhibits the action of the antimitotic agent vinblastine. It should be noted that the conditions of autoclaving can have important effects on experimental results, and once a set of conditions have been established they should be repeated faithfully.

Media can be prepared in large amounts, autoclaved, and stored until needed in the dark, preferably in a refrigerator or cold room.

Growth of contaminants are usually easy to detect and seldom a problem. It is also possible to add a volatile preservative to the medium which is then stored without autoclaving. Such a preservative is described by Mitchison (1970) and shown below. Mitchison recommends bringing the medium to a boil prior to autoclaving to ensure that all of the preservative is removed.

1-Chlorobutane, 2 parts by volume
Chlorobenzene, 1 part by volume
1,2-Dichloroethane, 1 part by volume

Another widely used organic medium is a mixture of bactotryptone, yeast extract, and glucose. Concentrations of bactotryptone may range from 0.25 to 2%, yeast extract from 0.1 to 0.5%, and glucose from 1.0 to 1.5%.

Frankel (1965) has modified this medium by replacing the yeast extract with a defined solution of vitamins and trace metals. As this formulation has been used fairly widely it will be described here (Table II). Frankel recommends dissolving the tryptone in hot glass-distilled H_2O (GDW). The vitamins, phosphates, and Mg^{2+} are added from separate concentrated stock solutions. Trace metals are added from a stock solution made up at a concentration of 200X (5 ml of stock/liter of

TABLE II
COMPOSITION OF A TRYPTONE MEDIUM

Component	Weight/100 ml medium
Bactotryptone	0.3 gm
Glucose	0.5 gm
KH_2PO_4	100 mg
K_2HPO_4	100 mg
$MgSO_4$	30 mg
Nicotinic acid	100 mg
Calcium pantothenate	150 mg
Thiamine-HCl	100 mg
Riboflavin monophosphate	90 mg
Pyridoxamine-2HCl	20 mg
DL-6-Thioctic acid	4 mg
Folic acid	4 mg
Biotin	0.4 mg
$Fe(NH_4)_2 \cdot (SO_4)_2 \cdot 6 H_2O$ ⎫	1.42 mg
$ZnSO_4 \cdot 7 H_2O$	0.45 mg
$MnSO_4 \cdot 4 H_2O$	0.16 mg
$CuSO_4 \cdot 5 H_2O$ ⎬ trace elements	30 μg
$Co(NO_3)_2 \cdot 6 H_2O$	50 μg
$(NH_4)_6Mo_7O_{24} \cdot 4 H_2O$ ⎭	1.2 μg

TABLE III
COMPOSITION OF NEFF'S MEDIUM

Component	Concentration
Ferric citrate	$0.1 \, \text{m}M$
KH$_2$PO$_4$	$2.0 \, \text{m}M$
CaCl$_2$	$0.05 \, \text{m}M$
MgSO$_4$	$1.0 \, \text{m}M$
Glucose	1.5%
Yeast extract (Difco)	0.75%
Proteose peptone (Difco)	0.75%

medium). Glass-distilled H$_2$O is added to make a final volume of 1 liter. The medium is then autoclaved at 15 psi for 15 minutes. The glucose is made up separately in a 40% solution, autoclaved separately, and added to the medium after sterilization. Frankel (1965) reports that this medium supports a generation time of about 3 hours at 28°C for the GL strain.

Another complex organic medium that has been reported to give increased yields of cells is described by Leick and Plessner (1968). This is given in Table III. Hjelm (1970) describes the preparation of this medium as follows. The ferric citrate is dissolved in 100 ml boiling GDW. The solution is then diluted with cold distilled H$_2$O to 90% of its final volume. The other components are added in the order listed without further heating. The medium is filtered and autoclaved for 20 minutes at 112°C. Hjelm does not specify the reason of autoclaving at this low temperature.

Hjelm has used this medium to grow *Tetrahymena* in his rotating bottle cultures (described in Section III,B). Under these conditions, in which gaseous exchange is optimal, higher densities of cells are obtained in Leick and Plessner's medium than in PPLS.

B. Defined Medium

For biochemical experiments in which specific labeling is required, or when the omission of specific precursors is desired to block macromolecular synthesis, it is necessary to use the chemically defined medium. This medium was originally worked out by Kidder and Dewey (1951). The most commonly used modification, and the one used in our laboratory, is that described by Elliott *et al.* (1954) (Table IV). Prescott has made a systematic study of the growth of the HSM strain in various media at its optimal growth temperature (32.5°C) and found that the

TABLE IV

Composition of *Tetrahymena* Synthetic Medium

Component	Amount (mg/liter)
L-Arginine-HCl	150
L-Histidine-HCl·H$_2$O	110
DL-Isoleucine	100
L-Leucine	70
L-Lysine-HCl·H$_2$O	35
DL-Methionine	35
DL-Phenylalanine	100
DL-Serine	180
DL-Threonine	180
L-Tryptophan	20
DL-Valine	60
Dextrose	1000
Sodium acetate	1000
Adenylic acid	25
Cytidylic acid	25
Guanylic acid	25
Uracil	25
Calcium pantothenate	0.1
Nicotinic acid	0.1
Pyridoxine·HCl	2.0
Riboflavin	0.1
Folic acid	0.01
Thiamine·HCl	1.0
Thioctic acid	0.001
K$_2$HPO$_4$	100.0
MgSO$_4$ · 7 H$_2$O	10.0
Zn(NO$_3$) · 6 H$_2$O	5.0
FeSO$_4$ · 7 H$_2$O	0.5
CuCl$_2$ · 2 H$_2$O	0.5

generation time in synthetic medium is rather long (7–12 hours) and variable. The addition of 0.5 mg/liter cholesterol eliminates the variability and reduces the generation time to a constant 5 hours and 20–30 minutes. The addition of a small amount of proteose peptone (0.4 gm/liter) causes a further reduction of the generation time to a constant 3 hours ± 2 minutes.

C.   Inorganic Media

Some workers report using simple saline solutions for washing *Tetrahymena*. These vary in concentration from 0.2 to 0.4% NaCl. Elliott and

Clemmons (1966) have found that the osmolarity of a soft pellet brei (homogenate) of *Tetrahymena pyriformis* strain E is 100 milliosmolar. An 0.4% NaCl solution is fairly close to this, being approximately 120 milliosmolar.

The most commonly used inorganic medium is that described by Hamburger and Zeuthen (1957). This is composed of:

KH_2PO_4	0.15 gm
K_2HPO_4	0.68 gm
NaCl	2.75 gm
$MgSO_4 \times 7 H_2O$	0.25 gm
Distilled H_2O to	1 liter

This composition is approximately 100 milliosmolar. However, in our experience with the GL strain the cells have a pronounced tendency to clump in this medium. Schmitt (1967) reports that there is also a size shift on transferring cells to this medium. It is possible that adjustment of the Mg^{2+} concentration or addition of a small amount of Ca^{2+} could correct this situation, as it is known that different strains have different ionic optima.

Tamura *et al.* (1969) have described a modified inorganic medium as follows:

NaCl	100 mg
KCl	4 mg
$CaCl_2$	6 mg
GDW to	1 liter

This medium appears to fall considerably below the osmolarity reported for *Tetrahymena* by Elliott and Clemmons (1966). It is similar to the classic Osterhaut's ciliate salt solution described below:

NaCl	0.105 gm
$MgCl_2$	0.0085 gm
KCl	0.0023 gm
$MgSO_4$	0.0040 gm
$CaCl_2$	0.0010 gm
Distilled water to	1000 ml
Adjust pH to 6.8 with Na_2HPO_4	

While falling below the osmolarity of the cells these media appear to satisfy the ionic requirements of the cell. *Tetrahymena* are very good osmoregulators and have been observed to stay alive in distilled water for several days. The ionic optima are determined empirically for this type of medium.

III.   Handling the Cells

A.   Stock Cultures

Stock cultures are normally maintained in test tubes. For this purpose Pyrex tubes covered with stainless steel tube closures are the easiest to handle. These may be obtained from Bellco (Vineland, New Jersey). One may also use screw-cap cultures tubes, but the handling of these in an aseptic manner is more difficult. The tubes are usually maintained at room temperature in an upright position. Better surface area to volume configurations can be obtained by incubating the tubes on a slant, but for the maintenance of stock cultures this is not necessary. Because photoperiodic effects have been demonstrated (Willie and Ehret, 1968) some workers feel that cultures should be kept in the dark. Light can also cause the destruction of essential nutrients, such as folic acid, and can thus limit growth (Kidder and Dewey, 1951).

Lee (1969) has demonstrated that the yield of cells grown in 1% tryptone–0.05% yeast extract was quite variable. This variation in the number of cells was found to depend on photolysis of some nutrient by visible light. Addition of solutions of flavin mononucleotide and thiamine caused an increase in number of cells to a final concentration corresponding to that obtained from cultures incubated in the dark. For this reason it is recommended that cells be grown in the dark.

Subculture can be accomplished in a number of ways. If it is desired that the stock cultures be kept in logarithmic growth one can either subculture fairly often or, by using smaller inocula, subculture less frequently. For this purpose a bacteriological loop of platinum or nichrome may be used. The volume of a loop with internal diameter of 4 mm is approximately 10 microliters and will contain hundreds of organisms. Since it has been demonstrated that there is no conditioning effect of the cells on the medium and that log phase cells transferred to fresh medium grow without a lag phase (Prescott, 1957b), there is no need to transfer a larger number of cells. Larger inocula of cells can be transferred with sterile, cotton-plugged Pasteur pipettes. One can avoid the necessity of frequent transfer by maintaining the cells at reduced temperatures. Several methods for freezing of *Tetrahymena* have also been described. For storage of cells over long period of time without subculture freezing is a useful technique.

It is necessary to maintain aseptic conditions during the routine subculture and inoculation of experimental cultures. Fortunately one has an advantage here in that *Tetrahymena* are bacteria feeders. Generally it is adequate to work over the flame of a Fisher burner. One might also take the precaution of washing off the working surface with a germicidal or surface-active agent prior to making transfers. The primary problem that is encountered is with fungi. Should stock cultures become contaminated there are several ways to decontaminate them. One is the use of penicillin, streptomycin, and antifungal agents such as Fungizone. Alternatively one can subculture frequently in the hope that the *Tetrahymena* will outgrow the contaminant. It is also possible to select a single cell with a braking pipette, wash it several times in sterile media, and then place it into a sterile growth medium. It is sometimes necessary to try a combination of these methods to eliminate contamination. Prevention is the best cure, and to this end it is wise to keep several replicate cultures going. When contamination becomes a problem one is generally aware of it as the contaminating organism outgrows the *Tetrahymena*. However, it is probably prudent to make periodic checks using standard bacteriological methods. A nutrient agar such as Difco stock culture agar can be used for the detection of aerobic contaminants. Difco thioglycolate broth medium is used to reveal the presence of anaerobic contaminants.

Differences in the phenotypic expression, and presumably in the genotypic constitution, of the cell do arise with time. For example Prescott reports (1957a) a shift in temperature optimum in the HS strain. Frankel (1964) also refers to the drift in physiological characteristics among different substrains of the GL strain maintained in different laboratories. For this reason maintenance of stock cultures in log phase under the same conditions of temperature, nutrition, lighting, etc., should favor maintenance of a constant cell type. Another reason for maintaining stock cultures in log phase is that when they are used for inoculation of an experimental culture log cells will grow up without a lag (Prescott, 1957b).

B. Growth of Experimental Cultures

There are a variety of ways in which the cells can be grown. Glassware varies from test tube cultures, to Erlenmeyer or low profile Fernbach flasks, to glass carboys. Two factors must be considered when deciding what culture conditions to use: generation time and maximal yield of cells desired.

In order to obtain the most rapid growth one must work at the optimal growth temperature, which is generally above room temperature. Optimal temperatures of various strains is considered in another section. To achieve temperatures above ambient one can use an incubator or water bath. Since it is also desirable to provide for maximal gas exchange, a combination of aeration and heating can be obtained in any of a number of types of shaker water baths. We have had good success with the Warner Chilcott water bath, which is quiet, can be run at slow to fast speeds, and maintains good temperature regulation. The Thyratron tube in older models should be replaced with a solid state plug-in available from the factory. This will eliminate frequent replacement of the tube.

The maximal cell yield is highly dependent upon the surface area to volume configuration in a standing culture. For this reason one should work with shallow layers, not more than 1 cm deep. The Fernbach type flask offers the optimal surface area to volume relationship for a standing culture.

Alternatives to growing cells in standing cultures involve various methods of aeration. One is described by Elliott et al. (1966) and involves placing a Teflon-covered stirring bar in the culture medium prior to autoclaving. The culture flask is then placed on a magnetic stirring motor. The stirring bar is rotated at a speed of 150 rpm.

Another variation is to bubble air into the cultures. This presents some problem with foaming, which can be overcome by the addition of an antifoaming agent such as Dow Corning AntiFoam B. Many workers have used this procedure (Watson and Hopkins, 1962; Leick and Plessner, 1968; Gorovsky, 1970). Watson and Hopkins (1962) recommend the addition of 2 ml antifoam emulsion per liter of culture.

It should be noted that Elliott et al. (1966) have demonstrated morphological differences between cells grown in standing cultures and cells grown in a number of other ways in which mechanical aeration was involved.

We have experimented with a method for growing *Tetrahymena* that has been used for growing tissue cells in culture. This procedure involves growing the cells in a rotating bottle. The cells are spun in a thin film which adheres to the inside surface of the bottle as it rotates. The cells in this film are thereby exposed to a maximal surface area for the diffusion of gases. It is easy to obtain high densities of cells under these conditions, but the nutrient supply soon becomes limiting. It is necessary to supplement the medium in order to get the best growth under these conditions. By doubling the concentrations of nutrients in our normal medium it was possible to get yields of log phase cells above 2×10^6 cells/ml.

The apparatus involved is shown in Fig. 2. The cells are grown in 2 liter volumes in a 5-gallon carboy. The carboy is rotated at about 66 rpm by an industrial jar mill. The one pictured is manufactured by U. S.

FIG. 2. Apparatus for growing *Tetrahymena* in a rotating bottle. A 5-gallon Pyrex glass carboy is rotated on an industrial jar mill at approximately 66 rpm. Provision is made for gassing the culture with pure oxygen through polyethylene tubing which passes through a length of glass in a one-hole rubber stopper. The stopper replaces a cotton plug which was removed after the culture had reached a density of 1×10^6 cells/ml.

Stoneware and was obtained from A. H. Thomas. We have experimented with the addition of vitamins, purines, and pyrimidines to these cultures in an attempt to stimulate higher densities of cells. Purines and pyrimidines had no effect. Concentrated solutions of vitamins seemed to prolong the increase in cell number, but at a very reduced rate. In pilot experiments in which medium was replaced without changing the concentration of cells we have obtained yields of cells greater than 6×10^6 cells/ml. However, this has not yet been accomplished on a practical scale. One can only increase the concentration of proteose peptone and liver fraction so far before some compound present in these substances becomes toxic. If the growth-limiting substance or substances were known these could be added to the culture at appropriate intervals. J. R. Sadler *et al.* (1971, personal communication) have used a similar procedure for growing bacteria cultures to high density. They have found it necessary to make periodic additions of glycerol to achieve the highest yields. Because of our success in pilot studies on a small scale we believe that it will be possible to grow *Tetrahymena* to much higher densities once the limiting factor is discovered.

In its present form this method provides yields of cells about five to ten times as great as those obtained by conventional procedures. In terms of preparation of medium, space requirements, and time involved in harvesting the cells, this is a considerable saving.

Hjelm (1970) has reported using a similar method to grow and synchronize *Tetrahymena*. His method of rotating is more involved but provides for temperature regulation. He investigated two different media and found that under these conditions nutrients did appear to be limiting and that higher yields of cells could be obtained in the enriched medium of Leick and Plessner (1968) (Neff's medium).

C.   Cell Counting

A number of methods for the determination of cell density exist. We will discuss three. Basically these involve (1) direct counting of fixed cells using a microscope and counting chamber, (2) electronic methods, and (3) turbidimetric methods. In most cases the latter two methods are calibrated by comparison with direct counts.

1.   DIRECT COUNTS

We have generally used one of two methods. The first is described by Scherbaum (1957). This method uses a Sedgwick–Rafter plankton counting chamber and a Whipple–Hauser ocular micrometer disc. The chamber holds a fixed volume of 1 ml and has a precise depth of 1 mm. The ocular micrometer is calibrated with a stage micrometer, and the area of a field under low power magnification can be calculated. We use a 15× wide field ocular and a 4× objective lens. Since the depth of the field is known to be 1 mm and the area is calculated the number of cells counted in a given field can be expressed per calculated volume. Enough fields are selected at random and counted to give a statistically significant result. We routinely count 400 cells using a hand tally to record cells.

A Whipple–Hauser ocular disc is shown in Fig. 3. Since some cells will fall on lines a method must be devised so that these cells are only counted once. The procedure we use involves scanning back and forth each row left to right, right to left, left to right, and so on. Cells falling on the top line of a row or the left line are counted, while cells falling on the bottom or right line are not counted.

The filling of the Sedgwick–Rafter counting chamber is as follows: A

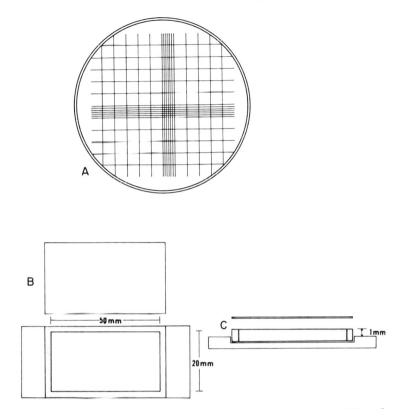

FIG. 3. (A) Drawing of an enlarged Whipple ocular micrometer. (B) Sedgwick–Rafter counting chamber seen from the top and (C) from the side.

1-ml sample of cells is placed in a small test tube. To this is added exactly 0.2 ml 3% Formalin. The 1.2 ml of fixed cells is swirled on a vortex mixer and rapidly introduced into the chamber. The cover slip is placed over the chamber and the excluded volume of cells blotted up with an absorbent tissue. With a little practice air bubbles can be avoided. Speed is of importance so that the volume of sample excluded contains the same density of cells as that remaining in the chamber. Since the fixed cells begin to settle as soon as the sample is introduced into the chamber, the need to work quickly is obvious. In theory it is possible to place a 1-ml sample in the chamber and apply the cover slip without obtaining air bubbles. In practice this is very difficult.

$$\frac{\text{(total no. of cells/number of fields)}}{\text{area of field (mm}_3\text{)}} \times 1000 \times \text{dilution} = \text{cell/ml}$$

We also have used the Levy counting chamber with Fuchs–Rosenthal rulings with good results. This method is better adapted for counting higher cell densities. The Sedgwick–Rafter chamber can be used with confidence at low densities and is therefore preferable for preparing growth curves. The Levy counting chamber can be obtained with two or four separate counting areas. This allows replicate counts to be made very quickly. This method is much faster and simpler than the Sedgwick–Rafter procedure. A small aliquot of cells is placed in a test tube and fixed with a known volume of fixative. A drop of this preparation is then applied with a Pasteur pipette to the space between the ground glass of the slide and the cover slip. The sample fills the count-

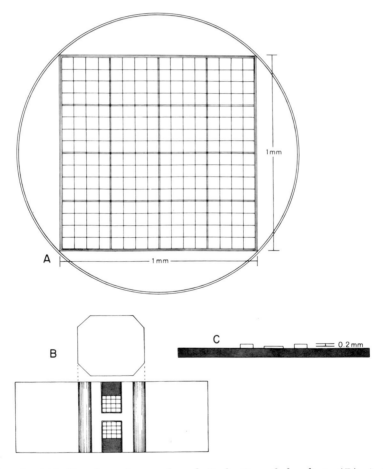

Fig. 4. (A) Drawing of an enlarged Fuchs–Rosenthal ruling. (B) A Levy counting chamber with cover slip as seen from above. (C) Levy counting chamber seen from the side.

ing area by capillarity. The calculation of cell density is similar to that used for the Sedgwick–Rafter procedure.

Drawings of the Levy counting chamber and the Fuchs Rosenthal rulings are shown in Fig. 4.

2. ELECTRONIC CELL COUNTING

A procedure for counting *Tetrahymena* in the Coulter counter model B electronic particle counter, using a 140 μ aperture, has been described by Schmidt (1967). This is the smallest aperture that can be used, since it is essential that the aperture be at least twice as large as the diameter of the cells. A *Tetrahymena* has an average length of about 50 μ. Cells must be suspended in a medium of high conductance, and any of the inorganic media can be used for this purpose. Since Schmidt was concerned with measuring cell volumes he did not fix the cells. However, in general, it is probably advantageous to include a fixative in the inorganic dilution medium to prevent the cells from swimming and to prevent growth in the counter sampling system. Since the cells are large and tend to settle quickly it is necessary to provide constant stirring during sampling. This is provided by a spark-free motor.

A procedure for counting *Tetrahymena* on the "Celloscope" electronic cell counter has also been described (Plessner *et al.*, 1964). Frankel (1965) used Coulter counter settings that were adjusted empirically to yield counts in agreement with parallel counts made using the Sedgwick–Rafter chamber. The cells were fixed in 1.8% Formalin in 0.5% NaCl.

The use of the Coulter counter to count other protozoa has been described (Fulton, 1970; Sonneborn, 1970). Fulton points out that, since the concentration of electrolyte determines the conductivity of the suspension medium and its resultant counting properties, it is important to reproduce the concentration exactly. Millipore filtering of the medium will give low backgrounds.

3. TURBIDIMETRIC ESTIMATION OF CELL DENSITY

A number of methods for turbidimetric estimation of cell density have been used. With this method one is measuring light scatter, and a number of factors can influence the results. Changes in cell size, shape, and in the absorbance of the culture medium during the growth cycle are among these. Thus in using this method to obtain growth curves one can get inaccurate pictures of what is happening. We have found for example that the time of entry into stationary phase as estimated by direct counts and by OD measurements differs by about 24 hours (Everhart and Ronkin, 1966).

However, this method can be very useful for estimating cell densities if one is dealing with populations of cells that are of the same age or if one is trying to use the same number of cells in a given experiment. Dilution of log phase cells to give a certain OD will probably be a satisfactory method.

We have used two different methods for measuring optical density of *Tetrahymena* cultures. The first involves the Bausch and Lomb Spectronic 20 at wavelength of 450 mμ. The second uses a Klett Somerson photoelectric colorimeter with a red filter having an absorbance maximum of 660 mμ. In general, light scatter is most efficiently measured at low wavelengths, but one must also pick a wavelength at which the medium does not absorb. A compromise wavelength is finally arrived at and this is best determined empirically depending upon the nature of the medium, cell density, etc. Optical density measurements of light scatter are relative, and measurements obtained on one instrument do not necessarily correspond to measurements obtained on another.

D.   Harvesting and Washing the Cells

The cells are chilled to slow down swimming by placing the culture vessel in a mixture of ice and water. They can then be pelleted in the centrifuge at 100–200g. We generally use 50-ml conical centrifuge tubes and spin the cells at 800 rpm for 3 to 5 minutes in an International refrigerated centrifuge PR-2 with the swinging bucket rotor No. 269. This method handles 400–600 ml of culture at a time. Additional medium can be poured in and the cells centrifuged without resuspending the pellets. In this manner several liters of cells can be harvested in about 20 minutes.

Some workers prefer to use a batch centrifuge for harvesting large cultures. Several commercial type centrifuges are available for this operation. Seaman (1960) recommends the following conditions for harvesting cells in the Sharples continuous-flow centrifuge: flow rate: 500 ml/min bowl speed: 3800 rpm. There is a continuous-flow attachment available for the Sorval RC-2 laboratory centrifuge also. This is the Blum–Szent Györgyi attachment. Seaman (1960) recommends a rotor speed of 1800 rpm for use with this apparatus. We have used the De-Laval cream separator for harvesting large cultures of *Tetrahymena*. The speed is reduced by running the centrifuge on a variable output transformer. The setting is determined empirically so that a minimum of cells are lost. However, the speed must not be so fast as to damage the cells. Preer (1959) has described a method for modifying the DeLaval centrifuge for the collection of *Paramecium*. This may be helpful for *Tetrahymena* also.

The harvested cells are generally washed free of all traces of growth medium before proceeding with fixation, homogenization, or whatever. This can be done with one of the inorganic growth media described in the previous section or with cold glass-distilled or deionized water. In our hands, cells grown in proteose peptone–liver fraction medium are stable in cold distilled water as long as they are in logarithmic growth. We have found that stationary phase cells are much more fragile and will lyse in distilled water washes.

Supernatant washing medium is removed from the cells most conveniently by a piece of large bore glass tubing attached to a water aspirator or other type of vacuum pump. This arrangement produces very little turbulance and consequently negligible loss of cells from the pellet.

Another commercially available centrifuge which can be used for collection of large volumes of *Tetrahymena* is the Foerst plankton centrifuge (manufactured by the Foerst Mechanical Specialties Company, 2407 N. St. Louis Ave., Chicago, Illinois). Conner *et al.* (1966) have reported having success with this centrifuge in harvesting *Tetrahymena*. However, two problems appear to arise with time: (1) corrosion and resulting toxicity of the centrifuge bowl and (2) contamination of the motor bearings with foam from the proteose peptone medium. Conner has described a modified bowl and cover that can be made of either stainless steel or a polycarbonate plastic to eliminate both problems. These adaptations are also available commercially from the Foerst Company.

With respect to the centrifugation of *Tetrahymena* two factors should be considered. First, Plessner (1964) reports that a single centrifugation of heat-synchronized *Tetrahymena* for 5 minutes at 100g delays the synchronous division for 30 minutes. Second, one must consider the effect of cold on the cells. Cold can delay and block division, as well as induce dissociation of microtubules (Frankel, 1967b). It is obvious that in certain types of experiments these factors must be considered.

E. Fixation

Fixation is necessary to preserve cells for later observation, such as for cell counting, autoradiography, and for general observation. Fixation for electron microscopy will not be discussed here. For methods used in fixation and for other electron microscope techniques the reader is referred to several excellent papers on the ultrastructure of *Tetrahymena*, for example, R. D. Allen (1967), Williams and Luft (1968), Pitelka (1968).

1. Acid Alcohol

Mixtures of acid and alcohol are often used as a fixative. The most common mixture is 3 parts of alcohol to 1 part of glacial acetic acid. Some workers use methanol, while others prefer ethanol. Similarly, 95% or absolute alcohol may be employed. Stone and Cameron (1964) recommend air drying the cells onto the slide prior to fixation. A small drop of medium containing cells is placed on a clean slide. As much medium as possible is removed with a braking pipette or other micropipette. The cells are then air dried, fixed for 10 minutes in 3:1, and dehydrated in several changes of 95% ethanol. These slides may be stored at room temperature indefinitely. Cells fixed in this manner are not spread out very well.

Flatter preparations of cells may be obtained in the following manner described by Prescott (1971, personal communication). A drop of cells is placed onto a drop of fixative on a clean microscope slide. As the cells begin to spread out the slide is flooded with additional fixative and then turned on edge to drain. For this purpose the ratio of alcohol to acid must be adjusted to about 5:1 to prevent lysis of the cells.

2. Formalin

Formalin is a useful general fixative for suspensions of cells for such purposes as cell counting. Cells can be killed with solutions as dilute as 1 drop/100 ml. However, for preservation the addition of several drops of a 3–5% (formaldehyde) solution per milliliter of cells is recommended. Glassware that has been used with Formalin and other fixatives should be kept separate from glassware used for growing cells. A solution of 37% formaldehyde equals 100% Formalin.

3. Lugol's Iodine

This is a classic fixative for ciliates and other protozoa composed of 6 gm KI and 4 gm I_2/100 ml of distilled water. A drop or two is added per milliliter of cell culture.

4. Osmium Vapor

A small amount of 4% osmium tetroxide is placed in the bottom of a scintillation vial or similar sized screw-cap container. A drop of cell culture is placed on a microscopic slide, and the slide inverted over the mouth of the bottle. This procedure should be carried out in a fume hood. The time required for fixation varies from 30 seconds to several minutes. After fixation the cells are allowed to air dry on the slide. This procedure is relatively slow if the time of death of the cells is critical, but it can be used for autoradiography.

5. GLUTARALDEHYDE

Several drops of 1–2% glutaraldehyde is added per milliliter of cell culture for short-term fixation. The preservation of surface detail and shape is superior with glutaraldehyde as compared to formaldehyde.

F. Simple Stains

1. METHYL GREEN–PYRONINE

Methyl green–pyronine is a useful stain for revealing nuclei. The nucleus stains green while the cytoplasm becomes a pink to red color. A drop of stain is mixed with a drop of cell culture on the microscope slide. Alternatively, a drop of stain can be placed next to the edge of the cover slip with the cells beneath. The stain can be drawn under the cover slip by placing a piece of absorbant paper at the opposite edge. In this manner a concentration gradient of stain is established across the slide and an optimally stained area can be found.

The formula for this differential stain is pyronine G (2.5% aqueous solution) 3.0 ml; methyl green (2.0% aqueous solution) 5.3 ml; 1 M acetate buffer, pH 4.8, 20.0 ml; distilled water to 100 ml. The stain may be stored for several months.

2. NIGROSIN

The nigrosin relief method is useful for demonstrating surface detail. A drop of 10% nigrosin (in water) is placed on the slide next to a drop of cell culture. The two are mixed together and spread into a thin film on the slide and air dried. A drop of mounting medium and a cover slip are applied for examination. For quantitative studies of surface structures other procedures such as Chatton–Lwoff silver staining are best used (see Section X,C).

3. FEULGEN NUCLEAR STAINING

Quantitative staining of nuclei may be achieved by the Feulgen procedure. This procedure is also used commonly with autoradiography for the demonstration of silver grains over the nucleus. Cells are fixed onto slides using 3:1, osmium vapor, or some other method. The slides are hydrated by passing them through a series of alcohols into distilled water for 2 minutes each. They are then hydrolyzed for 8 minutes in 1 N HCl at 60°C. The slides are then removed directly to Schiffs reagent at room temperature for 20 minutes. They are washed twice in 0.5% sodium metabisulfite for 2 minutes each and then dehydrated 5 minutes each through 50, 95, and 100% ethanol.

G.   Observation of Living Cells

Dr. Hidemi Sato has described the following procedure for the light microscope observation of living *Tetrahymena* (1971, personal communication). In this method the cells are immobilized by flattening them between the cover slip and the microscope slide. This obviates the need for a fixative and allows one to observe surface structures and cytoplasmic organelles in great detail with phase-contrast optics. A drop of cell culture is placed on the slide and covered with a cover slip. As one observes the preparation through the microscope, fluid is withdrawn from

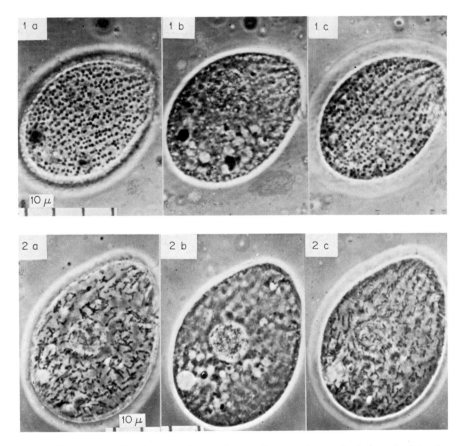

FIG. 5.   Living *Tetrahymena* flattened according to the method described in the text and viewed by phase-contrast microscopy. Upper series of photographs (1a–c) show a heat-synchronized cell at the end of the last heat shock at three different levels of focus. The lower series of photographs (2a–c) is a cell at EH + 45 minutes. Mitochondria are transformed from their normal spherical stage to a threadlike rod shape at this point. (Kindly provided by Dr. Hidemi Sato.)

beneath the cover slip by applying a wick of absorbant paper to the edge of the cover slip. Sufficient fluid is removed to flatten and immobilize the cells. The edges of the cover slip are then sealed with a mixture of melted Vaseline, lanolin, and paraffin (2:2:1), which is applied with a brush, pipette, or other instrument. The cells can be observed in this condition for some time. The proper amount of flattening is achieved with experience. Photographs of cells "fixed" in this manner are shown in Fig. 5.

Stone and Cameron (1964) have described another procedure for observing living *Tetrahymena*. A drop of culture is placed in a microchamber made by supporting a chip of cover slip on glass wool fibers which are held in place by cellophane tape. The depth of the chamber is determined by the thickness of the glass fibers, which have an average diameter of 8 μ. The flattening achieved by this procedure is inferior to that described above.

IV. Effects of Environmental Factors

In order to obtain maximal yields of cells, to plan experiments on the cell cycle, and to manipulate experimental conditions it is important to understand the effect of various conditions on the growth of the cells. A number of studies have been performed investigating the effects of various environmental parameters on the cell growth. Some of these will be considered here. This is not intended to be a comprehensive review on the subject, which has been considered in more detail elsewhere (Prescott and Stone, 1967).

A. Temperature

Prescott (1957a) has studied the effect of temperature on the multiplication rate of two strains of *Tetrahymena*-GL and -HS. The results are summarized in Fig. 6. The optimal temperature for GL is 29°C and for HS 32.5°C. Variation of temperature between 26°–32°C produces only small changes in generation time for GL. Similarly variation between 26°–34°C causes small changes for HS. Above these temperature ranges increased generation time and death occur for both strains. Generation time also increases rapidly below 26°C. MacKenzie *et al.* (1966) have examined the durations of various phases of the cell cycle in strain HSM at different temperatures. Their results indicate that the relative proportion of the cell cycle occupied by each of the 3 phases—G1 + ½D,

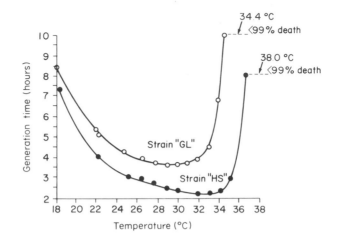

Fɪɢ. 6. The relationship between temperature and generation time is shown for two different strains of *Tetrahymena pyriformis*. Both strains were grown in 1% proteose peptone buffered at pH 7.35 with 0.01 M phosphate according to the method of Prescott (1957b). (Redrawn from Prescott, 1957a, with the permission of the Society of Protozoologists.)

S, G2 + ½D—is the same at the six temperatures studied between 17° and 32°C.

Cameron and Nachtwey (1967) also examined the effect of temperatures between 20°–29°C on the various phases of the cell cycle in strain HSM. In contrast to the findings of MacKenzie *et al.*, these workers report that the absolute amount of time spent in S remains constant while the proportion of the cell cycle spent in S decreases with decreasing temperature.

Cleffmann (1967) performed the same experiments on *Tetrahymena pyriformis* HSM and obtained results in agreement with MacKenzie *et al.* (1966), that is, he found that as the temperature changes, the fraction of the cell cycle occupied by each phase remains the same.

To explain the differences in these results Cameron and Nachtwey propose that increased nutritional value of their medium has changed the kinetics of DNA synthesis. While Cameron and Nachtwey use proteose peptone–liver fraction medium, both MacKenzie *et al.* and Cleffman use enriched synthetic medium. The effect of the nutritional state on DNA synthesis will be discussed later.

B. Nutritional State

The effect of the nutritional state on the maximal yield of cells has already been pointed out in the section on growth media. Prescott (1958,

TABLE V

GENERATION TIMES OF *Tetrahymena pyriformis* HS IN VARIOUS MEDIA

Medium	GT (in hours)
Proteose peptone —1.5%	2.08
Proteose peptone —1.5% + 0.1% liver fraction	1.85
Tetrahymena extract	1.65
Synthetic medium	7–12
Synthetic medium plus chloresterol	5
Synthetic medium plus 0.03% proteose	3

1959) investigated the growth rate of *Tetrahymena* HS in different media at the optimal pH and optimal temperature. The results are presented in Table V. The more enriched the medium, the shorter the generation time. Cameron and Nachtwey (1967) have found that not only is the length of the generation time a function of the nutritional state but also that the length and position of the phases of the cell cycle are also greatly influenced by this variable. Their results, as well as the results of others, indicate that growth on a less rich medium increases the duration of the S period and shifts it from early interphase toward the middle of interphase. This relationship is shown in Fig. 7 for the HS strain grown in three different media.

C. pH

The effect of pH on the growth rate of *Tetrahymena* has been studied with respect to yield. Prescott (1958) has also demonstrated the effect of pH on generation time. The cells appear to be capable of growth over a range from pH 5 to about 9. The optimum for the maximal growth rate

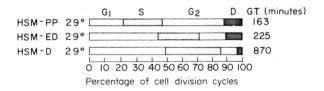

FIG. 7. The effect of different growth media on the position and duration of different phases of the cell cycle. *Tetrahymena pyriformis* HSM was grown at 29°C in three different media: PP, 1.5% (w/v) proteose peptone plus 0.1% liver extract; ED, synthetic medium (Elliott *et al.*, 1954) enriched with 0.04% proteose peptone; D, synthetic medium minus proteose peptone. The generation time in each medium is shown at the far right. (Redrawn from Cameron and Nachtwey, 1967, with the permission of Academic Press.)

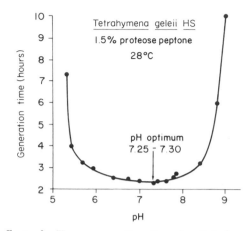

FIG. 8. The effect of pH on generation time for *Tetrahymena pyriformis* HS is shown. Cells were grown in 1.5% proteose peptone adjusted with concentrated NaOH to 15 different pH values over a range of 5.3 to 9.0. The cells were grown in pipette cultures according to the method of Prescott (1957b). (Redrawn from Prescott, 1958; with the permission of the University of Chicago Press.)

of the HS strain has been shown to be pH 7.25–7.30. Between the pH values of 5.65 and 8.40 the effect of pH on growth is not pronounced. These results as described by Prescott (1958) are shown in Fig. 8.

The study of DNA synthesis under varying environmental parameters may lead to information on the factors involved in the regulation of this process as suggested by Prescott and Stone (1967).

V.   Synchronization

Tetrahymena were first synchronized by Scherbaum and Zeuthen (1954). This experimental system has provided much information into the mechanisms of and preparations for cell division. Although the mechanisms of induction of synchrony have so far remained elusive, the system has stimulated many models and much interest. The mechanisms involved in the synchronization of *Tetrahymena* have been reviewed in many places. Most recently, Zeuthen and Rasmussen (1971) have summarized this subject in great detail. Earlier books and reviews that include studies on *Tetrahymena* are Zeuthen (1964), Scherbaum (1960, 1964), Cameron and Padilla (1966), and Zeuthen and Williams (1969). Although the heat-shock system of Scherbaum and Zeuthen (1954) has been most widely studied, many other methods for synchronizing *Tetra-*

hymena now exist. Many of these appear to act through the same mechanism as heat shock (Zeuthen and Rasmussen, 1971).

A. Heat Shocks

Plessner *et al.* (1964) have described much of the methodology involved in working with heat-synchronized cells. To effect heat synchronization one exposes a logarithmically growing culture to heat shocks at 34°C alternated with periods of growth at the optimal temperature of 28°C. As originally described the heat shocks were of 20-minute durations and were alternated with 40-minute periods at 28°C. In its simplest form this method can be carried out manually by transferring a flask between two water baths set at the appropriate temperatures. However, since six to seven shocks are generally applied, this involves a 6- to 7-hour wait before an experiment can be started.

To expedite this process a number of automatic synchronizing devices have been designed. Because the homemade device used in our laboratory incorporates some novel features, it will be described here in detail. The description of such a device will also serve to provide a starting point for the researcher who has never designed such an apparatus. A schematic representation of this synchronizer is shown in Fig. 9.

The controlling unit of this apparatus is an electronic timing device manufactured by Coulter Electronics. This has a 240-minute timing cycle which can be set to activate a microswitch for 1-minute to 240-minute intervals. When the microswitch is activated, a light is normally lit. We have replaced this light with a two-pole magnetic relay such that 120 V ac is carried by one outlet when the switch is closed and by a second outlet when the circuit is open.

During one 30-minute interval the timer activates an electronic relay controlled by a mercury thermoregulator. This relay activates a heating coil whenever the temperature in the water bath falls below 34°C. During the next 30-minute interval this unit is switched off and a Yellow Springs thermistor thermoregulator is switched on. The Yellow Springs unit regulates the lower temperature 28°C. If this unit is activated and the temperature is above 28°C cold water flows through a copper coil in the bottom of the water bath. The "break" circuit of the Yellow Springs thermoregulator controls the flow of water by means of a solenoid valve attached to the cold water tap. The "make" circuit of the Yellow Springs thermoregulator is attached to a 125-W immersion heater and is activated whenever the temperature falls below 28°C.

At the end of the 30-minute period controlled by the Yellow Springs thermistor unit, the unit is shut off and the electronic relay is activated.

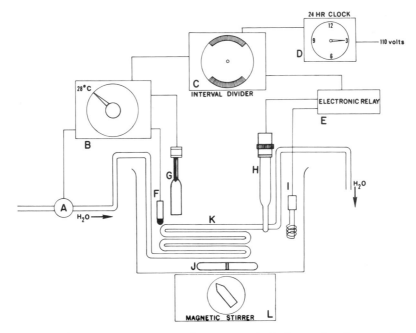

FIG. 9. Schematic representation of an automatic device used to heat synchronize *Tetrahymena*. A, solenoid water valve; B, Yellow Springs thermister temperature control unit; C, Coulter electronics interval divider; D, Montgomery Ward 24-hour timer; E, electronic relay; F, thermister probe; G, blade heater; H, mercury thermoregulator; I, heating coil; J, Teflon-coated magnetic stirring bar; K, copper coil carrying cold water; L, magnetic stirring motor. Elements F–K are contained within a water bath.

Whenever the temperature is below 34°C it "calls" for heat and a 250-W coil heater is activated. When the temperature reaches 34°C it is shut off. The 34°C temperature is opposed by ambient temperature.

The electronic timer is plugged into a Montgomery Ward 24-hour timer that can be set to turn the whole apparatus on at some time in the early morning. In this manner cells can be placed in the synchronizer at night and a synchronized population is ready the next morning.

The cells are grown in a covered Revere Ware stainless steel 2-quart pot. The transfer of heat across the metal is much faster than across glass and this facilitates rapid temperature changes between 28° and 34°C. The stainless steel exerts no adverse effect on the cells. Exchange of gases as well as heat exchange are both promoted by stirring the cells inside the pot with a Teflon stirring bar. The whole apparatus sits on a magnetic stirring motor.

Heat synchronization cannot normally be carried out on populations of *Tetrahymena* with a density greater than 100,000 cells/ml. The reason for this is not clear, but it may be related to inadequate gaseous exchange at high densities. In this respect Hjelm (1970) notes that a maximum population density of 2×10^5 cells/ml results in satisfactory synchronization of division in his system. It is also of interest that in Hjelm's rotating bottle cultures maximal synchronization of division occurs after only four shocks as compared to six to seven in the normal case. Hjelm suggests that the factors allowing growth at higher concentrations in the rotating bottle may also be responsible for producing maximal synchrony with a fewer number of shocks.

It is also important to note that heat synchronization of cell division cannot be accomplished in the defined medium, but only in complex organic medium. In fact Rasmussen and Zeuthen (1962) have noted that synchronization is improved if the normal 2% proteose peptone–0.1% liver fraction medium is supplemented to include 2% proteose peptone–0.4% liver fraction.

The various strains of *Tetrahymena pyriformis* have different temperature optima of growth as demonstrated by Prescott (1957a). Consequently different temperature regimen are required to bring about heat synchronization. Table VI shows the times and temperatures required for a few strains.

In the process of heat synchronization growth and division are separated. During the process cell size increases three- to fourfold. Protein, RNA, and DNA increase by a similar amount (Scherbaum, 1964). Because inhibitors of protein synthesis block the first synchronous division the concept of "division protein" has been developed (Rasmussen and

TABLE VI

CONDITIONS FOR HEAT SYNCHRONIZATION OF VARIOUS
STRAINS OF *Tetrahymena pyriformis*

Strain	Length of heat shock (min)	Length of intershock (min)	Upper temperature (°C)	Lower temperature (°C)	Number of shocks	Reference
GL	30	30	34	28	7	Zeuthen (1964)
HSM	25	25	39	29	5	Cameron *et al.* (1966)
W	30	30	34	28	10	Nachtwey and Giese (1968)
WH-6	30	15	42.8	35	3–5	Holz *et al.* (1957)
WH-14	30	15	43	34		Byfield and Lee (1970)
S	30	30	34	28	7	Randall and Disbrey (1965)

Zeuthen, 1962). This hypothesis states that the synthesis of division-related protein or proteins is disrupted during the process of heat synchronization. Synthesis of this protein must therefore occur during the recovery period after the last heat shock (EH) and before the first synchronous division.

Morphologically, other processes besides cell division are synchronized by heat treatment. At the end of the last heat shock all cells are found to be in the same stage of oral morphogenesis, called the anarchic field stage. While duplication of kinetosomes and cilia has proceeded through the process of heat synchronization, the organization of the kinetosomes that will give rise to the oral apparatus of the posterior daughter cell has been blocked (Holz et al., 1957; Williams and Scherbaum, 1959; Frankel, 1962). For the first time at EH + 40 minutes single and bundled microtubular and microfilamentous structures appear between the kinetosomes of the anarchic field, and development of the daughter oral apparatus resumes (Williams and Zeuthen, 1966). Inhibition of protein synthesis blocks this development (Frankel, 1967a, 1969a). It is proposed that the events that occur in the oral apparatus reflect processes occurring elsewhere in the cell, including macronuclear division and cytokinesis. Thus, the effect of heat shocks is to disrupt the organization and assembly of fibrous structures involved not only in oral morphogenesis but also in macronuclear division and cytokinesis (Zeuthen and Williams, 1969).

B. Cold

There are two methods of synchronization which use cold as the disruptive agent. The first is the multiple-cold shock method (Zeuthen, 1964) and the second is the repetitive temperature cycling method in which the cells are held in the cold for most of their generation time and then allowed to warm up and divide (Padilla and Cameron, 1964).

In the cold shock system of Zeuthen (1964) the cells are kept at 10°C for 2-hour periods interspersed with 40-minute periods at 28°C. After a series of six shocks the cells are allowed to develop at 28° or 22°C (DeBault and Ringertz, 1967).

The division index reaches a maximum of about 80% at 1.4 hours after the last cold shock. Cells synchronized by cold shocks are similar to heat-synchronized cells in that they have a higher mass and greater DNA content than normal cells. Cold-synchronized cells by the multiple-shock system are also similar to heat-synchronized cells in their sensitivity to inhibitors (Moner and Berger, 1966). In fact, Zeuthen and

Rasmussen (1971) have proposed that the effect of heat and cold is on the same cellular mechanism.

Padilla and Cameron (1964) described a method for synchronization that involved exposure to a single cold "shock." In this procedure strain HSM was used. The culture was held at 12°C for 9.5 hours and then the temperature was raised to 27°C for 2.5 hours. This 12-hour cycle is repeated over and over, and the cell density is maintained by dilution every 24 hours with fresh medium during a cold period.

In this system there is no cell division during the first 1.5 hours in the warm period; 90% of the cells then divide in the last hour. The division index reaches a maximum of only 40%. Micronuclear DNA synthesis appears to be synchronized as a consequence of the synchronization of cell division. Macronuclear DNA synthesis is shut off during the warm period at some point prior to the beginning of cell division.

C. Centrifugation

Corbett (1964) has described the centrifugal separation of a synchronous subpopulation from cultures of *Tetrahymena*. Log phase cultures at densities around 100,000 cells/ml are centrifuged for 6 minutes at 550g. The supernatant cells were shown by Corbett to represent a distinct population. The cells divide with some synchrony about 2 hours after the centrifugation with a maximum division index of 30% at 2 hours. All cells divide between 1 and 3 hours. Corbett shows that the synchrony is not a condition that preexisted in the culture or is it a result of the centrifugation itself.

In theory it should be possible to isolate a synchronous population of cells in this manner since, while later in the cell cycle they are larger and pyriform, these cells just after division are small and round. Corbett has also shown that density differences exist. Separation on the basis of sedimentation velocity has been achieved for other cell types including bacteria, yeast (Mitchison and Vincent, 1966), and mammalian cells (Fox and Pardee, 1970).

However, the situation in *Tetrahymena* is complicated by a number of factors, a major one being that they swim. Dr. John Bollinger (1971, personal communication) has spent a great deal of effort trying to select a subpopulation of synchronous cells. He has been unable to repeat Corbett's results and has had no success with Mitchison and Vincent's (1966) method in either sucrose or in Ficoll.

One must also keep in mind the effect of centrifugation on delays in cell division (Plessner, 1964) as mentioned earlier.

D.   Colchicine

Colchicine and colcemid are plant alkaloids with rather specific anti-mitotic activities. In eukaryotic cells they inhibit the process of mitosis by causing disruption of the mitotic spindle. Colchicine is thought to exert its action by a direct binding to the microtubule protein subunits. Because the growing spindle is in equilibrium with a pool of micro-tubular subunits the binding of colchicine to the subunit pool effectively depletes the pool and the assembled mitotic apparatus falls apart (Borisy and Taylor, 1967; Wilson and Friedkin, 1967).

Colchicine has been used successfully to accumulate mitotic mammalian cells grown *in vitro*. These arrested cells are then used to produce a synchronous population (Stubblefield, 1968). Although *Tetrahymena* do not have a spindle associated with the macronucleus, intramacronuclear microtubules have been demonstrated (Ito *et al.*, 1968; Tamura *et al.*, 1969).

Wunderlich and Peyk (1969) have shown that colchicine and colcemid can inhibit macronuclear division. However, after a period of time the cells overcome the inhibition and subsequently divide with a burst of synchrony. Wunderlich and Peyk used the GL strain grown in 2% proteose peptone–0.4% liver fraction L. Colchicine was used at 1.5 mg/ml and colcemid at 0.5 mg/ml. The cells were grown at 28°C. After 2.5–3 hours there is a burst of synchronous division with a maximum division index of 60%. A second burst of synchronous division occurs after 2–2.5 hours with a peak division index of 30–50%. The authors report that colcemid is more effective and repeatable than colchicine. The time interval between the two bursts of division is approximately equal to the natural generation time.

The authors further report that morphogenesis of the daughter oral apparatus is not affected. It is known that protein and RNA metabolism are not affected by these concentrations of colchicine (Rosenbaum and Carlson, 1969). One must conclude therefore that the targets of colchicine synchronization are far fewer and less complex than seen in heat-synchronized cells.

It will be interesting to examine the pattern of DNA synthesis in cells synchronized in this manner. The effect of both heat and cold synchronization of cell division is to synchronize the initiation of DNA synthesis (Hjelm and Zeuthen, 1967a). Moreover since colchicine-synchronized cells have a normal generation time between the first and second synchronous divisions they appear to be in a more balanced state of growth. Although the cells recover in the presence of the drug, the period of inhibition is short and excess amounts of DNA are less likely to have been accumulated as is the case in other systems.

E. Hypoxia

Rasmussen (1963) has shown that treatment of *Tetrahymena pyriformis* GL with two 50-minute exposures to pure nitrogen produces good synchronization of cell division. For 90 minutes after the second exposure no cells are seen in division. This is followed by a period of about 100 minutes during which the population exactly doubles. The maximum division index of 60% is reached about 2½ hours after the last exposure to anoxia.

This type of synchronization was accomplished by Rasmussen on strain GL grown in 2% proteose peptone–0.1% liver fraction plus salts. The density of the culture exposed to hypoxic shocks was about 30,000 cells/ml. The culture was transferred to a sterile glass cylinder 240 mm high and 28 mm in diameter. The cylinder was plugged with cotton and a wide glass tube passed through to the bottom. Nitrogen containing less than 0.02% O_2 was passed through the tube at a rate of 3 ml/second.

Rooney and Eiler (1967) modified this procedure by using hypoxia rather than anoxia, multiple shocks, enriched medium, and decreased duration of shocks. In their system the division maximum develops earlier and is more complete.

Subsequently Rooney and Eiler (1969) have described an automatic apparatus for hypoxic synchronization of cell division in *Tetrahymena*. This apparatus is somewhat technically involved, but produces a high degree of reproducible synchrony. In this apparatus the first synchronous division does not begin until 100 minutes after the termination of the last hypoxic shock. Ninety-eight percent of the cells divide whereas Rooney and Eiler (1969) report that only 86% of the cells divide after hyperthermic synchronization.

F. Starvation

Cameron (1965) found that upon replacement of pyrimidines to pyrimidine-starved cells there was a more or less synchronous entry into DNA synthesis by the G1-arrested *Tetrahymena*. He suggested that this might be modified for use as a synchronization technique.

Recently Cameron and Jeter (1970) have modified this procedure to produce synchronized mass cultures. This involves starving the cells in an inorganic medium followed by refeeding in an enriched nutritional growth medium. The first cells start to divide 240 minutes after refeeding; by 390 minutes all the cells have divided. A 50% increase in cell number occurs at about 310 minutes. The division index never exceeded 20%. Although this is rather poor synchronization of cell division, the system may prove useful for the examination of preparations necessary for the initiation of DNA synthesis. Pulse labeling of the cells with thymidine-^{3}H revealed that no cells incorporated thymidine-^{3}H into DNA until 150 minutes after refeeding. After 150 minutes the percentage of cells labeled in a 25-minute pulse rose to 40% at 300 minutes. Continuous labeling revealed that no cells divided without first synthesizing DNA. The authors point out that this method is easy and inexpensive. Moreover, it is readily adaptable to mass cultures. The procedure is as follows.

Cells were grown in 2% proteose peptone–0.1% liver extract, and harvested at densities between 10–100,000 cells/ml by repeated centrifugation in 50-ml plastic tubes at 2000 rpm for 2-minute intervals. The cells were then washed two to three times in sterile inorganic phosphate buffer (0.6 gm KH_2PO_4, 0.15 gm K_2HPO_4, 0.25 gm $MgSO_4$/liter, pH 6.5 with NaOH). The washed cells were inoculated into sterile 2.5-liter low form flasks containing 500 ml medium at a concentration of 10,000–25,000 cells/ml. After starvation for 24 hours, refeeding was accomplished by adding 0.5 liter of two times concentrated sterile enriched proteose peptone directly to the starved cells.

G. Vinblastine

The use of vinblastine to synchronize *Tetrahymena* has been described in detail by Stone (1968a,b). Strain WH6 can be inhibited from cell division by concentrations of vinblastine in the range 20–25 μg/ml. Strain GL requires about 50 μg/ml. Cells are inhibited for several hours. They are then released by centrifugation and replacement with fresh warm medium lacking vinblastine. They can also be released in inorganic medium.

The percentage increase in cell number is a function of the length of time spent in the presence of inhibitor. Inhibition for 4 hours leads to an 82% increase. Inhibitions for 6 hours results in a doubling or 100% increase. Inhibition for 8 and 12 hours leads to 194 and 225% increases, respectively. These latter results are a little difficult to explain.

Cultures inhibited for 6 hours show a division index peak of 60% at 65 minutes after release and a second division maximum of 38% at 140 minutes. In the inorganic medium there is only one division burst which shows a division index of about 80% at 85 minutes.

Both micronucleate (WH-6) and amicronucleate (GL) strains can be synchronized by this method.

Stone (1968b) points out that vinblastine is known to act as an inhibitor of macromolecular synthesis and may therefore have sites of nonspecific action in addition to the specific inhibition of division. Furthermore, the one to four division of many daughter cells after 12 hours of inhibition is rather unusual. While investigation of the peculiarities of this system may well lead to information on the process of cell division, it does not seem that this is a useful system for the study of the normal cell cycle.

DNA synthesis following vinblastine synchronization has been examined by Sedgley and Stone (1969). The cells appear to continue to synthesize DNA during the period of inhibition and even to accumulate excess amounts adequate to carry them through two subsequent divisions. There does not appear to be any synchronization of DNA synthesis after the first division as is seen in heat-synchronized cells (Hjelm and Zeuthen, 1967a,b).

H. Synchronous Rounding

The phenomenon of synchronous rounding was described by Tamura *et al.* (1966). To produce this effect, in which the cells do not divide but periodically lose their pyriform shafe and round up, the cells are washed several times in an inorganic medium. They are then transferred axenically to a complete synthetic medium minus amino acids. The cells are kept in this amino acid-free medium at 26°C for about 9 hours and then they are subjected to the standard cycling treatment for heat synchronization. They are given eight heat shocks at 34°C for 30-minute durations interspersed with 30-minute periods at 26°C. At 80 minutes after the final heat shock the cells are found to be in a state of synchronous rounding rather than undergoing division. This pattern of synchronous rounding is repeated twice more at times when heat-synchronized cells would normally divide. It is proposed that a common mechanism of phasing exists for both heat-synchronized cell division and for synchronous rounding. To study the underlying events responsible for synchronization the latter system is simpler since it has been separated from the more complicated process of cell division.

I. Synchronization of DNA Synthesis

Hjelm and Zeuthen (1967a,b) have shown that cells enter DNA synthesis synchronously following the first synchronous division induced by heat shocks. This appears to be a consequence of the synchronization of cell division rather than a cause. The synchronous first division maximum occurs at 70–80 minutes. Pulse labeling with tritiated thymidine shows that the curve for labeled macronuclei increases from 0 to 70 minutes to 80% at 110 minutes, and back down to 70% at 160 minutes. These events repeat themselves during the second division cycle.

It has been assumed that because heat-synchronized *Tetrahymena* have a two- to fourfold excess of DNA (see Scherbaum, 1964) and can carry out two synchronous divisions in an inorganic medium (Hamburger and Zeuthen, 1957), DNA synthesis is not necessary in these cells. For this reason one is curious about the nature of the DNA being synthesized in Hjelm and Zeuthen's (1967a,b) system.

One method that has been used to synchronize mammalian cells is to interfere with thymine metabolism. Since *Tetrahymena* have access to thymine from the medium as well as from synthetic pathways it is necessary to block both of these processes to bring about effective thymine starvation.

Zeuthen (1968) has described a method using methotrexate plus uridine which has the effect described above. Methotrexate is a folic acid analog and inhibits thymine synthesis. Uridine apparently acts to interfere with transport mechanisms that bring thymine compounds into the cell (Zeuthen, 1968).

Zeuthen recommends the use of 0.05 mM methotrexate (amethopterin, Lederle Laboratories) and 5 mM uridine to bring about the inhibition of cell division and the inhibition of incorporation of thymidine-^{3}H into DNA in synchronized cells. Concentrations of thymidine above 0.05 mM specifically release cells from the effects of M + U inhibition. Zeuthen suggests that it will be possible to apply synchronization procedures for cell division and DNA synthesis to the same culture.

VI.   Isolation of Organelles

A.   Methods of Lysis

Isolation of most cellular components or organelles begins with lysis or homogenization of the cells. The following general kinds of methods can be employed: (1) detergent lysis, (2) osmotic shock, (3) mechanical breakage (shear), and (4) ultrasonic disruption. Many of the isolation procedures involve the addition of a small amount of a nonionic detergent such as Nonidet P-40 (Shell Oil Corp.) or Triton X-100 (Rohm & Haas). The cells will usually spontaneously lyse in detergent. Osmotic methods do not always produce lysis because *Tetrahymena* are quite good osmoregulators. Swelling in distilled water will cause weakening and lysis of some cells. Addition of saturated indole (Lyttleton, 1963) is said to poison the contractile vacuole, with the resulting osmotic swelling in hypotonic medium producing lysis. A number of mechanical means may be employed. The Logeman press (a cream homogenizer) is often used because it is very fast. One may also use hand-held homogenizers of the Potter–Elvehjem type or shear produced by forcing a cell suspension through a glass syringe and needle. Finally ultrasound is a very effective means of breaking up cells. We allow cells to swell in cold glass-distilled water and then apply a 15-second pulse with a Branson sonicator model S-75 at No. 2 setting. This is the lowest effective setting that will produce breakage of >95% of the cells.

B.   Isolation of Macronuclei

Many methods for the isolation of *Tetrahymena* macronuclei have been described. In choosing a procedure one must keep a number of factors in mind. For example, some methods give very clean nuclei free of cytoplasmic contamination, but these may not have active DNA polymerase. Another method may give clean nuclei but with considerable loss of nuclear proteins. Other methods may not facilitate the isolation of histones. One must select a method of isolation to fit the requirements of the experiment.

Basically there are three procedures for isolating nuclei. These include aqueous media, aqueous media plus detergent, and nonaqueous or organic media. Examples of each of these types will be described. One must always bear in mind that a method worked out for one strain of *Tetrahymena* may not work for another strain.

Lee and Scherbaum (1966) reported the use of an isolation medium containing the nonionic detergent Triton X-100. This method has been modified by Byfield and Lee (1970) and it is this modification that is described here. Cells are washed in a medium containing 10 mM tris, 2 mM $CaCl_2$, and 1.5 mM $MgCl_2$, pH 7.4. They are lysed in the same buffer with sucrose added to a final concentration of 0.25% and Triton X-100 added to a final concentration of 0.1%. After lysis polyvinylpyrrolidone (PVP) is added at a final concentration of 2% to assure the nuclear integrity. This suspension is filtered through cotton filters and the filtrate layered over an equal volume of 0.5 M sucrose. Sedimentation of nuclei is brought about by successive 5-minute acceleration to 70, 250, and 800g. This procedure reduces cytoplasmic contamination.

Nuclei isolated in this manner are reported to have an active DNA-dependent RNA polymerase, to show little damage, and to be free of cytoplasmic contamination. They have been used for the isolation of histones with a recovery of 92–98%.

Mita *et al.* (1966) have described a procedure using a different non-ionic detergent—Nonidet P-40. Cells are washed in a medium containing 10 mM tris, 1 mM $MgCl_2$, and 3 mM $CaCl_2$, pH 7.5. To one volume of packed cells is added nine volumes of the sucrose buffer. Nonidet P-40 made up of sucrose buffer is added to a final concentration of 0.25%. This suspension is shaken by hand for several minutes or passed through a Potter–Elvehjem homogenizer with a loosely fitting Teflon pestle. One volume of lysate is layered over two volumes of sucrose buffer containing 0.33 M sucrose. Nuclei are pelleted by centrifuging 5 minutes at 1200g in a swinging bucket rotor. These nuclei are reported to have an active RNA polymerase. The Mita method has been used for isolation of nuclei for ultrastructural studies of nuclei by Nilsson and Leick (1970). They show the attachment of nucleoli to the inner nuclear membrane. In nuclei isolated from stationary phase cells there is a tendency for the membrane and this layer of nucleoli to pull off from the isolated nucleus. No structures are seen on the outer nuclear membrane, which is not well resolved and may have been removed in the isolation process.

Prescott *et al.* (1966) have described a similar procedure using Triton X-100 and spermidine. The cells are washed in 0.15 M KCl and suspended to 0.1% Triton, 0.001% spermidine, and 0.25 M sucrose. They are allowed to stand for 5 minutes and are then lysed by (repeated) forceful expulsion through a Pasteur pipette. The nuclei are pelleted at 700g for 30 minutes and washed with 0.25 M sucrose.

Gorovsky (1970) has reported a method that uses octanol and gum arabic. The cells are washed twice in "medium A" that contains 0.1 M sucrose, 4% gum arabic, and 1.5 mM $MgCl_2$, pH 6.75. They are then resuspended in the same medium containing 0.63 ml n-octanol/100 ml medium and homogenized in a Waring blendor for two 6-second periods. The nuclei are pelleted at 1000g, resuspended in medium A, and centrifuged 15 minutes at 250g. Cleaner nuclei can be obtained by spinning through 0.5–1.0 M sucrose. Greater yields can be obtained by recentrifuging the supernatant from the original homogenate, which contains many unsedimented nuclei.

Bollinger (1971) reports that neither the Mita nor Gorovsky methods yield nuclei which contain an active DNA polymerase. He suggests that in the case of the Gorovsky method this may be due to inactivation of the enzyme rather than to its leaking out, since the isolation medium was shown to inhibit active DNA polymerase from sea urchin embryos. Considerable loss of proteins from amoeba nuclei have been reported by Goldstein and Prescott (1967) using the Triton–spermidine method.

Bollinger (1971) has worked out an aqueous procedure for isolating macronuclei that does not use a detergent or stabilizing agent. Cells are washed in 10% v/v glycerin in 10 mM 2[N-morpholino]ethane sulfonic acid (MES) buffer containing 2 mM $CaCl_2$. The glycerin causes the cells to swell and also weakens the membrane. The cells are then suspended in 10 volumes of 2 mM $CaCl_2$ with 10 mM MES and centrifuged at 1800g for 10 minutes. The pellet is resuspended to 10 ml in 0.25 M sucrose MES–Ca and forced once through a No. 20 hypodermic needle. To the homogenate is added 45 ml 2.44 M sucrose in MES–Ca. The suspension is centrifuged 20 minutes at 20,000g. The pellet is resuspended in 2.44 M sucrose in MES–Ca and centrifuged 40 minutes at 20,000 rpm in a SW25 rotor. The resulting pellet is suspended in MES–Ca and centrifuged 10 minutes at 600g. The final pellet contains purified nuclei.

These nuclei have an active DNA polymerase and are free of cytoplasmic contamination. The method reduces aggregation of cytoplasmic components and is simple.

C. Micronuclei

Muramatsu (1970) describes a procedure for the isolation of micronuclei from *Tetrahymena*. This is based on lysis of cells in Nonidet by the method of Mita, differential centrifugal separation of the macronuclei from the micronuclei, and elimination of contaminating macronuclei by sonication.

A much more involved procedure is described by Gorovsky (1970). Gorovsky's procedure is similar in that both macro- and micronuclei are isolated together using the octanol procedure described earlier. The macronuclei are then destroyed by homogenization in a Waring blendor. Because there is no previous differential centrifugation step this homogenization step is not as effective in removing contaminating macronuclei and many subsequent centrifugation and homogenization steps are repeated.

D. Mitochondria

The following procedure for the isolation of mitochondria from the ST strain has been described by Suyama (1966). Washed cells are resuspended to 10 volumes of 0.2 M raffinose, 1 mM potassium phosphate buffer, and 0.25% bovine serum albumin, pII 6.2, and passed through a cream homogenizer. The homogenate is centrifuged at 5000g for 6 minutes, and the resulting supernatant removed by suction and discarded. The upper layer of pellet is resuspended in the same volume of buffer and care is taken not to include the bottom layer. This same

centrifugation step is repeated twice. The final pellet contains mitochondria with some contamination of cilia. The author states that this
method does not work well on other strains tested due to cytoplasmic
aggregation. Ribosomes, several types of RNA, and DNA have been
characterized from isolated *Tetrahymena* mitochondria (Chi and
Suyama, 1970).

Bollinger (1971) has described the following procedure for isolating
mitochondria from strain GL: Two liters of log phase cells are harvested
and the cell pellet washed twice by resuspension in 40 volumes of cold
glass-distilled water. The washed cells are suspended in 20 volumes of
0.2 M sucrose containing 5 mM N-tris(hydroxymethyl)methyl-2-amino-
ethane sulfonic acid (TES) at pH 7.0. The suspension is forced four
times through a No. 20 hypodermic needle fitted to a 20-ml glass syringe.
The homogenate is centrifuged at 500g for 6 minutes. The supernatant
is carefully removed and centrifuged at 6000 g for 6 minutes. The pellet
is resuspended in 40 ml 0.2 M sucrose–5 mM TES and centrifuged at 500
and 6000g as above. The pellet is resuspended and subjected to another
differential spin. The final pellet that contains mitochondria is suspended
in 100 mM potassium phosphate buffer, pH 7.0, or other buffer. Buetow
(1970) has discussed the isolation of *Tetrahymena* mitochondria and
their properties in Volume IV of this series.

E.   Pellicle

The term pellicle is used to describe the conglomeration of surface
structures of ciliates. This includes cilia, surface membranes, kinetosomes,
and tubular and fibrillar components. A number of workers have isolated
these structures usually in an attempt to get a fraction enriched for
kinetosomes. Most of the methods described involve fixation with ethanol.
Ironically, beautifully preserved "ghosts" are often seen in cell lysates
in the absence of any fixative. However, these structures are very fragile
and therefore difficult to isolate. The trick of isolating pellicles is to
stabilize them without aggregating everything else.

Seaman (1960) first described the isolation of pellicles from *Tetrahymena* as a source of kinetosomes. His method uses 40% ethanol at
−15°C to fix the pellicles. After at least 3 hours at −15°C the suspension was centrifuged at 250g for 10 minutes. The pellet contained
pellicles and was washed once with 1% digitonin in 0.4 M KCl and then
was allowed to sit in digitonin at 4°C with periodic examination until
only pellicles remained. When only pellicles remained they were centrifuged out at 1000g for 10 minutes.

Randall and Disbrey (1965) isolated pellicle fragments from *Tetra-*

hymena in the following manner: 3–10 ml of logarithmic phase culture are centrifuged down in a conical 12-ml centrifuge tube, and washed twice with glass-distilled water. The washed cells are resuspended in 0.6 ml H_2O and chilled 10 minutes on ice. Three milliliters of 12% cold ethanol is added, the contents are stirred gently with a pipette, and after 2 minutes transferred to a 37°C H_2O bath for 4 minutes. The cells are disrupted mechanically by passing them through a syringe needle. The pellicle fragments are collected by centrifugation at 800g for 5 minutes.

F. Kinetosomes

Although *Tetrahymena* pellicles should provide an excellent source for kinetosomes, these centriole-like structures have yet to be isolated in a sufficiently purified state to allow chemical characterization. A number of questions concerning their structure and function need to be examined. Seaman (1960), Argetsinger (1965), Hoffman (1965), and Satir and Rosenbaum (1965) have attempted to isolate kinetosomes from pellicles using methods modified from those of Seaman (1960). All of these procedures involve the homogenization of isolated pellicles, but each uses a different method. Satir and Rosenbaum (1965) are the only workers to present electron microscope evidence of a structurally intact isolated kinetosome fraction. Their preparation contains in addition to kinetosomes, the kinetodesmal fibers, striated fibers that arise at the proximal end of each kinetosome. It seems likely that the kinetosomal components could be selectively dissolved away from the kinetodesmal fibers, which are rather resistant to dissolution.

Satir and Rosenbaum's (1965) procedure can be described as follows. Isolated pellicles are resuspended in 0.25 M sucrose and washed twice. They are then disrupted in 0.25 M sucrose by passing through a Logeman emulsion homogenizer (available from most laboratory supply houses) or by sonication. The homogenate is centrifuged 15 minutes at 4000g (5750 rpm in a Sorvall SS-34 rotor) to sediment large fragments. The supernatant contains kinetosomes, which are brought down at 6000g for 15 minutes. They are then washed twice in glass-distilled water.

Whitson *et al.* (1966) reported the isolation of kinetosomes from digitonin-dissolved pellicles by zonal centrifugation. However, no characterization of this fraction in terms of purity or chemical composition was made. Their procedure is described in the section on isolation of the oral apparatus (Section VI,J).

Wolfe (1970) has recently used isolated oral apparatus as a source of kinetosomes. While these are free of kinetodesmal fibers they are held together by a microfilamentous meshwork. Wolfe predicts that a purification of kinetosomes will not follow from isolated oral structures.

G. Cilia

Cilia were first isolated from *Tetrahymena* by Child (1959), using a method derived from the earlier work of Child and Mazia (1956), which involved fixation with cold alcohol followed by digitonin solubilization as described in Section VI,E. This method gave a very few, impure cilia after tedious differential centrifugation. A second method, described in the same paper, used 20% glycerin at −8°C to harden the cells. The cilia are sheared off from the cells and separated from whole cell bodies by a simple differential centrifugation step. This procedure is said to be based on a localized cytolytic weakening of the pellicle in the area of the basal body.

Watson and Hopkins (1962) used a similar procedure for the isolation of *Tetrahymena* cilia, substituting 10% ethanol in place of glycerol. Addition of ethylenediaminetetraacetic acid (EDTA) prevented the release of cilia. However, addition of Ca^{2+} after EDTA treatment caused detachment of cilia and also stabilized the cell bodies and facilitated their separation from cilia.

The procedure of Watson and Hopkins was further modified by Gibbons (1965) and this is the procedure that will be described here. Cilia isolated by this method retain their intact morphology, are free from other cellular contaminants, can be isolated in mass, and are readily solubilized. We follow Gibbons' "ethanol–sucrose" modification directly for isolating cilia. Gibbons' procedure was worked out for the W strain, but we have used it with excellent success on the GL and the HS strain.

(1) Collect 50–100 ml of log phase cells grown in an enriched organic medium by centrifugation in the cold at 100g for 3–5 minutes.

(2) The cells are washed once with a solution containing 30 mM NaCl–0.2 M sucrose and resuspended in the same solution to a final volume of 200 ml in a 1500-ml beaker.

(3) The contents of the beaker are brought to 4°C in an ice-water bath.

(4) The beaker in ice water is placed on a stirring motor and a Teflon stirring bar placed in the beaker.

(5) Eight-hundred milliliters of a solution containing 11% ethanol, 2.5 mM EDTA, and 15 mM tris-thioglycolate buffer, pH 8.3, at 0°C is added to the suspension of cells.

(6) The mixture is stirred rapidly as 9.6 ml of 1 M $CaCl_2$ is quickly pipetted in. An almost immediate detachment of cilia results.

(7) The progress of cilia release is followed in the phase-contrast microscope. The cells are stirred until all somatic cilia are released. Sometimes the oral cilia remain behind after all somatic cilia are detached.

(8) The suspension is sedimented at 1000g for 5 minutes. The supernatant cilia are removed by aspiration using a 50-ml volumetric pipette with wide bore attached to a propipette. They can also be carefully decanted.

(9) The cilia are pelleted by centrifugation at 16,000 rpm (Sorvall SS-34 rotor) for 20 minutes in 30-ml Corex centrifuge tubes.

(10) The pellet is resuspended in 25 ml of 30 mM tris-HCl, pH 8.3, at 0°C, 2.5 mM MgSO$_4$, 0.2 M sucrose, and 25 mM KCl (TMSK) solution and centrifuged at 1000g for 5 minutes to remove any contaminating cell bodies. The cilia can be stored in TMSK for several days without change in morphology.

Because we are interested in microtubule proteins of cilia we normally process the isolated cilia according to either of two procedures: (1) that of Gibbons (1965) for the isolation of central pair microtubules or (2) that of Renaud *et al.* (1968) for isolation of acetone powders of cilia.

The first procedure involves removing the membranes of the cilia. This can be done by a wash in nonionic detergent such as digitonin (Gibbons, 1965) or Triton X-100 (Stephens, 1968). Alternatively the cilia can be dialyzed directly against 0.1 mM EDTA, 1 mM tris-HCl, pH 8.3, and 0.1 mM dithiothreitol (or 1 mM mercaptoethanol). After dialysis the suspension is brought to 10 mM tris to facilitate sedimentation of the intact outer fibers and membrane fragments. The effect of dialysis against EDTA at low ionic strength is to disrupt the membrane into small fragments and selectively solubilize the arms and central pair microtubules. The structures remaining intact are sedimented out at 16,000 rpm for 20 minutes. Occasionally the pellet is seen to consist of two layers. The upper fluffly layer is enriched in membrane fragments. Recently Witman (1970) has reported that membrane fragments can be separated from other ciliary components by isopycnic centrifugation. Outer fibers prepared by the dialysis procedure are shown in Fig. 10.

The second procedure involves washing the isolated cilia with 1 mM tris-HCl, pH 8.3, containing 0.1 mM EDTA, and 30 mM KCl. They are then washed twice with 5 ml of acetone at 0–5°C in a glass tube. The suspension is centrifuged, the acetone removed, and the pellet dried under a stream of nitrogen gas. The pellets are then stored at −20°C in a closed container containing silica gel or other desiccant.

Microtubule protein—presumably outer fiber but perhaps also including central pair—is extracted from acetone powders by two 30-minute extractions with 1 mM tris-HCl pH 7.8, at 0°C with stirring. The extracts

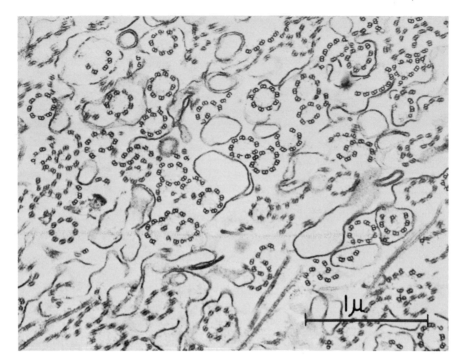

Fig. 10.   Thin section of an isolated cilia preparation that has been dialyzed for 48 hours against 0.1 m*M* EDTA, 1 m*M* tris-HCl, pH 8.3. Membranes have been removed from the axonemes but remain as vesicles. Central pair microtubules and arms have been solubilized. Some of the outer doublets have begun to break down but most remain intact and associated together in groups of nine. (Micrograph kindly provided by Dr. N. B. Gilula.)

are pooled and centrifuged at 16,000 rpm for 20 minutes. This yields a homogenous preparation of 6 S protein (Renaud *et al.*, 1968). Reduction and acetylation, followed by acrylamide gel electrophoresis, reveals two similar proteins. These two proteins have been characterized by Stephens (1970) and Everhart (1971).

H.   Ribosomes

Lyttleton (1963) described a procedure for the isolation of ribosomes from indole-lysed cells. A preliminary sedimentation of insoluble residue was pelleted at 10,000g for 15 minutes. The supernatant from this lysis is centrifuged at 105,000g for 15 minutes. The pellet was considered to be the ribosome fraction.

Leick and Plessner (1968) isolated ribosomes by a more involved procedure that probably yields cleaner and more intact ribosomes. Cells are washed twice in 0.01 M tris-HCl (pH 7.2) containing 0.15 M sucrose, 0.1 mM magnesium acetate, and 0.01 M KCl. The pellet is frozen at $-15°C$, thawed, and homogenized in a Potter–Elvehjem type homogenizer. The homogenate is spun at 15,000g for 15 minutes. The supernatant from this preliminary spin is sedimented at 100,000g for 60 minutes. The pellet contained ribosomes.

Kumar (1969) uses a similar procedure, washing the cells in the buffer described below. Washed cells are homogenized in 10 volumes of this buffer. Kumar emphasizes the importance of accomplishing this step quickly and in the cold. For this purpose he uses a Logeman press and estimates that the homogenization procedure takes 5 seconds.

We use the Kumar method in our laboratory for isolating ribosomes from strain HS. The slightly modified procedure is given in detail below.

TKM buffer
15 mM tris-HCl, pH 7.5
1.5 mM MgCl$_2$
5 mM KCl
0.1 mg/ml spermidine
0.25 M sucrose

(1) Harvest 1 liter of log phase cells. Wash three times in TKM buffer.

(2) Kill cells by freezing pellet in dry-ice acetone mixture.

(3) Homogenize at 0–4°C in a glass tissue grinder in a volume of 10 ml of TKM buffer plus spermidine and sucrose.

(4) Centrifuge at 11,500 rpm for 20 minutes in the Sorvall SS-34 rotor (16,000g).

(5) Remove supernatant and centrifuge for 2 hours at 40,000 rpm in Spinco No. 40 rotor (100,000g) or at 36,000 rpm on the Spinco SW 50-L.

(6) Resuspend pellet by vortexing in 0.01 M phosphate buffer, pH 7.5, 0.10 mg/ml spermidine.

For the examination of ribosomal subunits by sucrose density gradient centrifugation 0.3 μM EDTA is added per milligram of ribosomal material. This preparation is layered on top of a linear gradient of 15–30% sucrose made up in TKM buffer. The gradients are spun at 25,000 rpm for 5 to 10 hours at 7°C in a Spinco SW 27 rotor.

I. RNA Extraction

The method of Kumar (1969) has been used in our laboratory for the isolation and extraction of whole cell and ribosomal RNA in both *Tetra-*

hymena and Chinese hamster ovary cells. It is also used for examination of high molecular weight RNA from isolated nuclei. This procedure avoids degradation by using neutral pH. The modification, which follows, was described by Prescott *et al.* (1971).

Cells are collected by centrifugation and washed twice in the TKM buffer used in the isolation of ribosomes. The cell pellet (volume 1–4 ml) is frozen in a dry ice–acetone mixture and kept frozen until homogenized at 0°C in five volumes of 0.15 M sodium acetate, 0.1 M sodium chloride, and 1% sodium lauryl sarcosinate (SLS). An equal volume of 6% (w/v) sodium 4-aminosalicylate was added. The homogenate was then extracted with two volumes of phenol–cresol (500 gm solid phenol, 70 ml redistilled *m*-cresol, 55 ml water, 0.5 gm 8-hydroxyquinoline) for 25 minutes at 5°C on a wrist-action shaker. The phases were separated by centrifugation at 27,000g, the aqueous phase removed and retained,

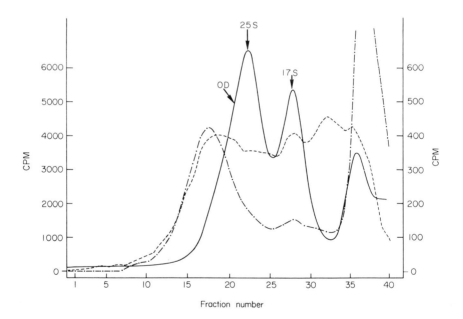

Fig. 11. Pulse-labeled *Tetrahymena* RNA. *Tetrahymena* were labeled for 5 minutes with both uridine-[14]C (- - -) and methyl methionine-[3]H (· - ·). The cells had previously been starved for 6 hours in synthetic medium minus pyrimidines to reduce pyrimidine pools. RNA was extracted as described in the text and run on a sucrose density gradient. The solid line represents the distribution of OD$_{260}$ absorbing material through the gradient as determined on a Gilford continuous recording spectrophotometer. The other lines show cpm in 1-ml fractions collected from the bottom of the gradient and analyzed as described in the text. (From Prescott *et al.*, 1971.)

and the phenol interphase reextracted for 10 minutes with one-half volume fresh aqueous solution. The aqueous phases were recombined, NaCl was added to a final concentration of 2%, and the mixture was reextracted with an equal volume of the phenol–cresol solution for 15 minutes. RNA was precipitated from the aqueous phase by the addition of NaCl to a final concentration of 0.5 M and of 2 volumes of ethanol. After 2 or more hours, the RNA was collected by centrifugation at 12,000g for 10 minutes, washed once with 95% ethanol, and dissolved in 0.1 M NaCl, 0.1 M tris, and 0.01 M sodium ethylenediaminetetraacetate (EDTA), pH 7.4, containing 0.2% SLS (NETS buffer).

The RNA was layered on a 40-ml sucrose gradient (15–30% w/v in NETS buffer) and centrifuged in a SW 27 rotor in a Spinco L-65 ultracentrifuge at 26,000 rpm for 16 hours at 5°C. Gradients were pumped from the bottom through a Gilford recording spectrophotometer. RNA extracted from whole *Tetrahymena* strain HS and fractionated in this manner is shown in Fig. 11.

J. Oral Apparatus

The ultrastructure of the isolated oral apparatus has been examined by Williams and Zeuthen (1966) and by Wolfe (1970). The oral apparatus is composed of kinetosomes, microtubules, and microfilaments. Isolated oral apparatus are therefore a potentially good source for these three structures. Isolated oral apparatus have also been used to investigate the relationship between oral morphogenesis and cell division (Williams et al., 1969).

Williams and Zeuthen's (1966) procedure involves lysis of the cells in butanol. Cells from 40 ml of culture are centrifuged into a pellet and the pellet resuspended in 12 ml of 1.5 M t-butanol by stirring on a vortex mixer. This results in cell lysis and the release of the oral apparatus. The lysate is transferred to a round-bottom tube, diluted with an equal volume of water, and centrifuged at 2000g for 10 minutes. The isolated oral structures are found in the pellet. Depending on the vigor with which the cells are lysed oral apparatus can be obtained in various states. Very gentle disruption leaves pieces of pellicle intact from which early morphogenetic stages of the developing mouth can be identified. More vigorous homogenization removes all traces of pellicle and shears off the deep fiber bundle (Williams and Zeuthen, 1966).

Whitson et al. (1966) have described a procedure for isolating mass amounts of oral structures from *Tetrahymena* by zonal centrifugation. Cells are washed in 0.01 M tris–HCl containing 5 mM Mg^{2+} at pH 7.5. They are lysed by the addition of an equal volume of saturated indole

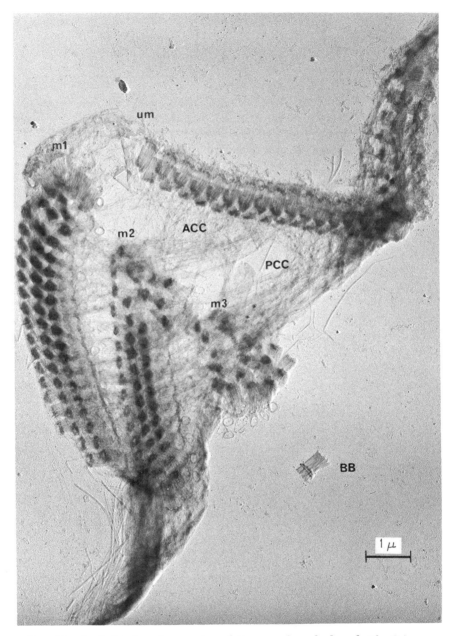

Fɪɢ. 12. An isolated oral apparatus, platinum–carbon shadowed. The four membranelles are joined by microfilamentous cross-connectives. UM, undulating membrane; M1, membranelle number 1, etc.; ACC, anterior cross connectives; PCC posterior cross-connectives; BB, basal body.

made up in 0.01 M tris buffer, pH 7.5. The indole lysate is treated with four times its volume of cold 95% EtOH for 2 to 3 hours and then centrifuged. The pellet is washed with 1% digitonin in 0.4 M KCl and then extracted for 18 to 24 hours in the cold digitonin solution. Oral structures and kinetosomes survive this treatment. The lysate is centrifuged at 2000 rpm for 10 minutes, and the resulting pellet is resuspended in 10% sucrose. This suspension is then separated by zonal centrifugation in an A-X11 rotor (the author's designation) containing 1300 ml of a linear 10–55% (w/w) sucrose, and the oral structures are sedimented to the 45–49% sucrose level. Zonal centrifugation probably represents an improvement in the isolation of clean kinetosomes and oral structures over conventional differential centrifugation.

Wolfe (1970) has described the following procedure for the isolation of oral structures. Cells from 10-liter cultures were collected and resuspended in 200 ml 0.12 M sucrose. To this was added 1 liter of 0.01 M tris, pH 9.3, containing 1 M sucrose, 1 mM EDTA, and 0.1% mercaptoethanol. After the cells swelled, 10 ml of 10% Triton X-100 was added. The cells immediately lyse. The lysate is centrifuged at 12,000g for 30 minutes. The pellet is resuspended in isolation medium without detergent and homogenized in a glass homogenizer with a Teflon pestle. Three 15-minute spins in the same buffer eliminate small contaminating particles.

Wolfe (1970) advises that if DNA were originally present in the basal bodies of the oral apparatus it would be lost in this procedure. An electron micrograph of an intact isolated oral apparatus is shown in Fig. 12.

VII.   Inhibitors

A.   Drug Penetration

In considering the use of inhibitors and other drugs for studies on *Tetrahymena* several factors must be kept in mind. First, is the question of penetration. For a number of drugs the concentrations required to achieve a given effect are much higher for *Tetrahymena* than for mammalian cells. The ciliate cortex is a complex surface structure composed in many places of three unit membrane layers in close approximation to one another. This may represent a substantial barrier to the entry of molecules from the outside. The controversy as to whether uptake of nutrients and other materials occurs across the surface membranes or only by means of the oral apparatus has continued for some time and is still unresolved. Seaman (1961) has proposed that most nutrient substances enter *Tetrahymena* through the surface membrane rather than by mouth. Byfield and Lee (1970) have confirmed Seaman's observations and suggest that nucleosides as well as actinomycin D enter the cell by general surface transport, not by food vacuole phenomena. In direct contrast, Nachtwey and Dickinson (1967) propose that food vacuole formation is the sole route of entry for actinomycin D. These workers found a strong correlation between food vacuole formation and actinomycin sensitivity. Because of this one must be suspicious of any negative results obtained from inhibitor studies. If inhibitors enter by food vacuoles one must consider the following factors: time of formation of the vacuole; time between formation of the vacuole and release of the drug to the cytoplasm; and effect of hydrolytic vacuole enzymes on the drug molecule.

B. Recovery

The phenomenon of recovery from drugs appears to be unique to *Tetrahymena*. *Tetrahymena* has the capacity to recover from the inhibitory effects of a wide variety of drugs including cycloheximide, actinomycin D, 5-fluorodeoxyuridine, mercaptoethanol, colchicine, and vinblastine (see Frankel, 1970).

This recovery has been examined in detail in only two cases. Rosenbaum and Carlson (1969) found that colchicine inhibited the regeneration of cilia in deciliated *Tetrahymena*. However, after a period of 3 to 5 hours in the presence of 4 mg/ml colchicine the cells apparently recovered from this inhibition and begin to regenerate new cilia. Addition of fresh colchicine at this point had no effect; furthermore, at the time of recovery the concentration of colchicine-^{3}H taken up by the cell began to decline. These observations suggest a twofold mechanism of recovery: (1) inhibition of further uptake related to the feeding process and (2) active excretion of the inhibitor.

Frankel (1970) studied the recovery of *Tetrahymena* from the inhibitory effects of cycloheximide on leucine-^{14}C incorporation. He found that cycloheximide in the medium of recovering cells was able to inhibit cells that had not previously been exposed to the drug, indicating that recovery was not the result of degradation. Furthermore, a specificity of the recovery process was demonstrated. Cells recovering from the effects of cycloheximide were insensitive to additional cycloheximide, but were fully inhibited by colchicine. Frankel suggests that a mechanism involving transport phenomena is more attractive than the alternative involving adaptation at the biochemical site of action.

If drugs enter the cells solely by food vacuole formation, there is a period during each cell cycle when the cells will not take up drugs. Nachtwey and Dickinson (1967) and Chapman-Andresen and Nilsson (1968) have demonstrated that food vacuole formation is dependent upon the stage of the cell cycle. Food vacuole formation in logarithmically growing cells ceases during cell division from the onset of macronuclear division up to about 5 minutes after the separation of the daughter cells. The rate of vacuole formation in heat-synchronized cells is constant for the first 60 minutes after EH and then decreases during the first synchronous division only to rise again thereafter. This behavior is the same as that seen in log phase cells.

C. Actinomycin D

Whitson and Padilla (1964) studied the effects of actinomycin D on *Tetrahymena* strain HSM synchronized by the repetitive cold cycle procedure described earlier. The inhibitor was added at 1 hour and 45 minutes before the beginning of the warm period. The cells normally divide during the last 60 to 90 minutes of the 2.5-hour warm period, but the addition of actinomycin D completely inhibits this division.

At a concentration of 8 μg/ml the inhibitor blocks cell division if added 90 minutes before the end of the cold period. If it is added later more and more cells divide. In control populations morphogenesis of daughter oral apparatus is inhibited during the cold period. However, 90% of the cells are in some stage of oral morphogenesis just prior to the peak of cell division. Actinomycin D added at 90 minutes before the end of the warm period completely inhibits oral development.

Lazarus *et al.* (1964) obtained similar results for heat-synchronized GL strain. Inhibition of the first synchronous division was dependent on the concentration of the drug and the time of its application. Complete inhibition is obtained if 20–50 μg/ml is added within the first 30 minutes after the last heat shock. The second synchronous division was inhibited completely at all concentrations studied (5–50 μg/ml) regardless of time of application. Incorporation of radioactive phosphate into RNA was inhibited 90% by 5 μg/ml of actinomycin D and 96% by 20 μg/ml immediately after addition, while 20 μg/ml added at 0 to 15 minutes blocks the first division but not at 30 minutes; 50 μg/ml blocks division if added at 15 or 30 minutes and at 45 minutes blocks only 64%. The inhibition of cell division in heat-synchronized *Tetrahymena* by actinomycin D has been studied with similar results by a number of workers (Frankel, 1965; Moner and Berger, 1966; Nachtwey and Dickinson, 1967).

Frankel (1965) found that actinomycin D concentrations down to 2.4

μg/ml ($2 \times 10^{-6} M$) almost completely prevent increase in cell number in a logarithmic population. After addition of actinomycin D to a log phase culture the cell number continues to increase for a period of time. There seems to be no correlation between the length of this period and the concentration. The average time required for cessation of culture growth at concentrations between 1 to 4×10^{-4} was 1.8 hours. Under the growth conditions employed in this experiment it was estimated that the log cells become insensitive to actinomycin D at a point corresponding to 0.58 of the cell generation time. This was verified using single cell experiments.

Satir (1967) applied actinomycin D to cultures of strain W over a range of concentrations from 5 to 30 μg/ml. When the drug was applied to cultures at the time of inoculation at concentrations of 5–20 μg/ml the following effects were observed: (1) the lag phase was lengthened, (2) the generation time was increased, (3) the period of growth deceleration was shortened, and (4) the cell concentration at which the culture entered the stationary phase was reduced. Concentrations of 30 μg/ml or higher inhibited cell division in most cases. The author proposes that the duration of the various phases of the culture cycle is a function of the amount of RNA per cell.

D. Colchicine

Colchicine can be used to delay division in *Tetrahymena* but the cells have the ability to recover rather quickly from this drug (Wunderlich and Peyk, 1969). The recovery leads to a synchronous division as described earlier (Section V,D). The inhibition of cell division by colchicine has been correlated with the absence of macronuclear microtubules (Tamura *et al.*, 1969; Ito *et al.*, 1968). At the concentrations used (1.5 mg/ml colchicine, or 0.5 mg/ml colcemid) by Wunderlich and Peyk (1969) to inhibit cell division, cell motility was unaffected in the GL strain. They report that concentrations of 7 mg/ml colchicine (or 2 mg/ml colcemid) killed most of the cells. Microtubules of the oral apparatus and other cytoplasmic microtubules were reported to be unaffected by 1.5 mg/ml colchicine.

Tamura *et al.* (1969) report the use of 30 mg/ml colchicine to reversibly inhibit the appearance of macronuclear microtubules in strain W. Microtubules of cilia, kinetosomes, and oral apparatus were unaffected. The need to use higher concentrations of colchicine than those used by Wunderlich and Peyk may represent strain differences. The addition of 30 mg/ml colchicine to heat-synchronized cells prior to 45 minutes after the last heat shock (EH + 45) blocked both the elongation and division of the macronucleus and cytokinesis. If added after EH + 45, macronuclear elongation was inhibited but cytokinesis proceeded and resulted in uneven macronuclear division.

In direct contrast to the findings of other workers, Kennedy (1969) reports that 2 mg/ml colchicine prevents the appearance of microtubules in the developing oral apparatus of heat-synchronized cells. The drug was applied at the end of the last heat shock and delayed the first synchronous division by 45 minutes.

Rosenbaum and Carlson (1969) used colchicine at a concentration of 4 mg/ml (0.5 mg/ml colcemid) to inhibit the regeneration of cilia from deciliated *Tetrahymena pyriformis* W. This concentration has no effect on the motility of normal cells, on the incorporation of uridine-^{14}C into RNA, or on the incorporation of ^{14}C-amino acids into TCA-insoluble material. In the absence of inhibitors 90% of the deciliated cells recovered their motility by 70 minutes after amputation. In the presence of colchicine deciliated cells start to regain their motility after 3 to 5 hours. Since colchicine has no effect on intact cilia it presumably must exert its action at the time of assembly. The question of recovery from the drug was referred to in Section VII,B.

E. Cycloheximide

The effect of cycloheximide on protein synthesis in logarithmically growing cells has been examined by Rosenbaum and Carlson (1969). In the presence of 1 μg/ml cycloheximide, protein synthesis continued at a very reduced rate, but at 10 μg/ml incorporation of ^{14}C-amino acids into acid- precipitable protein was completely inhibited.

Byfield and Scherbaum (1968) report that a concentration of 5 μg/ml cycloheximide immediately inhibits precursor uptake by more than 95%. They also add that cycloheximide has no effect on RNA synthesis for at least 30 minutes.

Frankel (1969a) investigated the effect of cycloheximide on protein synthesis, oral development, and cell division in heat-synchronized *Tetrahymena pyriformis* GL. The incorporation of leucine-^{14}C into hot TCA-insoluble material was inhibited by 85% at a concentration of 1 μg/ml. This inhibition occurs immediately upon addition of the drug as shown in Fig. 13. Lower concentrations of the drug produce correspondingly lower levels of inhibition. At concentrations of 1 μg/ml and less the cells eventually regain their original rate of protein synthesis. At 0.2 μg/ml protein synthesis is inhibited by about 80% for the first hour. How-

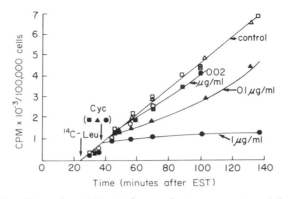

FIG. 13. The effect of cycloheximide on the incorporation of leucine-¹⁴C into hot TCA-insoluble material in heat-synchronized *Tetrahymena pyriformis* GL. Leucine-¹⁴C was added at 25 minutes after the last heat shock and cycloheximide was added at 40 minutes. The concentrations of cycloheximide used were 1.0 μg/ml (●), 0.1 μg/ml (▲), and 0.02 μg/ml (■). A control was run for each dose in which no cycloheximide was added. Incorporation by control cultures is shown by open symbols. (From Frankel, 1969a, with the permission of the Wistar Institute Press.)

ever, at about 60 minutes after the addition of the drug there is a rapid increase in the rate of synthesis with eventual recovery to the rate of control cells. This is shown in Fig. 13. This recovery from the effects of cycloheximide has been referred to earlier (Section VII,B) and has been studied in detail by Frankel (1970).

If cycloheximide is added to heat-synchronized cells less than 55 minutes after the last heat shock at concentrations of 10 μg/ml and higher, the first synchronous division is blocked. Lower concentrations delay but do not block division. This delay is a function of the concentration of inhibitor and the time after the last heat shock of its addition.

Cycloheximide inhibits the development of the oral apparatus and can cause the resorbtion of already differentiated primordia. The delay of division referred to above is correlated with the blockage of development. Resumption of oral morphogenesis is temporally correlated with a recovery of protein synthesis. Frankel concludes that continued protein synthesis is necessary to maintain oral development. The relationship between oral development and cell division is not clear.

Gavin and Frankel (1969) performed a similar study on heat-synchronized cells of the WH-6 strain. It is interesting to note that in this strain the initiation of oral development requires protein synthesis but the later phases of oral development, nuclear division, and cytokinesis proceed in the absence of concurrent protein synthesis.

F. Puromycin

The effect of puromycin on protein synthesis in heat-synchronized *Tetrahymena* was studied by Frankel (1967a). The rate of incorporation of histidine-^{14}C into hot TCA-insoluble material in the presence of 43 μg/ml puromycin was roughly 70% of the control rate while at 340 μg/ml incorporation was reduced to 10–20% of the control rate. The effect of puromycin on heat-synchronized cell division shows a transition point at about 45 minutes after the last heat shock. Puromycin (43 μg/ml) will inhibit the first synchronous division if applied before that time, while after that time a concentration of 43 μg/ml prevents only 20% of the cells from dividing. If puromycin is added before the transition point, oral development is arrested. If it is added after the transition point there is a slight delay at low concentrations (43 μg/ml) and considerable delay at high concentrations (420 μg/ml).

Puromycin appears to be one of the few drugs from which *Tetrahymena* cannot recover (Frankel, 1970). Although the concentration required to inhibit protein synthesis is higher than required for mammalian cells, the specificity of the drug as an inhibitor of protein synthesis is indicated by Frankel (1967a). Respiratory rates are essentially unaffected even by 430 μg/ml.

G. Mercaptoethanol

The effects of mercaptoethanol on heat-synchronized cell division have been studied by Gavin and Frankel (1966) and Mazia and Zeuthen (1966). Gavin and Frankel found that mercaptoethanol inhibits division in the two strains studied (GL and Wh-6, amicronucleate and micronucleate, respectively). In exponentially growing cells, increase in cell number ceases after a 30-minute lag. In heat-synchronized cells, division is prevented or greatly delayed if the mercaptoethanol is added before a transition point that differs for the two strains.

For strain WH-6 the addition of $4 \times 10^{-2} M$ mercaptoethanol during the first 25 minutes after the last heat shock was effective in blocking the first synchronous division. If added at EH + 35 a normal synchronous division occurs. For the GL strain $1.5 \times 10^{-2} M$ mercaptoethanol will block or greatly delay the first synchronous division if added before EH + 50 minutes. After 60 minutes it has no effect. For both strains the inhibition of cell division was correlated with the arrest of oral development.

Micronuclear and macronuclear division are inhibited in WH-6 if the mercaptoethanol is added before the transition point. After the transition point macronuclear division in the WH-6 strain is normal. For the GL strain $1.5 \times 10^{-2}\,M$ mercaptoethanol causes macronuclear division to be blocked or highly abnormal even though cytokinesis proceeds normally.

Mazia and Zeuthen (1966) have performed a similar study on the GL strain. They found that the concentration–time relationship was quite complex. Using short-term exposures they established that both the amount of delay and its time relationship depend upon concentration. The length of delay is also a function of time spent in the presence of inhibitor.

H.   Ultraviolet Irradiation

Division of heat-synchronized *Tetrahymena pyriformis* strain W was studied in the presence of low doses of ultraviolet light (UV) by Nachtwey and Giese (1968). The division is delayed if the irradiation is administered in the first 30 minutes after the last heat shock. After that time most cells become insensitive. The interesting finding here is that the transition to insensitivity to UV (30 minutes) comes much earlier than (for example) transition to insensitivity to heat, which occurs for this strain at 60 minutes. This suggests that two different processes are involved in the preparation for cell division.

I.   Vinblastine

The effects of vinblastine on log phase *Tetrahymena* have previously been discussed in the section on synchronization (Section V,G). Concentrations of 25 γ/ml for strain WH-6 and 50 γ/ml for strain GL are effective in inhibiting cell division. The site of action is believed to be rather specific because cells that are released from vinblastine inhibition by washing divide with a high degree of synchrony (Stone, 1968a,b). Addition of 20 γ/ml to a log phase population of strain WH-6 causes a slight inhibition in the rate of incorporation of thymidine-^{3}H, but it is unlikely that this is the cause for its effect on cell division (Sedgley and Stone, 1969).

VIII.   Radiochemical Procedures

A.   Filter Disc Assay

The rate of synthesis of various cellular macromolecules can be studied using radioactive precursors. Byfield and Scherbaum (1966) have described a simple method for measuring protein and nucleic acid metabolism using whole cells and filter paper discs. This procedure was adapted from the method of Bollum (1959) for measuring *in vitro* DNA synthesis and the method of Mans and Novelli (1961) for measuring *in vitro* protein synthesis.

Cells are labeled with a ^{14}C-amino acid mixture to study protein synthesis or with uridine-^{14}C or thymidine-^{14}C to study RNA or DNA, respectively. The radioactive material can be added directly to the standard growth medium but reduction of the concentration of nutrients will result in a greater incorporation of labeled material. For example, Byfield and Scherbaum (1966) use a medium containing 0.25% proteose peptone. At various times after the addition of label 75 μ aliquots are removed and placed on 2.3 cm filter discs of Whatman 3 MM chromatography paper. The discs can be numbered with a pencil and are held on stainless steel straight pins. The pins facilitate handling and prevent adherence of the discs during washing. The discs are placed directly into trichloroacetic acid (TCA) at 0°–4°C. Five percent TCA is used for experiments involving the labeling of RNA; 10% is used for labeling protein. A volume of 10 ml is allowed for each disc. Mans and Novelli (1961) report that the discs can be stored in the initial TCA wash for up to 4 days without loss of radioactivity. After all time points are collected the discs are washed in three separate changes of 5% TCA for 15 minutes each. They are then washed twice in ethanol:ether 1:1 and twice with anhydrous ether. Washing is accomplished by swirling in a flask or beaker containing 5 ml of the washing solution per disc. The pins are removed and the discs are dried and then counted in 10 ml of a toluene-based scintillation cocktail.

Ether washing is used to remove all traces of water, which can lead to significant quenching. An alternative procedure is to wash twice with 95% ethanol. The discs are then dried under an infrared heating lamp. If all the water is removed the disc will "clear" or become transparent on placement into the scintillation fluid.

High backgrounds sometimes result from the binding of unincorporated label to the filters. This can be reduced by presoaking the filter discs in a solution containing unlabeled precursor so that these binding sites are occupied. The filters are then blotted dry before applying the sample.

A similar procedure involving suction filtration can also be used. We use a stainless steel filter chimney for this purpose (available from W. A. Shaerr, 356 Gold St., Brooklyn, New York). An aliquot of cell culture is applied to the filter, which has been prewashed with unlabeled precursor. The cells are washed several times with 5- to 10-ml amounts of inorganic medium containing an excess of unlabeled precursor. They are

then washed with 5% TCA followed by 95% ethanol. The wash solutions
are delivered from polyethylene squirt bottles. The discs are dried under
an infrared lamp and counted in 10 ml of Omnifluor (New England
Nuclear).

Either ³H- or ¹⁴C-labeled precursors may be used in this type of experiment. However, if cells from different stages of the culture or cell
cycle are compared one must be alerted to the possibility of differential
quenching especially with ³H-labeled compounds.

B.   Autoradiography

Incorporation of radioactive precursors into macromolecules and their
subsequent cellular localization can be determined using autoradiography. The procedure as used for *Tetrahymena* has been described by
Stone and Cameron (1964) and Prescott (1964).

This method is particularly useful for studying the incorporation of
thymidine-³H into DNA. One can determine the number of cells synthesizing DNA at any given time. For example, Hjelm and Zeuthen
(1967b) detected DNA synthesis in heat-synchronized cells using 10
μCi/ml thymidine-³H given in 5-minute pulses. The slides were exposed
for 3 days prior to development. In general, levels of thymidine between
2.5–10 μCi/ml are used. Pulses range from 5 to 15 minutes and slides are
developed for 1 to 2 weeks. The desirability of using low doses of
radioactivity is discussed by Nachtwey and Cameron (1968). These
dosages are, of course, dependent upon the specific activity of the
thymidine.

Autoradiographic determination of rates of protein or RNA synthesis
is not of great use with the advent of more sensitive and simpler methods
for measuring their synthesis. Grain counts are necessary in order to
quantitate rates of macromolecular syntheses, and these are subject to
considerable experimental variability. Autoradiography has been performed on isolated structures such as the oral apparatus (Williams *et al.*,
1969) and pellicle (Randall and Disbrey, 1965). Higher resolution can
be obtained using sectioned material and electron microscope techniques
as described by Stevens (1966). Murti and Prescott (1970) have used
these techniques to investigate micronuclear RNA in *Tetrahymena*.

C.   Labeling RNA

RNA can be labeled with uridine-³H, uridine-¹⁴C, or methyl methio-
nine-³H. Prior to labeling with uridine Prescott *et al.* (1971) collected
log cells by centrifugation, resuspension in pyrimidine-free synthetic

medium, and incubation for 6–15 hours to reduce pyrimidine pools. *Tetrahymena* continue to grow and divide for at least 24 hours in pyrimidine-free medium.

After isolation of labeled RNA as described earlier (Section VI,I) and separation by sucrose density gradient centrifugation, 1-ml fractions are collected, precipitated with an equal volume of cold 20% TCA, collected on Whatman GF/A glass fiber filters, washed with cold 5% TCA, cold 95% ethanol, and dried under an infrared lamp. The filters are then counted in Omnifluor on a scintillation counter.

IX. The Cell Cycle

A. Markers of the Cell Cycle

During balanced growth there is a doubling of all cellular components in the interval between the end of one cell division and the beginning of the next. This doubling results from numerous processes and reactions that are repeated in each cell generation. These sequential events together constitute the cell cycle.

The replication of DNA is restricted in most cells to a discrete part of the cell life cycle. This restriction has formed the basis for subdividing the cell cycle into four parts. These periods were designated by Howard and Pelc (1953) as G1, the interval between the end of nuclear division and the start of DNA synthesis; S, the period of DNA synthesis; G2, the period following DNA synthesis and preceding nuclear division; D, the period of division.

The cell cycle is commonly subdivided on this basis because of the ease with which these periods can be determined by autoradiography. However, other markers of the cell cycle are readily defined in *Tetrahymena*. These include the duplication of somatic cilia, oral morphogenesis, and the periodic cessation of food vacuole formation.

The duplication of cortical structure in *Tetrahymena* appears to occur in a well-ordered sequence. The data of Randall and Disbrey (1965) suggest that somatic kinetosome duplication preceeds oral morphogenesis (specifically the duplication of oral kinetosomes and formation of the anarchic field). Frankel (1960) has clearly demonstrated that somatic kinetosome doubling precedes oral development in the closely related ciliate *Glaucoma pyriformis*.

In a similar manner oral morphogenesis always takes place before cell division. Frankel (1964, 1967a,b, 1969a) has shown that any treatment that disrupts the development of the daughter oral apparatus will also block cell division. The stages in the morphogenesis of the oral apparatus and their position in the cell cycle have been described by Frankel (1967a).

The formation of food vacuoles has been shown to cease for a period in each cell cycle extending from the beginning of macronuclear division to just after separation of the daughter cells (Chapman-Andresen and Nilsson, 1968). The nuclear events of the cell cycle have been discussed by Prescott and Stone (1967). The placement of the period of DNA synthesis in the *Tetrahymena* life cycle is extremely variable and dependent upon such factors as temperature and nutritional state. However, under all conditions it has been shown to be noncontinuous. Under a given set of conditions the length and position of the S period is very reproducible.

B.   Methods for Cell Cycle Analysis

Cell cycle analysis requires a population of cells that are in the same stage of the cell cycle. For morphological and autoradiographic studies only a small number of cells are needed and these can be hand selected. However, for biochemical studies large numbers of cells are required and one must resort to the use of artificial methods of synchronization.

1.   HAND SELECTION

The selection of dividing cells for use in cell cycle studies has been described in detail by Stone and Cameron (1964). A culture of cells is poured into a petri dish and after a few minutes dividers settle to the bottom. A dissecting microscope is used while the dividing cells are picked up with a braking pipette made according to the instructions of Stone and Cameron (1964). The dividing cells are inoculated into a small volume of medium contained in a capillary pipette, sealed at one end with an agar plug and a cotton plug at the other. These pipette cultures are observed under the dissecting microscope as described by Prescott (1957b).

Another method for observing hand-selected cells has been described by Nachtwey and Dickinson (1967). Approximately 1 μl drops of culture medium are prepared by tapping a vertically held 1-mm capillary tube containing medium on the bottom of a Falcon Plastics tissue culture petri dish. These drops are covered with a thin layer of paraffin oil to prevent evaporation. Individual cells are then introduced into the drops with a micropipette and their progress through the cell cycle observed through the dissecting microscope.

Hand-selected cells give rise to a naturally synchronous population. Because dividers in almost precisely the same stage can readily be recognized, the first division is undergone almost in unison and the second division shows very good synchrony (Stone and Cameron, 1964). These populations have the advantage over cells synchronized artificially in that their synchrony has not been imposed on them by some metabolic insult.

Hand-selected cells have been used for a variety of purposes, including growth studies (Prescott, 1957b), autoradiographic studies (Prescott, 1960), inhibitor studies (Nachtwey and Dickinson, 1968), electron microscopy (Flickinger, 1965), and determination of the distribution of generation times (Nachtwey and Cameron, 1967).

2. ARTIFICIAL SYNCHRONIZATION

Methods for artificial synchronization have been discussed in Section V. Some of these methods lend themselves to the study of particular stages of the cell cycle better than others. For example the system of Cameron and Jeter (1970) that produces G1 arrest by starvation seems applicable for the study of G1 events, particularly the initiation of DNA synthesis. After refeeding starved cells there is a lag period which precedes DNA synthesis. Presumably, during this period the cells are involved in preparations that are specific for the initiation of the S period.

G2 is the period in which specific preparations for cell division may be made. Heat-synchronized cells are good material for an investigation of these preparations. Although heat-synchronized cells are in a state of unbalanced growth the period between the last heat shock and the first synchronous division is unquestionably a time in which division associated processes take place. The normal sequence of events in oral morphogenesis, macronuclear division, and cytokinesis occur.

X. Special Techniques

A. Emacronucleation

Nachtwey (1965) has described this procedure for emacronucleation of single cells. Individual cells are isolated into small drops of proteose peptone medium. These are then drawn up into a fire-polished Pasteur pipette with a diameter of 12–15 μ. Forceful expulsion of the cell ruptures the posterior end and allows the macronucleus to float free. This procedure results in the loss of some 20–75% of the cytoplasm.

Nachtwey (1965) employed this procedure to detect the role of the macronucleus in heat-synchronized cell division. No cells enucleated before 42 minutes after the last heat shock ever divided. These results support those obtained with chemical "enucleators" such as actinomycin D. However, due to problems with penetration of and recovery from the drugs the results with enucleated cells strengthen the argument.

One can think of other experiments in which this procedure might be useful. For example can deciliated cells regenerate their cilia in the absence of a macronucleus? This type of experiment might more easily be conducted on the larger ciliates such as *Blepharisma* or *Stentor,* but there is less biochemical literature on these.

Enucleation might also be useful for the isolation of small numbers of macronuclei for ultrastructural or radiochemical study. The nuclei could be extruded from their cytoplasm as described above but in a suitable nuclear isolation medium such as the Triton–spermidine medium described by Prescott *et al.* (1966). Radioactive labeling experiments have been performed on nuclei isolated from individual *Amoeba proteus* by Goldstein and Prescott (1967).

B. Deciliation

Rosenbaum and Carlson (1969) have described a procedure for de-ciliation of *Tetrahymena* that allows for subsequent viability of the cells and the regeneration of cilia. This procedure is adapted from the isolation procedure of Watson and Hopkins (1962) mentioned earlier. Its main modification is the omission of ethanol.

The cells may be grown in PPL or a defined medium. They are harvested and resuspended to a final concentration of 2×10^6 cells/ml in their original growth medium. The cells are chilled to 0–4°C in a mixture of ice and water. At zero time, 2.5 ml of concentrated cells are added to 5 ml of medium A (10 mM disodium EDTA; 50 mM sodium acetate buffer, pH 6). The contents is mixed by swirling. At 30 seconds 2.5 ml cold distilled water is added. At 90 seconds 0.25 ml 0.2 M CaCl$_2$ is added and the suspension mixed by inversion of the tube. At 3 minutes 30 seconds the suspension is forced three to four times through a 10-ml glass syringe fitted with a No. 18 gauge needle. One volume of deciliated cell suspension is then immediately pipetted into 20 volumes of recovery medium (fresh defined growth medium). The deciliated cells are then incubated at 25°C with gentle shaking on a Dubnoff shaker.

Rosenbaum and Carlson (1969) worked out this procedure for *Tetrahymena pyriformis* strain W. We have modified the procedure slightly for strain HS in order to get suitable results. The concentration of Ca^{2+} was found to be very critical. It was necessary to reduce the final concentration of Ca^{2+} to 1 mM which is 1/5 that used by Rosenbaum and Carlson. Higher concentrations were lethal to the cells. We also centrifuged the cells after deciliation and resuspended them in fresh medium at densities of 100,000 cells/ml. Dilution wastes large amounts of medium and also requires excessive amounts of isotope in radioactive labeling experiments. Centrifugation for 2 minutes at 100g does not seem to have any adverse effect on regeneration, though this point may require more careful consideration.

C. Silver Staining of Cortical Structures

Silver staining has been used for some time to bring out the so called "silver line system" in ciliates. Recently Dippel (1962) has demonstrated that for paramecium silver staining results in a nonspecific deposition of silver in depressions on the cell surface. Frankel (1969b) has shown that in *Tetrahymena* there is a good correspondence between the black dots revealed by silver staining and kinetosomes demonstrated by other methods.

The original methods of Chatton and Lwoff (1930) has been modified by Corliss (1953) and Frankel and Heckmann (1968). The method described below is taken from the latter two references.

(1) The cells are centrifuged in a 10–15 ml conical centrifuge tube and the supernatant medium discarded. The pellet is resuspended in the small volume of remaining fluid and fixed with 5 volumes of Champy's solution for 3 minutes.

(2) The fixative is withdrawn and replaced with two changes of Defano fixative. The cells are allowed to sit in a third change for at least 2 hours.

(3) The cells are allowed to settle out. A concentrated drop of cells is placed on a warm slide and mixed with a drop of saline gelatin at 35–45°C. The cells are mixed with the tip of a warm needle and spread into a thin film. The thickness of the film should be about 50–100 μ or not much greater than the diameter of the cells.

(4) The gelatin is hardened by placing the slide in a petri dish on ice. The petri dish is kept moist with a piece of wet paper toweling.

(5) The slide is placed in 3% $AgNO_3$ at 5–10°C for 10–20 minutes in a darkened container.

(6) The slide is washed with cold distilled water and placed in a white porcelain pan containing cold distilled water at a depth of 3–4 cm. The slide is exposed to ultraviolet irradiation of 254 mμ for about 30 minutes. The light source is maintained about 30 cm from the slide. The time of exposure must be determined by trial and error.

(7) When the cells have acquired a yellow–brown color they are transferred to cold 70% ethanol, through several changes of 100% ethanol, cleared in cold xylene, and mounted in Permount. The composition of the solutions used in this procedure are given below:

Champy's fixative
    3% Potassium dichromate   7 parts
    1% Chromic acid   7 parts
    2% Osmic acid   4 parts
De Fano fixative
    Cobalt nitrate   1 gm
    100% Formalin   10 ml
    Sodium chloride   1 gm
    Distilled water   90 ml
Saline gelation
    Powdered gelation   10 gm
    Sodium chloride   0.05 gm
    Distilled water   100 ml

ACKNOWLEDGMENTS

I wish to thank the following people who have contributed in various ways to the preparation of this chapter: Drs. J. Bollinger, C. J. Bostock, I. Cameron, J. Frankel, K. G. Murti, D. M. Prescott, J. Wolfe, and Misses Marlene Lauth, Norma Sattler, and Catherine Verhulst. I especially thank Dr. R. R. Ronkin who introduced me to *Tetrahymena*.

REFERENCES

Allen, R. D. (1967). *J. Protozool.* 14, 553.
Allen, S. L. (1967). *In* "Chemical Genetics of Protozoa" (G. W. Kidder, ed.), Vol. I, pp. 617–694. Academic Press, New York.
Argetsinger, J. (1965). *J. Cell Biol.* 24, 154.
Bollinger, J. (1971). In preparation.
Bollum, F. J. (1959). *J. Biol. Chem.* 234, 2733.
Borisy, G., and Taylor, E. W. (1967). *J. Cell Biol.* 34, 535.
Buetow, D. E. (1970). *In* "Methods in Cell Physiology" (D. M. Prescott, ed.), Vol. IV, pp. 83–115. Academic Press, New York.
Byfield, J. E., and Scherbaum, O. H. (1966). *Anal. Biochem.* 17, 434.
Byfield, J. E., and Scherbaum, O. H. (1968). *Exp. Cell Res.* 49, 202.
Byfield, J. E., and Lee, Y. C. (1970). *J. Protozool.* 17, 445.
Cameron, I. L. (1965). *J. Cell Biol.* 25, 9.
Cameron, I. L., and Jeter, J. R., Jr. (1970). *J. Protozool.* 17, 429.
Cameron, I. L., and Padilla, G. M. (eds.) (1966). "Cell Synchrony," Academic Press, New York.
Cameron, I. L., and Nachtwey, D. S. (1967). *Exp. Cell Res.* 46, 385.
Cameron, I. L., Padilla, G. M., and Miller, O. L., Jr. (1966). *J. Protozool.* 13, 336.
Chapman-Andresen, C., and Nilsson, J. R. (1968). *C. R. Trav. Lab. Carlsberg* 36, 405.
Chatton, E., and Lwoff, A. (1930). *C. R. Soc. Biol.* 104, 834.
Chi, J. C. H., and Suyama, Y. (1970). *J. Mol. Biol.* 53, 531.

Child, F. M. (1959). *Exp. Cell Res.* **18**, 258.

Child, F. M., and Mazia, D. (1956). *Experientia* **12**, 161.

Cleffmann, G. (1967). *Z. Zellforsch. Mikrosk. Anat.* **79**, 599.

Connor, R. L., Cline, S. G., Koroly, M. J., and Hamilton, B. (1966). *J. Protozool.* **13**, 377.

Corbett, J. J. (1964). *Exp. Cell Res.* **33**, 155.

Corliss, J. O. (1953). *Stain Technol.* **28**, 97.

DeBault, L. E., and Ringertz, N. R. (1967). *Exp. Cell Res.* **45**, 509.

Dippel, R. V. (1962). *J. Protozool. Suppl.* **9**, 24.

Elliott, A. M. (1959). *Ann. Rev. Microbiol.* **13**, 77.

Elliott, A. M., and Clemmons, G. L. (1966). *J. Protozool.* **13**, 311.

Elliott, A. M., Brownell, L. E., and Gross, J. A. (1954). *J. Protozool.* **1**, 193.

Elliott, A. M., Travis, D. M., and Work, J. A. (1966). *J. Exp. Zool.* **161**, 177.

Everhart, L. P. (1971). *J. Mol. Biol.* (in press).

Everhart, L. P., and Ronkin, R. R. (1966). *J. Protozool.* **13**, 646.

Flickinger, C. J. (1965). *J. Cell Biol.* **27**, 519.

Fox, T. O., and Pardee, A. B. (1970). *Science* **167**, 80.

Frankel, J. (1960). *J. Protozool.* **7**, 362.

Frankel, J. (1962). *C. R. Trav. Lab. Carlsberg* **33**, 1.

Frankel, J. (1964). *J. Exp. Zool.* **155**, 403.

Frankel, J. (1965). *J. Exp. Zool.* **159**, 113.

Frankel, J. (1967a). *J. Cell Biol.* **34**, 841.

Frankel, J. (1967b). *J. Protozool.* **14**, 639.

Frankel, J. (1969a). *J. Cell Physiol.* **74**, 135.

Frankel, J. (1969b). *J. Protozool.* **16**, 26.

Frankel, J. (1970). *J. Cell. Physiol.* **76**, 55.

Frankel, J., and Heckmann, J. (1968). *Trans. Amer. Microsc. Soc.* **87**, 317.

Fulton, C. (1970). *In* "Methods in Cell Physiology" (D. M. Prescott, ed.), Vol. IV, pp. 341–476. Academic Press, New York.

Gavin, R., and Frankel, J. (1966). *J. Exp. Zool.* **161**, 63.

Gavin, R., and Frankel, J. (1969). *J. Cell. Physiol.* **74**, 123.

Goldstein, L., and Prescott, D. M. (1967). *J. Cell Biol.* **33**, 637.

Gibbons, I. R. (1965). *Arch. Biol.* **76**, 317.

Gorovsky, M. (1970). *J. Cell Biol.* **47**, 619.

Hamburger, K., and Zeuthen, E. (1957). *Exp. Cell Res.* **13**, 443.

Hjelm, K. K. (1970). *Exp. Cell Res.* **60**, 191.

Hjelm, K. K., and Zeuthen, E. (1967a). *C. R. Trav. Lab. Carlsberg* **36**, 127.

Hjelm, K. K., and Zeuthen, E. (1967b). *Exp. Cell Res.* **48**, 231.

Hoffman, E. J. (1965). *J. Cell Biol.* **25**, 217.

Holm, B. J. (1968). *Exp. Cell Res.* **53**, 18.

Holz, G. G., Scherbaum, O. H., and Williams, N. (1957). *Exp. Cell Res.* **13**, 618.

Howard, A., and Pelc, S. R. (1953). *Heredity suppl.* **6**, 261.

Ito, J., Lee, Y. C., and Scherbaum, O. H. (1968). *Exp. Cell Res.* **53**, 85.

Kennedy, J. R., Jr. (1969). *In* "The Cell Cycle: Gene-Enzyme Interactions" (G. M. Padilla, G. L. Whitson, and I. L. Cameron, eds.), pp. 227–248. Academic Press, New York.

Kidder, G., and Dewey, V. C. (1951). *In* "Biochemistry and Physiology of Protozoa" (A. Lwoff, ed.), Vol. I, pp. 323–400. Academic Press, New York.

Kumar, A. (1969). *Biochim. Biophys. Acta* **186**, 326.

Lazarus, L. H., Levy, M. R., and Scherbaum, O. H. (1964). *Exp. Cell Res.* **36**, 672.

Lee, D. (1969). *J. Cell. Physiol.* **74**, 295.

Lee, Y. C., and Scherbaum, O. H. (1966). *Biochemistry* **5**, 2067.

Leick, V., and Plessner, P. (1968). *Biochim. Biophys. Acta* **169**, 398.

Lwoff, A. (1923). *C. R. Acad. Sci. (Paris)* **176**, 928.

Lyttleton, J. W. (1963). *Exp. Cell Res.* **31**, 385.

MacKenzie, T. B., Stone, G. E., and Prescott, D. M. (1966). *J. Cell Biol.* **31**, 633.

Mans, R. J., and Novelli, G. D. (1961). *Arch. Biochem. Biophys.* **94**, 48.

Mazia, D., and Zeuthen, E. (1966). *C. R. Trav. Lab. Carlsberg* **35**, 341.

Mita, T., Shiomi, H., and Iwai, K. (1966). *Exp. Cell Res.* **43**, 696.

Mitchison, J. M. (1970). In "Methods in Cell Physiology" (D. M. Prescott, ed.), Vol. IV, pp. 131–165. Academic Press, New York.

Mitchison, J. M., and Vincent, W. S. (1966). In "Cell Synchrony" (I. L. Cameron and G. M. Padilla, eds.), pp. 328–331. Academic Press, New York.

Moner, J. G., and Berger, R. O. (1966). *J. Cell. Physiol.* **67**, 217.

Muramatsu, M. (1970). In "Methods in Cell Physiology" (D. M. Prescott, ed.), Vol. IV, pp. 195–230. Academic Press, New York.

Murti, K. G., and Prescott, D. M. (1970). *J. Cell Biol.* **47**, 460.

Nachtwey, D. S. (1965). *C. R. Trav. Lab. Carlsberg* **35**, 25.

Nachtwey, D. S., and Cameron, I. L. (1968). In "Methods in Cell Physiology" (D. M. Prescott, ed.), Vol. III, pp. 213–259. Academic Press, New York.

Nachtwey, D. S., and Dickinson, W. J. (1967). *Exp. Cell Res.* **47**, 581.

Nachtwey, D. S., and Giese, A. C. (1968). *Exp. Cell Res.* **50**, 167.

Nanney, D. L. (1968). *Science* **160**, 496.

Nilsson, J. R., and Leick, V. (1970). *Exp. Cell Res.* **60**, 361.

Padilla, G. M., and Cameron, I. L. (1964). *J. Cell. Comp. Physiol.* **64**, 303.

Phelps, A. (1936). *J. Exp. Zool.* **72**, 479.

Pitelka, D. (1968). In "Research in Protozoology" (T. T. Chen, ed.), Vol. III, pp. 280–388. Pergamon, London.

Plessner, P. (1964). *C. R. Trav. Lab. Carlsberg* **34**, 1.

Plessner, P., Rasmussen, L., and Zeuthen, E. (1964). In "Synchrony in Cell Division and Growth" (E. Zeuthen, ed.), pp. 543–563. Wiley (Interscience), New York.

Preer, J. R., Jr. (1959). *J. Protozool.* **6**, 88.

Prescott, D. M. (1957a). *J. Protozool.* **4**, 252.

Prescott, D. M. (1957b). *Exp. Cell Res.* **12**, 126.

Prescott, D. M. (1958). *Physiol. Zool.* **31**, 111.

Prescott, D. M. (1959). *Exp. Cell Res.* **16**, 279.

Prescott, D. M. (1960). *Exp. Cell Res.* **19**, 228.

Prescott, D. M. (1964). In "Methods in Cell Physiology" (D. M. Prescott, ed.), Vol. I, pp. 365–370. Academic Press, New York.

Prescott, D. M., and Stone, G. E. (1967). In "Research in Protozoology" (T. T. Chen, ed.), Vol. II, pp. 118–146. Pergamon Press, Oxford.

Prescott, D. M., Rao, M. V. N., Evenson, D. P., Stone, G. E., and Thrasher, J. D. (1966). In "Methods in Cell Physiology" (D. M. Prescott, ed.), Vol. II, pp. 131–142. Academic Press, New York.

Prescott, D. M., Bostock, C. J., Gamow, E., and Lauth, M. (1971). *Exp. Cell Res.* **67**, 124.

Randall, J., and Disbrey, C. (1965). *Proc. Roy. Soc.* **B162**, 473.

Rasmussen, L. (1963). *C. R. Trav. Lab. Carlsberg* **33**, 53.

Rasmussen, L., and Zeuthen, E. (1962). *C. R. Trav. Lab. Carlsberg* **33**, 333.

Renaud, F. L., Rowe, A. J., and Gibbons, I. R. (1968). *J. Cell Biol.* **36**, 79.

Rooney, D. W., and Eiler, J. J. (1967). *Exp. Cell Res.* **48**, 649.
Rooney, D. W., and Eiler, J. J. (1969). *Exp. Cell Res.* **54**, 49.
Rosenbaum, J. L., and Carlson, K. (1969). *J. Cell Biol.* **40**, 415.
Satir, B. (1967). *Exp. Cell Res.* **48**, 253.
Satir, B., and Rosenbaum, J. L. (1965). *J. Protozool.* **12**, 397.
Scherbaum, O. H., and Zeuthen, E. (1954). *Exptl. Cell Res.* **6**, 221.
Scherbaum, O. H. (1957). *Acta Pathol. Microbiol. Scand.* **40**, 7.
Scherbaum, O. H. (1960). *Ann. Rev. Microbiol.* **14**, 283.
Scherbaum, O. H. (1964). *In* "Synchrony in Cell Division and Growth" (E. Zeuthen, ed.), pp. 177–194. Wiley (Interscience), New York.
Schmidt, P. (1967). *Exp. Cell Res.* **45**, 460.
Seaman, G. R. (1960). *Exp. Cell Res.* **21**, 292.
Seaman, G. R. (1961). *J. Protozool.* **8**, 204.
Sedgley, N. N., and Stone, G. E. (1969). *Exp. Cell Res.* **56**, 174.
Sonneborn, T. M. (1970). *In* "Methods in Cell Physiology" (D. M. Prescott, ed.), Vol. IV, pp. 241–339. Academic Press, New York.
Stephens, R. E. (1968). *J. Mol. Biol.* **32**, 277.
Stephens, R. E. (1970). *J. Mol. Biol.* **47**, 353.
Stevens, A. R. (1966). *In* "Methods in Cell Physiology" (D. M. Prescott, ed.), Vol. II, pp. 255–310. Academic Press, New York.
Stone, G. E. (1968a). *In* "Methods in Cell Physiology" (D. M. Prescott, ed.), Vol. III, pp. 161–170. Academic Press, New York.
Stone, G. E. (1968b). *J. Cell Biol.* **39**, 556.
Stone, G. E., and Cameron, I. L. (1964). *In* "Methods in Cell Physiology" (D. M. Prescott, ed.), Vol. I, pp. 127–140. Academic Press, New York.
Stubblefield, E. (1968). *In* "Methods in Cell Physiology" (D. M. Prescott, ed.), Vol. III, pp. 25–43. Academic Press, New York.
Suyama, Y. (1966). *Biochemistry* **5**, 2214.
Tamura, S., Toyoshima, Y., and Watanabe, Y. (1966). *Jap. J. Med. Sci. Biol.* **19**, 85.
Tamura, S., Tsuruhara, T., and Watanabe, Y. (1969). *Exp. Cell Res.* **55**, 351.
Watson, M. R., and Hopkins, J. M. (1962). *Exp. Cell Res.* **28**, 280.
Whitson, G. L., and Padilla, G. M. (1964). *Exptl. Cell Res.* **36**, 667.
Whitson, G. L., Padilla, G. M., Canning, R. E., and Cameron, I. L. (1966). *Nat. Cancer Inst. Monogr.* **21**, 317.
Wille, J. J., Jr., and Ehret, C. F. (1968). *J. Protozool.* **15**, 785.
Williams, N. E. (1964). *In* "Synchrony in Cell Division and Growth" (E. Zeuthen, ed.), pp. 159–175. Wiley, New York.
Williams, N. E., and Luft, J. H. (1968). *J. Ultrastruct. Res.* **25**, 271.
Williams, N. E., and Scherbaum, O. H. (1959). *J. Embryol. Exp. Morphol.* **7**, 241.
Williams, N. E., and Zeuthen, E. (1966). *C. R. Trav. Lab. Carlsberg* **35**, 101.
Williams, N. E., Michelson, O., and Zeuthen, E. (1969). *J. Cell Sci.* **5**, 143.
Wilson, L., and Friedkin, M. (1967). *Biochemistry* **6**, 3126.
Witman, G. B. (1970). *J. Cell Biol.* **47**, 229a.
Wolfe, J. (1970). *J. Cell Sci.* **6**, 679.
Wunderlich, F., and Peyk, D. (1969). *Exp. Cell Res.* **57**, 142.
Zeuthen, E. (ed.) (1964). "Synchrony in Cell Division and Growth." Wiley (Interscience), New York.
Zeuthen, E. (1968). *Exp. Cell Res.* **50**, 37.
Zeuthen, E., and Williams, N. E. (1969). *In* "Nucleic Acid Metabolism, Cell Differ-

entiation, and Cancer Growth" (E. V. Cowdry and S. Seno, eds.), pp. 203–217. Pergamon, Oxford.

Zeuthen, E., and Rasmussen, L. (1971). *In* "Research in Protozoology" (T. T. Chen, ed.), Vol. IV, in press. Pergamon, London.

Chapter 8

Comparison of a New Method with Usual Methods for Preparing Monolayers in Electron Microscopy Autoradiography[1]

N. M. MARALDI, G. BIAGINI, P. SIMONI, AND R. LASCHI

Centro di Microscopia Elettronica, Universita di Bologna, Italy

I. Introduction

At present two techniques are used most frequently to prepare monolayers of nuclear emulsion for electron microscope autoradiography. The first one, described by Granboulan and Granboulan (1964), is a variation of the "dropping" method (Salpeter and Bachmann, 1964) and consists of dipping, into the emulsion, the collodion-coated slides that support the sections. The use of a mechanical device (Kopriwa, 1966; Vrensen, 1970) for dipping the slides into the emulsion allows the highest standardization of the process. The method is simple and quick, although a loss of specimens may occur when they are taken off the slides and placed on the grids. Monolayers obtained in this way are uniform and

[1] This work was supported by Grant No. 69.02310/115.3608 from C.N.R.

highly reproducible. At first, this technique was used with the Kodak NTE emulsion (Salpeter and Bachmann, 1964). It was later used mainly with the Gevaert NUC 307 (Granboulan and Granboulan, 1964), while most work employing the Ilford L4 emulsion was done by using the loop method.

The loop method, first described by O'Brien and George (1959) and later modified by several authors (Revel and Hay, 1961; Caro and van Tubergen, 1962; Moses, 1964), has a lower degree of reproductiveness compared with the dipping method. On the other hand, it has the advantage of being directly applicable to sections collected on grids, which makes the staining and carbon coating easier. Variations of the loop method have been increased by the introduction of various types of expandable loops (Eheret *et al.*, 1964; Karasaky, 1965; Maraldi and Di Caterino, 1968). Such systems allow one to obtain films of emulsion of the required thickness by varying the loop circumference. However, the expandable loop easily becomes misshapen, and thus uniform monolayers are not obtained. Other techniques such as the dipping of grids attached to a slide (Hay and Revel, 1963; Young and Kopriwa, 1964) are used infrequently at present. In this paper the results obtained using the above-mentioned methods are compared with those achieved by means of a simple device, of new design, already described in a preliminary paper (Maraldi and Di Caterino, 1968).

II. Methods

All of the following methods were performed with Ilford L4 nuclear emulsion. In fact, with both the Gevaert NUC 307 and the Kodak NTE emulsion the dipping method is the only one that can be used. The loop method, owing to the extreme fragility of the supporting gelatin, is not at all reliable (Granboulan, 1965). Tests performed by us with these nuclear emulsions to obtain monolayers with the "emulsion film drawer" device confirmed that techniques requiring the loop cannot be used.

A. Preparation of Monolayers

1. DIPPING

Carefully cleaned 2.5 × 7.5 cm slides are coated by immersion to a depth of 6 cm in 2% collodion dissolved in isoamyl acetate. The emulsion is melted in the original bottle in a water bath for 1 hour at 38°C. One part of melted emulsion is added to four parts of double-distilled water at 38°C and mixed in a mixer at a low speed in a water bath at 38°C for 30 minutes. The collodion-coated slides are dipped 5 cm into the emulsion after it has remained at room temperature for a few minutes. The lifting of slides from the emulsion is performed as evenly as possible. The slides are set with a 20° inclination in plastic boxes and are dried in a humidified room (relative humidity about 70%).

2. Loop

The emulsion is melted as in the dipping technique, but the dilution is one part of emulsion to two parts of water. After mixing at 38°C for 30 minutes the emulsion is chilled in an ice bath for about 2 minutes and then kept for 10 minutes at room temperature. Stainless steel wire (Ø 1 mm diameter) loops, 4 cm in diameter, are dipped into the emulsion. The excess emulsion is removed with filter paper. After the emulsion has gelled, the film is applied by gentle contact onto the collodion-coated slides. The slides are placed horizontally, sheltered from dust, and dried in a humidified room.

3. Emulsion Film Drawer

This simple device consists of a rectangular stainless steel frame provided with a movable arm that makes it possible to obtain films of nu-

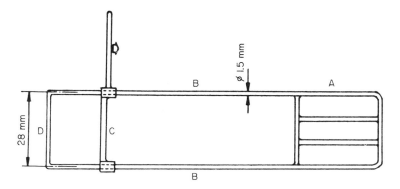

Fig. 1.   Scheme of the emulsion film drawer device. A, Handle; B, fixed guides; C, movable arm; D, fixed terminal arm. A drop of melted emulsion is placed between the fixed terminal arm (D) and the movable arm (C) tightly joined. Then the movable arm is slowly shifted in the direction indicated by the large arrow until the film displays the required interference color. From Maraldi and Di Caterino (1968).

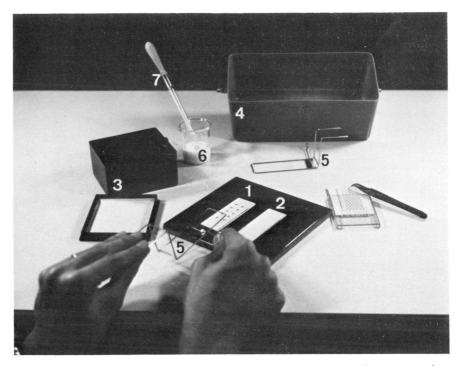

FIG. 2. Photograph of the basic equipment for preparing monolayers of nuclear emulsion with the emulsion film drawer device. 1, 2.5 × 7.5 cm slide supporting several Formvar-coated grids; 2, 2.5 × 7.5 cm collodion-coated slide with ultrathin sections deposed on the marks; 3, plastic black box for storage of slides; 4, plastic box for drying the emulsion-coated slides; 5, emulsion film drawer devices; 6, melted nuclear emulsion; 7, Pasteur pipette. In this figure the operator applies a film of nuclear emulsion directly on the grids.

clear emulsion of such dimensions as to completely coat a 2.5 × 7.5 cm slide (Fig. 1). The method we have been using in order to obtain monolayers of nuclear emulsion follows. The emulsion is melted as already described and is diluted 1:3. After shaking for 30 minutes in a water bath at 38°C, the emulsion is kept at room temperature for about 10 minutes. An emulsion drop is placed with a Pasteur pipette between the terminal arm (Fig. 1,D) and the movable arm (Fig. 1,C) of the emulsion film drawer with the movable and terminal arms tightly opposed to one another. The movable arm is then moved away from the terminal arm until the film shows the required interference color. The film of emulsion thus obtained is applied by gentle contact on the collodion-coated slide or directly on Formvar-coated grids (Fig. 2). The drying is the same as described for the loop method.

B.   Electron Microscope Observations

After an overnight drying of the films of emulsion obtained according to the different techniques, the film of collodion plus emulsion is removed from the slides by floating it on a water surface. The monolayers are selected according to the interference color. The glossy side of grids rinsed in carbon tetrachloride is placed on the floating film. The grids are then collected by means of a moist filter paper applied over a porous funnel connected with a water pump aspirator (Salpeter and Bachmann, 1964). These grids (noncarbon-coated) were observed in a Siemens Elmiskop I electron microscope. Areas of 3 μ^2 were isolated in equally magnified micrographs of monolayers of emulsion obtained according to the different techniques. In these areas, the number of silver halide grains/μ^2 was carefully determined.

C.   Determination of Gelatin Content

Carefully cleaned, noncollodion-coated slides were covered with monolayers of nuclear Ilford L4 emulsion according to the various above-mentioned techniques. At the same time, controls were carried out by electron microscopy in order to estimate the quality of monolayers undergoing chemical analysis. Identical surfaces (300 cm^2) of monolayers obtained according to different techniques were removed from the slides with 10 N HCl. It is necessary to dip the slide in HCl for 10 minutes to detach the emulsion completely from the slides. Subsequently, the solution is brought to 5 N HCl with H$_2$O. After hydrolysis at 100°C for 24 hours the amino acid content of the different specimens was determined by the ninhydrin reaction spectrophotometrically (λ 570 mμ) using a leucine solution as a standard.

III.   Results and Discussion

The monolayers of Ilford L4 emulsion obtained with the emulsion film drawer device look quite different from those obtained by means of other techniques when examined in the electron microscope. In fact, the silver halide grains are clear-cut and separated from each other (Fig. 5) with the drawer device, while in monolayers obtained with dipping, the grains are fuzzy and often closely joined with one another (Fig. 3). An intermediate aspect is that shown by monolayers obtained with the loop method (Fig. 4). In order to ascertain whether, in spite of their different

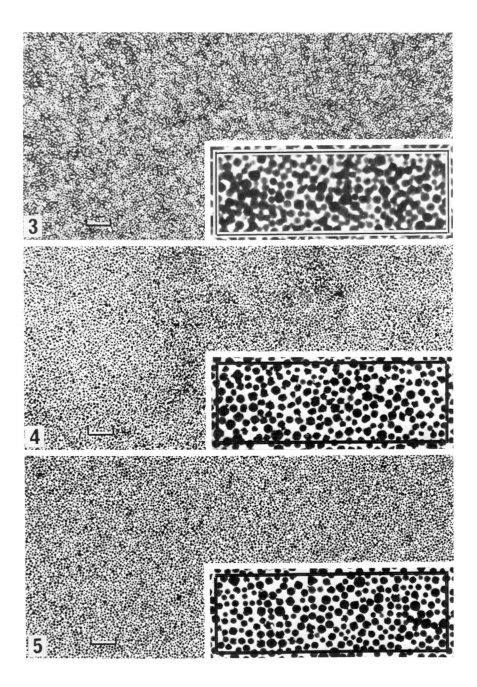

TABLE I

CHARACTERISTIC DATA OF ILFORD L4 MONOLAYERS

Preparation techniques of monolayers	Dilution	Interference color	Thickness (Å)	Number of grains per square micron	AA/cm² of monolayers (μg)
Dipping	1:4	Purple	1500	71 ± 3	9
Nonexpandable loop	1:2	Purple-copper	1400	68 ± 5	8
Emulsion film drawer	1:3	Copper	1200	72 ± 2	6

appearance, the films obtained with different techniques form real mono-layers of grains, closely packed but not superimposing, we worked out statistically the number of grains per surface unit (Figs. 3, 4, and 5 inserts). The number of grains is, within the limit of error, the same in the three cases (Table 1). These data agree with the theoretical number of silver halide grains, 1200 Å in diameter, monolayered and closely packed, that may occupy a given area. The various films examined thus represent real monolayers.

Previous papers have reported that the very best Ilford L4 monolayer obtainable with the dipping method has a purple interference color and has a 1500 Å thickness (Salpeter and Bachmann, 1965), while the emulsion grains have an average size of 1200 Å (Caro, 1964). The monolayers obtained by us during the present study with the dipping method are also purple in color. Those obtained with the loop have a purple-copper interference color and a thickness of about 1400 Å. When the emulsion

---

FIG. 3. Dipping method. Interference color, purple; thickness, 1500 Å. The distribution of the grains is uniform, but frequently the outlines of the grains are not clearly defined; they apparently sometimes fuse together. *Insert:* In the area limited by the continuous line (3 μ^2) 208 grains are present.

FIG. 4. Loop method. Interference color, purple-copper; thickness, 1400 Å. The distribution of the grains is not completely uniform, and the outline of each grain is fuzzy and out of focus. *Insert:* In the limited area there are 200 grains.

FIG. 5. Emulsion film drawer method. Interference color, copper; thickness, 1200 Å. The distribution of the grains is uniform, and the outline of the single grain is well defined and completely in focus. *Insert:* In the limited area the number of grains is 212.

FIGS. 3–5. Monolayers of Ilford L4 nuclear emulsion. Magnification: 6000 ×. *Inserts:* The area limited by the continuous line is 3 μ^2. The number of grains present in equal areas is determined in the different photographs. Theoretically, in a 3 μ^2 area of a perfect monolayer 209 grains, each 1200 Å in diameter, may settle. Magnification: 20,000 ×.

film drawer device is used, monolayers are regularly obtainable, showing at an equal number of silver halide grains per surface unit, a copper interference color and a thickness of about 1200 Å.

The autoradiographic emulsions contain two kinds of gelatin; one adsorbed onto the grains that is indispensable to their own stability and sensitivity (Mees, 1954). This gel layer has a thickness of about 50 Å (Pouradier and Roman, 1952). The other gelatin is in solution and is necessary for the cohesion of the grains (Granboulan, 1965). Of course, the less gelatin in solution, the better the characteristics of the monolayer of emulsion will be. In a 1500 Å thickness, typical of monolayers obtained with dipping, the grains of the Ilford L4 emulsion, whose average diameter is 1200 Å, possibly settle at different levels in the gelatin in solution (Fig. 6, top). The monolayers obtained with the emulsion film

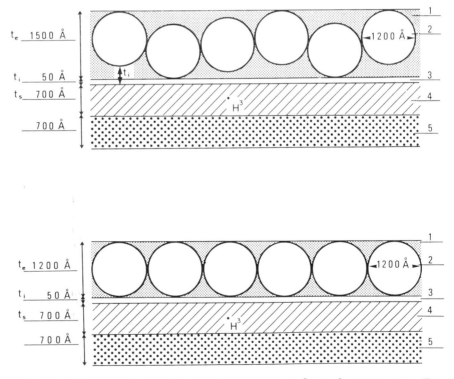

FIG. 6. Diagram of two electron microscope autoradiograph preparations. *Top:* preparation coated by a monolayer obtainable with the dipping method. *Bottom:* monolayer obtainable with the emulsion film drawer method. The value of $t_e + t_i + t_s$ is 2150 Å (top) and 1950 Å (bottom). Arrows (top) show the possible variation of t_i (see text). [3]H, Radioactive source in the section; 1, gelatin; 2, grains of Ilford L4 nuclear emulsion; 3, carbon layer; 4, section; 5, supporting film (collodion).

drawer, having an average thickness of 1200 Å, show silver halide grains all lying on the same level and closely packed by a very small quantity of gelatin in solution (Fig. 6, bottom). Owing to their great instability under the electron beam, monolayers of Ilford L4 emulsion, obtained with common techniques, may be easily examined only if negatively stained and previously fixed (Granboulan, 1965). This instability is revealed by the slightly out-of-focus aspect of the grains in the electron microphotographs of monolayers obtained by us with ordinary techniques (Figs. 3 and 4). This aspect is mainly due to the sublimation of the excess of gelatin in solution. Monolayers obtained with the emulsion film drawer reveal instead a high stability under the electron beam due to their very low gelatin content. This is the reason why the grains have such a clear and defined shape even in noncarbon-protected preparations (Fig. 5). In order to determine quantitatively the amount of gelatin present in each type of monolayer, we measured the quantity of amino acids obtainable from equal surfaces of hydrolyzed monolayers of emulsion. Of course, owing to the thickness variations with the dipping technique between one end of the slide and the other, we only isolated the part nearest to the area in which the sections are placed (about 2 cm from the slide's distal end where the film interference color is that typical of the monolayer). This contrivance had to be employed also with films obtained with the loop, since the peripheral areas displayed a greater thickness. The whole film obtained with the emulsion film drawer may be utilized since its thickness is nearly the same throughout. The results drawn by these analyses are reported, together with other characteristic data, in Table I.

The data reported in Table I show that the thickness of the monolayer obtainable with different techniques is reflected by the amino acid content, that is, the amount of gelatin, while the number of halide silver grains remains constant, within the limit of error.

It is well known that the resolving power of an electron microscope autoradiography depends on the following factors: (a) diameter of the silver halide crystals of the nuclear emulsion; (b) photographic processes employed; (c) physical characteristics of the radiation emitted by the radioisotope; and (d) "source-detector" geometry. The value of the "source-detector" geometry is expressed in the following equation:

$$r = d \tan \theta \tag{1}$$

in which d is the distance between the source and the latent image plane, or more simply between the source and a point in the center of the silver halide crystal, and θ is the radiation angle of incidence. The distance d is taken as

$$d = t_s/2 + t_i + t_e/3 \qquad\qquad (2)$$

where t_s is the section thickness, t_i the thickness of any kind of material interposed between the section and the silver halide crystal, and t_e is the thickness of the nuclear emulsion (which is generally assumed to be equal to the average diameter of the silver halide crystals). From Eq. (1) it follows that the more the distance d decreases, the more the resolving power increases. The thickness of the sections being equal, the resolving power may be improved by reducing the value of t_i and t_e, which affect it by a factor of 1 and a factor of $\frac{1}{3}$, respectively.

The data in the present paper show how, by means of the emulsion film drawer device, monolayers of Ilford L4 emulsion can be obtained which are much thinner, that is 1200 Å against the 1500 Å in the dipping technique and the 1400 Å in the loop technique, while the same number of silver halide crystals per square micron is still achieved. The resolving power is thus improved by this method owing to the decrease in the t_e value (thickness of the emulsion) as well as a decrease in t_i. In fact, t_i includes any material interposed between the radiation emission point and the silver halide crystal, for example the carbon layer. In the instance of a monolayer of emulsion having an excess of gelatin in solution, the gelatin itself may be responsible for the increase of t_i (Fig. 6, top, arrows) in various areas, owing to the unevenness of the distribution of the silver halide crystals. On the other hand, in the case of a monolayer almost clear of gelatin in solution, as that obtained with the emulsion film drawer, the t_i value is limited to the thickness of the carbon layer (Fig. 6, bottom). Therefore, since the t_e value is lower in the monolayer obtained with the emulsion film drawer than in any other instance, the resolving power of the autoradiographs thus prepared turns out to have been remarkably improved.

Monolayers obtained with the emulsion film drawer may be directly placed on the grids. A large number of grids all arranged on a single slide may be coated with the emulsion at the same time. The monolayer of emulsion keeps the grids on the slide during the exposure time, but cannot be relied upon during the photographic process. For this reason we made a small basket in which many grids at a time may undergo photographic processing after they have detached from the slide (Fig. 7). However, our experience has brought us to regard as more convenient the coating of the monolayer, obtained with the emulsion film drawer, on the collodion-coated slides bearing the sections. In fact, in placing the monolayer on the grid, the give of the supporting film where the holes of the grid are found may cause an uneven distribution of the halide grains.

Many authors have suggested that the excess of gelatin be digested in

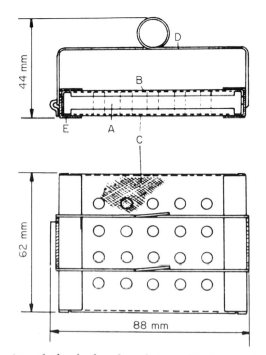

FIG. 7. Drawing of the basket for photographic treatment of the grids. A, Perspex grid holder with 20 holes; B and C, stainless steel wire gauzes; D and E, removable holdfasts. Depressions in the gauze prevent the grids from escaping. From Maraldi and Di Caterino (1968).

order to sharpen the contrast and to lessen the opacity of the electron autoradiographs. Proteolytic digestion (Comer and Skipper, 1954; Przybylski, 1961), alkaline digestion (Revel and Hay, 1961), acid digestion (Granboulan *et al.*, 1962), warm water melting (Silk *et al.*, 1961), and irradiation with an electron beam (Meek and Moses, 1963) have been tried. However, these treatments, and particularly the chemical and enzymatic ones, have drawbacks such as the removal of grains or their dislocation. The use of the electron beam to clarify the image may cause the sublimation of some of the grains, while a protracted heating of the specimen causes the dislocation of the whole supporting film as well as a progressive deterioration of the biological material and, consequently, a loss in definition of the ultrastructural details. On the other hand, the removal of the excess gelatin is necessary in order to obtain fine electron micrographs suitable for reproduction (Granboulan, 1965).

Autoradiographs obtained with the emulsion film drawer do not require gelatin removal. In fact, these monolayers contain, almost exclusively, the gelatin adsorbed onto the grains and a very small quantity of gelatin in solution. The monolayer has therefore a higher stability under the electron beam and a lower electron density thus allowing very good results to be reached in the definition of the ultrastructural details.

ACKNOWLEDGMENT

The technical assistance of Mr. U. Serra is gratefully acknowledged.

REFERENCES

Caro, L. G. (1964). *In* "Methods in Cell Physiology" (D. M. Prescott, ed.), Vol. I, pp. 327–363. Academic Press, New York.

Caro, L. G., and van Tubergen, R. P. (1962). *J. Cell Biol.* **15**, 173.

Comer, J. J., and Skipper, S. J. (1954). *Science* **119**, 441.

Eheret, C., Savage, N., and Albinger, J. (1964). *Z. Zellforsch. Mikrosk. Anat.* **64**, 129.

Granboulan, N., and Granboulan, P. (1964). *Exp. Cell Res.* **34**, 71.

Granboulan, P. (1965). *In* "The Use of Radioautography in Investigating Protein Synthesis" (C. P. Leblond and K. B. Warren, eds.), pp. 43–63. Academic Press, New York.

Granboulan, P., Granboulan, N., and Bernhard, W. (1962). *J. Microsc. (Paris)* **1**, 75.

Hay, E. D., and Revel, J. P. (1963). *Develop. Biol.* **7**, 152.

Karasaky, S. (1965). *J. Cell Biol.* **26**, 937.

Kopriwa, B. M. (1966). *J. Histochem. Cytochem.* **14**, 923.

Maraldi, N. M., and Di Caterino, B. (1968). *Sci. Tools* **15**, 34.

Meek, G. A., and Moses, M. J. (1963). *J. Royal Microsc. Soc.* **81**, 187.

Mees, C. E. K. (1954). "The Theory of the Photographic Process." Macmillan, New York.

Moses, M. J. (1964). *J. Histochem. Cytochem.* **12**, 115.

O'Brien, R. T., and George, L. A., II. (1959). *Nature (London)* **183**, 1461.

Pouradier, J., and Roman, J. (1952). *Sci. Ind. Photogr.* **23**, 4.

Przybylski, R. J. (1961). *Exp. Cell Res.* **24**, 181.

Revel, J. P., and Hay, E. D. (1961). *Exp. Cell Res.* **25**, 474.

Salpeter, M. M., and Bachmann, L. (1964). *J. Cell. Biol.* **22**, 469.

Salpeter, M. M., and Bachmann, L. (1965). *In* "The Use of Radioautography in Investigating Protein Synthesis" (C. P. Leblond and K. B. Warren, eds.), pp. 23–41. Academic Press, New York.

Silk, M. H., Hawtrey, A. O., Spence, I. M., and Gear, J. H. S. (1961). *J. Biophys. Biochem. Cytol.* **10**, 577.

Vrensen, G. F. J. M. (1970). *J. Histochem. Cytochem.* **18**, 278.

Young, B. A., and Kopriwa, B. M. (1964). *J. Histochem. Cytochem.* **12**, 438.

Chapter 9

Continuous Automatic Cultivation of Homocontinuous and Synchronized Microalgae[1]

HORST SENGER, JÜRGEN PFAU, AND KLAUS WERTHMÜLLER

Botanisches Institut der Universität Marburg, Marburg, Germany

I. Introduction

Mass cultures of microorganisms have been the basis for many different kinds of physiological studies. As experiments became more sophisticated,

[1] This investigation was supported by the Deutsche Forschungsgemeinschaft.

more attention was paid to the conditions of the cultures. Among these conditions, the developmental factor was the latest to be recognized; today, we are well aware of how important the developmental stage and the age of a culture are for the results of an experiment. Therefore, the possibility of reproducing cultures with identical distribution of developmental stages is of greatest importance. The most reproducible material is yielded by homocontinuous (Pfennig and Janasch, 1962; Herbert, 1960) and synchronous cultures (Tamiya, 1966; Pirson and Lorenzen, 1966; Lorenzen, 1970; Senger and Bishop, 1971). Much consideration has been given to the continuous cultivation of microorganisms, and several symposia with a wealth of literature have been dedicated in recent years to this special topic (Burlew, 1953; Continuous Cultivation of Microorganisms, 1959; Continuous Fermentation and Cultivation of Microorganisms, 1960; Continuous Culture of Microorganisms, 1960; Malek *et al.,* 1964; Malek and Fenel, 1966; Watanabe and Hattori, 1966; Kontinuierliche Züchtung von Mikroorganismen, 1967; Powell *et al.,* 1967; Malek *et al.,* 1969).

Growth factors of continuously growing cultures are complex and not all their changes can sufficiently be compensated for by the traditional, time-consuming methods. Therefore full automation of such cultures becomes increasingly desirable.

In heterotrophically growing cultures the essential growth factors are temperature, gassing, and nutrient medium. Temperature and gassing are easily regulated in continously growing cultures, whereas, automatic constancy of the nutrient medium (including pH) can be achieved by dilution, controlled via a chemostat (Myers and Graham, 1959). Photoautotrophic microorganisms, however, like unicellular algae, need light as an additional and most important factor for growth and development. In this regard, it is more important to keep constant the average light intensity provided per cell rather than the intensity of the light source itself. When cultures become denser during growth, the cells progressively shade themselves. This requires a density control of the culture by dilution, which might either be continuous (homocontinuous cultures, cf. Senger and Wolf, 1964) or periodic (synchronous cultures, cf. Pfau *et al.,* 1971). Thereby the nutrient medium factor is stabilized simultaneously. An apparatus to control such a dilution photoelectrically and fully automatically (turbidostat, cf. Pfennig and Janasch, 1962; Phillips and Myers, 1954) was designed and built by the authors (Senger and Wolf, 1964; Pfau *et al.,* 1971). Its application for permanent cultivation of homocontinuous or synchronous cultures of microalgae will be described in this chapter and its operation evaluated.

II.   Characterization of Different Types of Mass Cultures

A.   Homocontinuous Cultures

Every cell in a mass culture of microalgae has its own individual life cycle. In a homocontinuous culture the start of each life cycle depends only on statistical laws. Assuming that the individual life cycle lengths do not deviate considerably from the average, the schematic diagram of Fig. 1 would represent such a culture. Theoretically, a homocontinuous culture is always in the logarithmic growth phase and at steady state. The distribution of all physiological or developmental stages is completely randomized and should be the same at any time. Thus, samples taken from a homocontinuous culture always represent the average of all possible developmental stages of the population and all parameters should be identical, including cell number (Fig. 2) and transmittance (Fig. 3).

To obtain a homocontinuous culture, and realizing the requirements listed, all growth conditions have to be kept constant. There is no difficulty in stabilizing the temperature as well as flow rate and composition of the gas. In addition, the culture must be continuously adjusted to a constant optical density to compensate for the decrease in light intensity provided for a single cell, caused by self-shading of the cells during growth. Dilution with fresh nutrient medium guarantees the constancy of the growth factor "nutrient medium" automatically.

B.   Synchronous Cultures

Synchronization of cells in a mass culture is achieved when all life cycles are initiated at the same time. This condition is shown by a diagram in Fig. 1. Such a synchronized culture never assumes a steady state and only temporarily remains in the logarithmic growth phase. Samples taken from a synchronous culture always represent the developmental stage of the life cycle of one cell. Thus, neither cell number (Fig. 2) nor transmittance (Fig. 3) of the synchronous culture remain constant. Deviations in any parameter in a synchronous culture from the mean value is caused by the individual life cycle length of each single cell. This deviation is inherent in the strain and should be identical for all cultures grown under the same conditions.

In order to obtain a synchronous culture at least one growth parameter has to be changed periodically. For cultures of microalgae the most

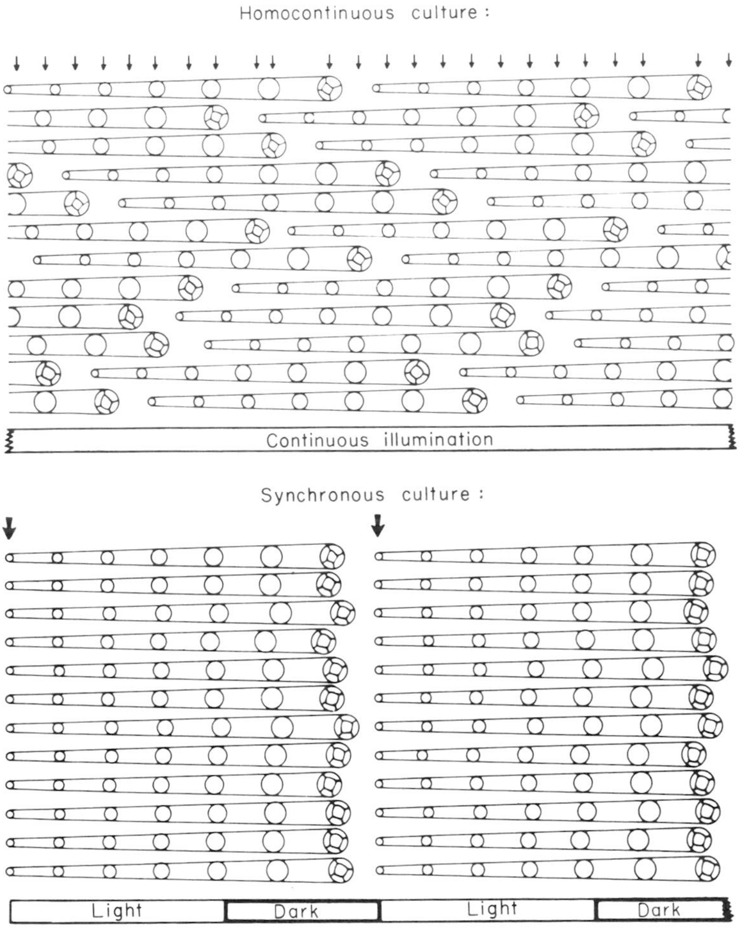

Homocontinuous culture:

Continuous illumination

Synchronous culture:

| Light | Dark | Light | Dark |

FIG. 1. Schematic representation of life cycle distribution for homocontinuous and synchronous cultures of microalgae. The arrows mark points of dilution. The mode of illumination is given below. In the scheme of the life cycle the number of daughter cells released was confined to one for clarity.

adequate parameter is light and, in some few cases, a change of temperature. If synchrony is to be maintained over repeated life cycles, the synchronizing change from dark to light must be accompanied by dilution to a constant density.

C. "Batch" Cultures

The completely randomized homocontinuous culture and the synchronous culture represent the two possible extremes in distribution of the

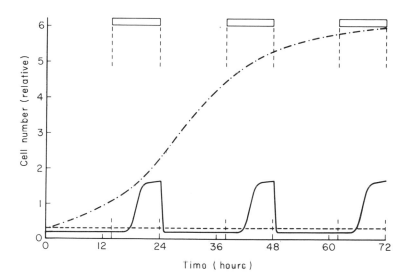

FIG. 2.   Cell number of a homocontinuous (- - -), synchronous (——), and batch culture (-•-•-) during 3 days of growth. The data were taken from *Scenedesmus* cultures but might, in principle, be applied to *Chlorella* cultures. The dark periods of the synchronous culture are marked on top.

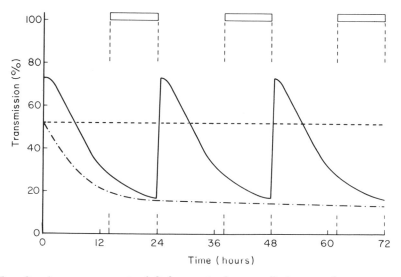

FIG. 3.   Average amount of light received per cell (expressed as percentage transmission) in a homocontinuous (- - -), synchronous (——), and batch culture (-•-•-) during 3 days of growth. Transmission of the cultures was measured by means of the photovoltaic cell (cf. p. 312). The data were taken from *Scenedesmus* cultures but might, in principle, be applied to a *Chlorella* culture.

TABLE I

Comparison of External and Internal Culture Conditions for
Homocontinuous, Synchronous, and Batch Cultures

External and internal culture conditions	Culture types		
	Homocontinuous culture	Synchronous culture	Batch culture
Illumination	Permanent	Periodical	Permanent
Dilution	Permanent	Periodical	None
Transmittance	Constant	Periodically changing	Changing
Steady state	Continuous	Never reached	Never reached
Logarithmic growth	Continuous	Temporary	Temporary
Cell number ⎫ Distribution of cell ⎬ parameters ⎭	Constant	Periodically changing	Changing toward saturation

developmental and physiological stages. The remaining types of cultures, with more or less undefined cell populations, will be summarized under the term "batch" cultures.

Batch cultures can be started with any kind of inoculum. Theoretically there are no restrictions on either the physiological conditions of the inoculum or on the amount of inoculum. Nevertheless, the most predictable way to grow a batch culture is to begin with a homocontinuous culture. The external growth conditions can be the same as for homocontinuous cultures, but self-shading and selective uptake of nutrients from the medium change in a batch culture since they are not compensated for by dilution. Thus, the logarithmic growth phase is variable and short. The cell population may be at any stage between a randomized and a synchronous culture, again depending on the inoculum and the period of growth. After the logarithmic phase, growth will approach stagnation more or less rapidly. Until stagnation all parameters of the culture continuously change without reaching a steady state. Changes in cell number and transmittance for a batch culture derived from a homocontinuous culture are demonstrated in Figs. 2 and 3.

External and internal conditions for the three culture types are summarized in Table I.

III.   Organisms and Growth Conditions

The organisms used in the experiments to be reported are the two unicellular green algae *Scenedesmus obliquus* (strain D_3) (Gaffron, 1939)

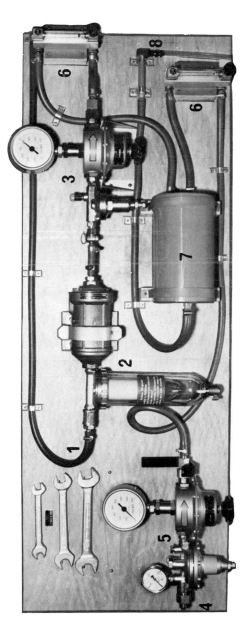

Fig. 4.   Gas mixing unit. (1) Compressed air inlet; (2) oil and water separator; (3) double-reducing valve for air; (4) CO_2 inlet; (5) double-reducing valve for CO_2; (6) flow meters; (7) cotton filter, sterilized gas mixing chamber; (8) outlet for sterile air/CO_2 mixture. (After plans of Dr. A. Kuhl, Göttingen.)

and *Chlorella pyrenoidosa* (strain 211-8b, algae collection, Göttingen). The course of the life cycles of both strains is straightforward. Young daughter cells grow into mother cells which again divide into daughter cells (autospores). No sexuality has been reported.

The algae are grown in an inorganic nutrient medium (Senger and Bishop, 1971) in glass culture tubes 42 cm in length and 3.7 cm in diameter. A glass capillary, through which the culture is aerated, is fused to the conical bottom of the tube. Thus, the cells are provided with CO_2 and kept in suspension. Culture tubes are maintained in a light thermostat (Kuhl and Lorenzen, 1964) at a temperature of 28°C. A combination of Osram 40 W/24-1 and Osram 40 W/15-1 fluorescent lamps provides an intensity of 10,000 lux at the surface of the culture tubes. By providing air and CO_2 of constant pressure via two flow meters (Fig. 4) a constant rate of 3% CO_2 in air is obtained. The mixture is filtered and led into the culture tubes at a rate of 20 liters/hour. The whole system can be kept sterile for several weeks.

To obtain homocontinuous cultures the cells are constantly illuminated and diluted to a preset density. Cultures become synchronized under light–dark regimes of 14:10 hours (*Scenedesmus*) or 16:12 hours (*Chlorella*) combined with dilution at the beginning of each light period.

IV.   Photoelectrically Controlled Automatic Dilution

A.   Experimental Setup of the Dilution Apparatus

The automatic dilution apparatus has been designed to fit into the commercially available light thermostat (Fig. 5). A schematic diagram for the experimental setup is given in Fig. 6. The transmission sensing photovoltaic cell is mounted behind the lower third of the culture tube. In this area the glass of the culture tube is frosted to cause complete light scattering, thus preventing the effect of selective scattering by different cell sizes. Signals from the photovoltaic cell operate the solenoid valve via the controller. The solenoid valve closes or opens the connection to the nutrient medium storage bottle by compressing a silicon rubber tube. This arrangement allows easy sterilization of the whole system (culture tube, storage bottle with nutrient medium, and connecting tubing). The fresh nutrient medium is fed during dilution into the aeration capillary, thus perfectly mixing with the gas stream and the culture suspension. The level of the nutrient medium has to be high enough to push the liquid

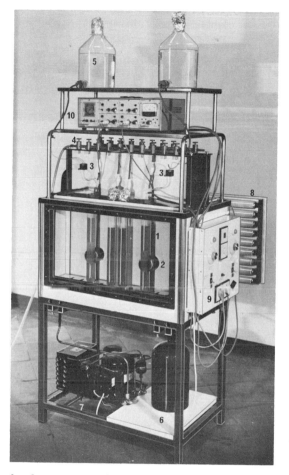

FIG. 5.   Light thermostat combined with dilution apparatus. (1) culture tube;
(2) photovoltaic cell housing; (3) solenoid valve; (4) gassing valves; (5) nutrient
medium storage bottle; (6) overflow storage bottle; (7) cooler assembly; (8)
fluorescent lamp unit (the front unit is removed); (9) light thermostat control
panel with dimmers for the fluorescent lamps; (10) electronic dilution and temperature control unit—left to right: timer, two double-dilution controllers, and temperature controller. (The light thermostat was built by Fa. Fritz Kniese, Marbach, Germany. The opto-electronic dilution apparatus with timer and temperature control unit
was designed and built by the authors and will now be manufactured and distributed
by the same company.)

into the gassing system with a gas pressure equal to that of a 1000 mm
water column. The culture tube is closed by a rubber stopper with two
outlets, one for the gas and the overflow of culture suspension during
dilution and the other for taking samples from the culture.

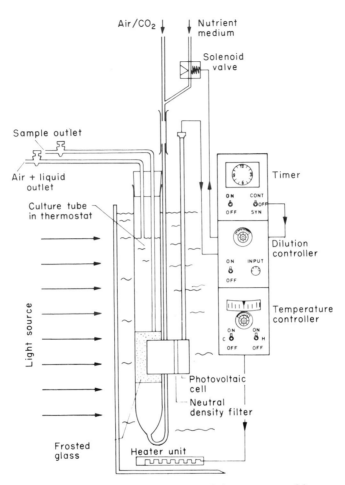

Fɪɢ. 6. Scheme for the experimental setup of the automatic dilution control apparatus, timer, and temperature controller. For further explanation see Fig. 5 and text.

B. Opto-electronic Controller of the Dilution Apparatus

A block diagram of the opto-electronic controller of the dilution apparatus as well as a detailed circuit diagram are given in Figs. 7 and 8. Transmission changes in the culture tube are sensed by a selenium photovoltaic cell (type 781 N, Electrocell, Berlin-Dahlem). Since such photovoltaic cells act as linear current generators at low levels of light intensity, a monolytic operational amplifier (A1) in the input circuit is wired in a current/voltage transducer configuration with a typical input impedance of less than 5 Ω. A general-purpose type of operational amplifier like the

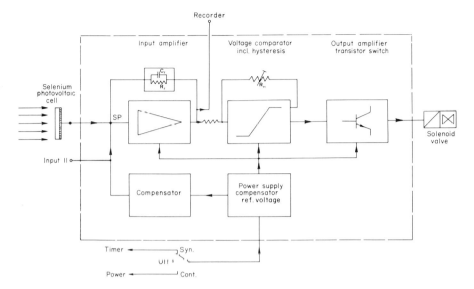

FIG. 7. Block diagram of the dilution control unit. For explanation see text. (From Pfau *et al.*, 1971.)

"709" would do in most similar applications. In critical cases, however, i.e., in precision measurements, at very low levels of light or under conditions where a photocell with considerably smaller area is used, an operational amplifier with very small input offset current and small input

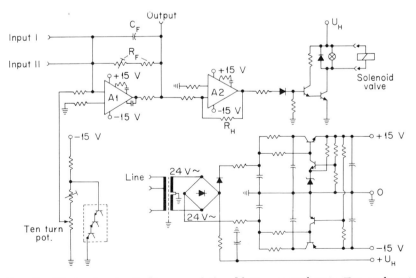

FIG. 8. Electronic circuit diagram of the dilution control unit. For explanation see text.

current drift (i.e., a FET-op.-amp.) should be selected. To maintain linearity and to prevent longtime wear-out the maximum photocurrent is limited to 660 μA, corresponding to 550 lux at a cell area of 12.5 cm², while higher intensities are reduced by Plexiglass neutral density filters. The amplifier gain is set to give 10 V into 5 KΩ for each 50 lux at the input, and a calibrated compensation current can be fed into the summing point (SP) of the op.-amp. (A1), allowing sensitive pickup of transmission changes at high levels of incident light. This compensation is linear and proportional to the transmission of the culture. If a direct relation to the packed cell volume is desired, a logarithmic compensation current is necessary.

The feedback resistor R_F and the filter capacitor C_F form a delay circuit necessary to compensate for short-term transmission changes caused by gassing bubbles in the culture tube.

A voltage comparator (A2), again a monolytic op.-amp. of the 709 type, with a differential input transistor pair for stability and with positive feedback over resistor R_H, senses the output voltage of the op.-amp. (A1) and activates the solenoid valve via a Darlington power amplifier stage as soon as the transmission in the culture tube is reduced by growth of the algae and drops under a preset level. Hysteresis of the switching levels is essentially determined by resistor R_H and can be as low as a few hundred microvolts, corresponding to differences of light intensities in the millilux level.

Power to the controller is either fed from the timer (switch position "SYN"), controlling only one dilution to a present density at the beginning of the life cycle (H. Pfau et al., 1971) or directly from the line (switch position "CONT"), allowing for continuous dilution to obtain homocontinuous cultures (H. Senger and Wolf, 1964).

In addition, the dilution control device is fitted with an analog output, which allows direct recording of the transmission changes in the culture sensed by the photovoltaic cell. The second input may be used for differential measurements of the growth of cultures in two separate culture tubes.

C. Optical Correlation between Light Source, Transmission of the Cultures, and Spectral Response of the Photovoltaic Cell

The same combination of fluorescent lamps used as a light source for photosynthetic growth of the cultures equally serves as incident light for the photovoltaic cell, enabling the operation of the dilution controller (cf. Fig. 6). The emission spectrum of the lamps derived from recordings of the spectroradiometer (ISCO, Model SR/SRR) is demonstrated in Fig. 9. In the same way, the transmission spectrum of this light passing through a culture (i.e., the light actually hitting the photovoltaic cell) is recorded (Fig. 9). The difference between the emission and the absorption spectra represent the actual absorption spectrum of the cells (Fig. 10). Comparison with an *in vivo* absorption spectrum for the same cells, measured with a dual-beam spectrophotometer (Shimadzu, MPS-50L), shows a divergence resulting from the different actinic light (Fig. 10). In the region of the emission lines of the fluorescent lamps an apparent enhancement in the absorption is notable, whereas the decline in energy output of the lamps above 660 nm causes a depression of the long-wave chlorophyll absorption.

The relative spectral sensitivity of the photovoltaic cell, applied with

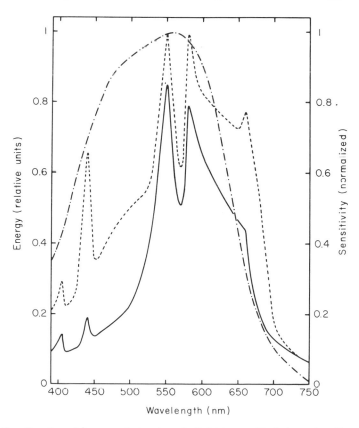

FIG. 9.   Spectra of lamp emission (- - - -), light transmitted through a *Scenedesmus* culture (——), and photovoltaic cell response (- • - • -). For experimental details see text. (Data for photovoltaic cell response from Electrocell, Berlin.)

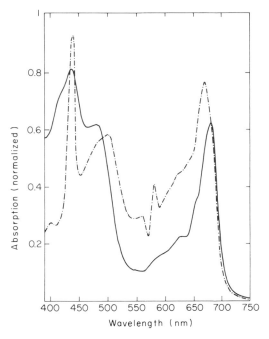

FIG. 10. Comparison of spectral difference between the lamp emission and culture transmission (-•-•-) and an *in vivo* absorption spectrum of *Scenedesmus* (——). For experimental detail see text.

the dilution controller, nearly covers the whole range of the emission spectrum of the lamps as well as the absorption spectrum of the cells with the exception of a small portion in the red region (Fig. 9). This type of selenium photovoltaic cell was especially chosen for its distribution of sensitivity although silicium photovoltaic cells are generally preferred because of their long-term stability and their relatively high output voltage. Regarding its sensitivity in the blue region, this type of selenium photovoltaic cell not only covers changes in chlorophyll but also changes in carotenoids or other pigments in blue or red microalgae, which also could be grown in this culturing and dilution apparatus.

Linearity of the response of the photovoltaic cell to different densities of the culture was determined by filling a culture tube with suspensions of known densities and recording the output (see p. 312). The results are given in Fig. 11. Density is expressed in PCV but could equally be presented in dry weight or cell number. The relation between transmission and density is logarithmic, thus following the Beer–Lambert law. In this way good accuracy and reproducibility in the optical system are achieved.

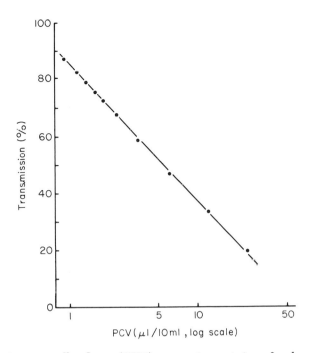

Fig. 11.   Average cell volume (PCV) versus transmission of a homocontinuous culture of *Scenedesmus obliquus.*

V.   Results and Accuracy in Achieving Permanent Homocontinuous Cultures

A.   Growth Parameters of Permanent Homocontinuous Cultures

As noted above, samples taken from an ideal homocontinuous culture should be identical at all times. Consequently, the best way to prove homocontinuity of a culture is to test several of its parameters with re-gard to their constancy for samples taken at various times.

Such experiments have been carried out with homocontinuous cultures of *Scenedesmus* and *Chlorella*. Samples were taken daily over a period of 8 days and different growth parameters determined on the basis of the same volume of culture liquid. Experimental details and methods have been reported earlier (Senger, 1970a). Data for cell number, PCV, dry weight, and total chlorophyll are presented in percentage of the mean

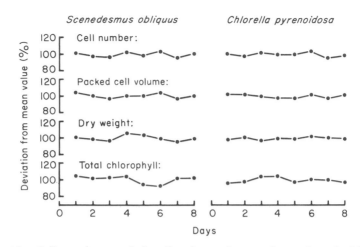

FIG. 12.   Cell number, packed cell volume, dry weight, and total chlorophyll measured on the basis of equal amounts of culture liquid for 8 subsequent days in homocontinuous cultures of *Scenedesmus obliquus* and *Chlorella pyrenoidosa*.

value (Fig. 12). The absolute data with standard deviations are compiled in Table II. Even the most extreme deviations do not exceed 9%, and the maximal standard deviation was 5.3%. In an earlier publication (Senger and Wolf, 1964) such data have been reported for samples taken at shorter intervals with equally small deviations. This proves that no daily fluctuations appear in the homocontinuous culture.

All the parameters mentioned above represent, in a strict sense, mean values of the whole culture. Theoretically, cultures with quite different distributions could yield the same mean values. The only way in which statements could be made on the distribution of the culture, is to determine parameters which can be measured on single cells. In this respect, cell size seems to be the best parameter that is experimentally accessible. We determined the cell sizes of representative samples of the culture at different times by measuring the longest diameter. The longest diameter is a sufficient parameter for the cell size, since we only want to prove its identity of distribution. Distribution curves of two samples each are given for *Scenedesmus* and *Chlorella* (Fig. 13). The pairs of distribution curves look almost identical with deviations of the mean values well within statistical limits (Table III).

B.   Photosynthetic Response of Permanent Homocontinuous Cultures

It is of great importance for the application of homocontinuous cultures in physiological experiments that, in addition to the growth param-

TABLE II

MEAN VALUES AND DEVIATIONS OF DIFFERENT GROWTH PARAMETERS FOR
HOMOCONTINUOUS AND SYNCHRONOUS CULTURES OF *Scenedesmus obliquus*
AND *Chlorella pyrenoidosa*

Parameter	Object	Culture type	Mean value and standard deviation (absolute)	Standard deviation (%)	Maximal deviation (%)
Cell number ($\times 10^4$/ml culture	Scen.	Homo.	520 ± 11.5	2.23	4.0
suspension)	Scen.	Syn.	318 ± 4.8	1.52	2.5
	Chl.	Homo.	593 ± 15.1	2.55	4.0
	Chl.	Syn.	398 ± 5.0	1.25	2.0
Dry weight (mg/20 ml culture	Scen.	Homo.	3.81 ± 0.13	3.41	6.0
suspension)	Scen.	Syn.	3.10 ± 0.07	2.37	5.0
	Chl.	Homo.	7.37 ± 0.13	1.77	3.0
	Chl.	Syn.	5.09 ± 0.15	2.95	5.0
PCV (μl/10 ml culture	Scen.	Homo.	7.50 ± 0.21	2.80	4.0
suspension)	Scen.	Syn.	6.04 ± 0.16	2.05	4.0
	Chl.	Homo.	14.70 ± 0.25	1.70	2.0
	Chl.	Syn.	10.15 ± 0.16	1.58	4.0
Total chlorophyll (mg/200 ml	Scen.	Homo.	1.09 ± 0.05	4.30	8.0
culture suspension)	Scen.	Syn.	0.85 ± 0.01	2.37	4.0
	Chl.	Homo.	1.68 ± 0.05	2.87	4.5
	Chl.	Syn.	1.08 ± 0.06	5.33	9.0

eters, the physiological activity of the cell samples remains constant. Thus,
experiments on several photosynthetic reactions in these cultures of
photosynthetic microorganisms, again based on identical amounts of culture liquid, were carried out over an 8-day period. Potential photosynthetic capacity and the Hill reaction with *p*-benzoquinone as Hill reagent
(a photosystem II reaction) were determined polarographically by oxy-

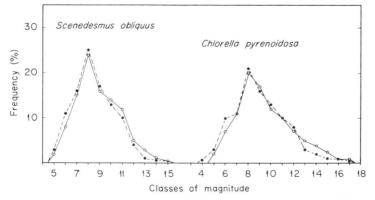

FIG. 13. Distribution curves of the cell sizes of *Scenedesmus obliquus* and
Chlorella pyrenoidosa measured at different times for homocontinuous cultures.

TABLE III

MEAN VALUES AND STANDARD DEVIATIONS OF THE DISTRIBUTIONS OF CELL SIZES OF
DIFFERENT CULTURE TYPES FROM *Scenedesmus obliquus* AND
Chlorella pyrenoidosa[a]

Object	Culture type; developmental stage	Mean values (standard deviation)	
		Sample I	Sample II
Scenedesmus	Homo.	8.53 ± 1.90	8.84 ± 1.92
	Syn.; 0 hour	7.33 ± 1.07	7.23 ± 1.05
	Syn.; 14th hour	12.97 ± 1.79	12.65 ± 1.65
Chlorella	Homo.	9.35 ± 2.56	9.08 ± 2.27
	Syn.; 0 hour	5.73 ± 0.85	5.94 ± 0.88
	Syn.; 16th hour	11.45 ± 1.10	11.53 ± 1.10

[a] The data are given in relative classes of magnitude. Multiplied by a factor of 1.11 they represent the longest diameter of the cells in micrometers (μm).

gen evolution. Photoreduction of carbon dioxide with hydrogen as the electron donor (a photosystem I reaction) was measured manometrically as carbon dioxide uptake. *Chlorella pyrenoidosa*, lacking hydrogenase, shows an inability to carry out photoreduction. The methods involved in these experiments have been reported in detail (Senger, 1970b) (Fig. 14).

The constancy of the measurements of photosynthetic capacity is excellent. Determinations of the Hill reaction and photoreduction are more complicated, and methodological pitfalls—especially in the case of the Hill reaction—may account for the largest deviations measured. Still, the most extreme deviation remains within 12% of the mean value and the highest standard deviation is 9.2% (Table IV).

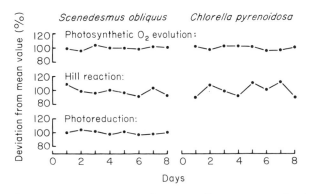

FIG. 14. Photosynthetic oxygen evolution, Hill reaction, and photoreduction of carbon dioxide measured for 8 consecutive days in homocontinuous cultures of *Scenedesmus obliquus* and *Chlorella pyrenoidosa*.

TABLE IV

MEAN VALUES AND DEVIATIONS OF PHOTOSYNTHETIC REACTIONS IN
HOMOCONTINUOUS AND SYNCHRONOUS CULTURES OF *Scenedesmus*
obliquus AND *Chlorella pyrenoidosa*[a]

Photosynthetic reaction	Object	Culture type	Volume (ml)	Mean values and standard deviation (absolute)	Standard deviation (%)	Maximal deviation (%)
Photosynthesis (O_2	*Scen.*	Homo.	40	310 ± 7.3	2.33	4
evolution in relative	*Scen.*	Syn.	50	384 ± 8.4	2.19	4
units)	*Chl.*	Homo.	20	227 ± 5.3	2.33	4
	Chl.	Syn.	30	280 ± 5.8	2.07	3
Hill reaction (O_2	*Scen.*	Homo.	40	179 ± 11.9	6.65	9
evolution in relative	*Scen.*	Syn.	50	222 ± 14.4	6.49	10
units)	*Chl.*	Homo.	20	49.3 ± 4.5	9.14	12
	Chl.	Syn.	30	60.0 ± 5.1	8.50	12
Photoreduction (μl CO_2	*Scen.*	Homo.	40	346 ± 7.4	2.14	4
uptake/10 minutes)	*Scen*	Syn.	50	470 ± 12.5	2.66	7

[a] In order to determine the photosynthetic reactions in samples of similar density, different but constant volumes of culture liquid have been used for the experiments with each type of culture (cf. column 4).

VI.   Results and Accuracy in Achieving Permanent Synchronized Cultures

A.   Growth Parameters of Permanent Synchronized Cultures

Synchronous cultures continuously undergo changes during their whole life cycle. Thus, permanently synchronized cultures cannot be tested for their accuracy without taking samples at fixed times after the start of the life cycle.

From permanent synchronized cultures of *Scenedesmus* and *Chlorella*, obtained as described above (see p. 308), samples were taken exactly at the 8th hour after the onset of illumination for eight subsequent life cycles. Cell number, dry weight, PCV, and total chlorophyll were determined on the basis of equal amounts of culture suspension. The results are shown in Fig. 15. The maximal deviation does not exceed 9% and the standard deviation remains within 5.3%. Considering that small differences in the developmental stage of subsequent synchronized cultures would cause marked changes in the growth parameters, the reproducibility of these permanently synchronized cultures is excellent.

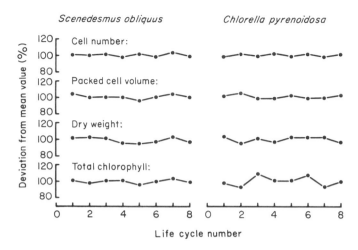

FIG. 15. Cell number, packed cell volume, dry weight, and total chlorophyll measured each at the 8th hour of eight consecutive synchronous life cycles of *Scenedesmus obliquus* and *Chlorella pyrenoidosa*. Synchronization was completely automized. (From Pfau *et al.*, 1971.)

Determinations of the distributions in cell size were carried out at the beginning (daughter cells) and the end of the light period (mother cells) for *Scenedesmus* (Fig. 16) and *Chlorella* (Fig. 17). The longest diameter of the cells, shown in the distribution curves, is only a rough

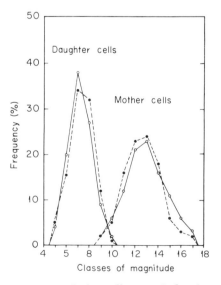

FIG. 16. Distribution curves of the cell size of daughter cells (0 hour) and mother cells (14th hour) from *Scenedesmus obliquus* measured for two consecutive life cycles of a permanently synchronized culture.

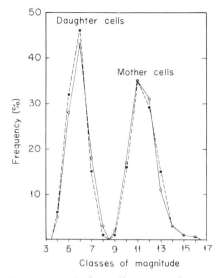

Fɪɢ. 17. Distribution curves of the cell size of daughter cells (0 hour) and mother cells (16th hour) from *Chlorella pyrenoidosa* measured for two consecutive life cycles of a permanently synchronized culture.

approximation of cell volume. It obviously has a broader spread in the case of the spindle-shaped *Scenedesmus* than the football- to sphere-shaped *Chlorella*. This is most probably the reason for the different shapes and the overlapping of the distribution curves for the two algae. It is understandable from the previous explanation on p. 303 and the illustration in Fig. 1, that mother cells generally demonstrate a broader distribution curve than daughter cells.

In each case, as illustrated in Figs. 16 and 17, the difference in the shape of the distribution curves of two subsequent samples and the divergence between the mean values (Table III) is so small that further statistical evaluations seem to be unnecessary. The results on the similarity of the distribution curves is additionally supported by the finding that the packed cell volume of samples taken at a fixed time from permanent synchronized cultures is constant (cf. Fig. 15).

B.   Photosynthetic Response of Permanent Synchronized Cultures

Experiments on photosynthetic reactions in the permanent synchronized cultures have been carried out in the same way as for the homocontinuous cultures (see p. 317). The results are shown in Fig. 18. The photosynthetic capacity with a peak value at the 8th hour of the life cycle (cf. Senger, 1970b), i.e., the time of measurement, shows a stand-

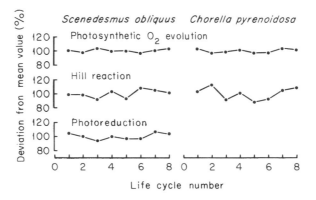

Fig. 18. Photosynthetic oxygen evolution, Hill reaction, and photoreduction of carbon dioxide measured each at the 8th hour of eight subsequent life cycles of *Scenedesmus obliquus* and *Chlorella pyrenoidosa*. Synchronization was completely automized. (From Pfau *et al.*, 1971.)

ard deviation of 2.3% only and a maximal deviation of 4% (Table III). Comparing this with the values obtained for the PCV (cf. Table II), the only direct accessible parameter, again demonstrates the accuracy of the dilution apparatus in producing permanent synchronized cultures with identical physiological reactions.

ACKNOWLEDGMENTS

We gratefully acknowledge the technical assistance of Mrs. Hedwig Werner, Mr. Hermann Becker, and Mr. Hans Dersch.

REFERENCES

Burlew, J. S. (1953). "Algal Culture, from Laboratory to Pilot Plant." Carnegie Inst. of Wash. Publ. 600. Washington, D. C.

Continuous Cultivation of Microorganisms. (1958). Publ. House, Czechoslavakia Acad. Sci., Prague.

Continuous Culture of Microorganisms. (1960). S.C.I. Monogr., No. 12, London.

Continuous Fermentation and Cultivation of Microorganisms. (1960). Pishche-promizdat, Moscow.

Gaffron, H. (1939). *Biol. Zentr.* **59**, 302–313.

Herbert, D. (1960). *S.C.I. Monogr.* No. 12, p. 21.

Kontinuierliche Züchtung von Mikroorganismen, Symposium Wien 1967. (1967). Vol. 21, Mitt. Versuchstat. Gärungsgew. Wien.

Kuhl, A., and Lorenzen, H. (1964). In "Methods in Cell Physiology" (D. M. Prescott, ed.), Vol. I, p. 159. Academic Press, New York.

Lorenzen, H. (1970). In "Photobiology of Microorganisms" (P. Halldal, ed.), p. 187. Wiley (Interscience), New York.

Malek, I., and Fenel, Z. (eds.). (1966). "Theoretical and Methodical Basis of Continuous Cultivation of Microorganisms." Academic Press, New York.

Malek, I., Beran, K., and Hospodka, J. (eds.). (1964). "Continuous Cultivation of Microorganisms." Academic Press, New York.

Malek, I. *et al.* (eds.). (1969). "Continuous Cultivation of Microorganisms." Academic Press, New York.

Myers, J., and Graham, I. R. (1959). *Plant Physiol.* **34**, 345–353.

Pfau, J., Werthmüller, K., and Senger, H. (1971). *Arch. Mikrobiol.* **75**, 338–345.

Pfennig, N., and Janasch, H. W. (1962). *Ergeb. Biol.* **25**, 93–135.

Phillips, J. N., and Myers, J. (1954). *Plant Physiol.* **29**, 148–152.

Pirson, A., and Lorenzen, H. (1966). *Ann. Rev. Plant Physiol.* **17**, 439–458.

Powell, E. O., Evans, C. G. T., Strange, R. E., and Tempest, D. W. (eds.). (1967). *Proc. 3rd Int. Symp., Microbial Physiol. Continuous Culture.* Her Majesty's Stationary Office, London.

Senger, H. (1970a). *Planta* **90**, 243–266.

Senger, H. (1970b). *Planta* **92**, 327–346.

Senger, H., and Bishop, N. I. (1971). *In* "Methods in Enzymology, (A. San Pietro, ed.), Vol. 23, Academic Press, New York.

Senger, H., and Wolf, H.-J. (1964). *Arch. Mikrobiol.* **48**, 81–94.

Tamiya, H. (1966). *Ann. Rev. Plant Physiol* **17**, 1–26.

Watanabe, A., and Hattori, A. (eds.) (1960). *Proc. U. S.–Jap. Conf. Cultures Collections Algae,* Jap. Soc. of Plant Physiologists, Kyoto, Japan.

Chapter 10

Vital Staining of Plant Cells[1]

EDUARD J. STADELMANN AND HELMUT KINZEL

Department of Horticultural Science, University of Minnesota, St. Paul, Minnesota and Pflanzenphysiologisches Institut der Universität Wien, Austria

[1] This work is supported by the Minnesota Agricultural Experiment Station and the Graduate School, University of Minnesota; Agricultural Experiment Station Scientific Journal Series, Paper No. 7679.

I.   Introduction

This chapter is based upon a survey on vital staining of Kinzel (1962).
Also, some of the material in this chapter is used as background reading
for the section on vital staining in the laboratory course Experimental
Protoplasmatology at the University of Minnesota. The experiments described in Section IV are exercises performed in this laboratory course,
many of which were adopted from Strugger (1949). The results described will be attained if a minimum of care is used in the experimental
procedure.

Since there are almost no English texts on vital staining of plant cells,
this chapter may serve as an introduction to stimulate interest on further
applications and new investigations in this field. Vital staining is one of
the few means available to investigate intact cells with little or sometimes almost no disturbance of cell functions. It is also possible that work
of this type may profit from recent progress in optical instrumentation
(see Runge, 1966; Ruch, 1968). These developments should contribute
significantly to the better analysis and interpretation of experimental
results in vital staining (e.g., Schwantes, 1965; Tsekos, 1970).

For more information on vital staining of plant cells and tissues the
outstanding compendium by Drawert (1968) should be consulted, since
it is the most up-to-date compilation in this field, giving a comprehensive
review of experimental results and theories of vital staining of plant
cells. A broader view of staining for microscopic work is discussed
thoroughly by Harms (1965) who explains the theoretical basis and the
practical aspects of many of the staining procedures used in biology.
More general information on biological stains may be found in Conn's
classic book now available in its eighth edition (Lillie, 1969).

A. General Remarks

Staining, in the broadest biological sense of the term, refers to treatment of a tissue or cell so that its components becomes distinguishable by the color they assume. Dyes used for such treatment are called biological stains. In some cases metal salts (e.g., silver nitrate) may be applied for staining (Jones, 1966, p. 28; Lillie, 1969, p. 1).

Dyes are substances that exhibit selective absorption in the visible light spectrum (Karrer, 1950, p. 484). However, not every substance that absorbs light selectively is a dye. For example, potassium permanganate has a very intense violet color, but it is not a dye. For a substance to be considered a dye, it must also possess the ability to become firmly attached to the structure that is to be stained.

In general, dyes are organic compounds. *Natural dyes* are extracted from plants and animals. Hematoxylin from *Haematoxylon campechianum* (Caesalpiniacae), and the carmines from *Coccus cacti,* a scale insect that lives on cacti in the warmer regions of America (Jones, 1966) are two commonly used natural dyes. *Synthetic dyes* usually are derivatives of substances found in coal tar.

For a substance to be considered a dye, its molecule must contain certain radicals or atomic groups. Two types of groups can be distinguished: (1) the chromophoric "color carrying" group and (2) the auxochromic "color deepening" group.

Chromophoric groups are unsaturated and include the azo group —N:N—, the nitroso group —N=O, the nitro group —NO$_2$, and the ethylene bond C=C. However, the presence of a chromophoric group does not make a substance a dye. Molecules with chromophoric groups occasionally give a color or stain tissues, but they can easily be removed again. A molecule with one or more chromophoric groups is called a *chromogene* (Harms, 1965, p. I/9).

A chromogene becomes a dye when an auxochromic group is added to its molecule. Auxochromic groups are polar and may exhibit hydrogen bonding and salt formation. Halogens may also be auxochromic groups (Harms, 1965).

Even though auxochromic groups are very simple and do not have double bonds, they have a considerable influence on the quality of the dye. In the amino group one or two hydrogen atoms are often replaced by methyl, acetyl, or benzylamino groups, as shown below.

$-N{\overset{CH_3}{\underset{CH_3}{\diagup\diagdown}}}$	$-N{\overset{H}{\underset{CO\cdot CH_3}{\diagup\diagdown}}}$	$-N{\overset{CH_2\cdot C_6H_5}{\underset{CH_2\cdot C_6H_5}{\diagup\diagdown}}}$
Dimethylamino group	Acetylamino group	Dibenzylamino group

The effect of these groups in dye molecules versus non-dye molecules can be illustrated as follows (Harm, 1965, p. I/13).

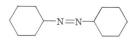

Azobenzene
[red, but not
a dye]

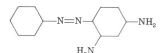

4-Aminoazobenzene
[dye (yellow), due
to the presence
of the auxochrome —NH₂]

[2, 4-Diaminoazobenzene
chrysoidin, dye (red brown),
two auxochrome groups]

Absorption of light in the visible spectrum is due to the presence of conjugate double bonds, which are formed when auxochromic and chromophoric groups are brought together (Harms, 1965). π-Electrons are important here, because they have great mobility. Oscillations of the π-electrons along a chain of conjugated double bonds cause the shift of the absorption maximum from the UV to the visible region.

B.   Staining in Microscopy

Stains were used very early in microscopy to increase visibility of organs, tissues, cells, and cell components. Van Leeuwenhoek (1714) used saffron tincture to stain muscle fibers, which were otherwise too translucent to observe (Harms, 1965). Since that time, stains have been applied in ever increasing numbers and with a variety of procedures to many specimens.

Three different types of staining can be distinguished for biological material (see Jones, 1966, p. 31).

a. Vital Staining or Intravitam Staining. Here the stain is applied to the living organism, cell, or tissue. In animals the stain may be injected or ingested (see Foot, 1950); in plants, the stain may be supplied via the root or other absorbing surfaces or tissue cuttings may be placed in the stain solution.

Vitally stained material is usually studied with the light microscope, which does not damage the specimen. Recently, staining with metal ions (accumulated by the living cell in the wall and other cell parts) has been used for electron microscopic observation (Lowenhaupt, 1970; Lowenhaupt and Lowenhaupt, 1966). Although the cells are necessarily killed during fixation, this type of staining may still be considered vital staining since the staining process was performed with the living cell.

b. *Supravital Staining*. This term was coined especially for certain staining procedures with zoological material, the stain being applied to living cells *in vitro* (not *in situ*). For example, mitochondria in leukocytes exposed to Janus green on a slide will be supravitally stained (cf. Doan and Ralph, 1950). Staining of cells immediately after their death has also been called supravital staining (Cappell, 1929, p. 595).

c. *Postmortem Staining*. Here the stain is applied after the cell or tissue is dead, e.g., after sectioning of animal or plant material, with killing before or by the sectioning procedure. In the first case staining may be done before sectioning. A great number of staining procedures are known for postmortem staining and are used in histological and cytological studies.

Since sections of tissues are often used in botanical studies involving vital staining, no sharp distinction between vital and supravital staining exists. In some instances, it might be difficult to differentiate between vital and postmortem staining. A clear understanding of the term *vital* is needed. No absolute criterion defines the exact time of death of an organism, a cell, or a cell organelle. Since a continuous transition between the living and dead state of the cell exists, it sometimes is difficult to decide whether a given cell is "still alive" or "already dead" (Drawert, 1968, p. 6f.).

Every stain introduced into a tissue or cell will cause physiological or morphological alterations and often, after a certain time, will result in a disturbance of the cell functions that may ultimately cause death. The best that may be expected in vital staining is a sufficiently long postponement of such alterations so that the specimen may be studied in a state approximating its natural condition.

Vital staining, therefore, may be defined as a method of applying stains to living cells, tissues, organs, or organisms for physiological or anatomical studies. After some (often a considerably long) time, disturbance or even death of the material may result. Therefore, only those staining procedures that result in immediate cell death should not be called vital staining methods (Drawert, 1968, p. 7).

The transition of the tissue from living to dead is gradual, and the time required for this transition differs considerably, depending on the specimen, the stain, and the procedure. Strugger (1937) set up a classification of stains based on their estimated degree of toxicity to living material. He distinguished four categories.

1. *Inturbant stains* (Latin: *inturbidus:* undisturbed, quiet) are those in which no toxic effects occur immediately after application of a stain, after an extended time, or after prolonged culture of the stained specimen. This is an ideal condition, one that is only rarely approached.

2. *Turbant stains* (*turbare:* to confuse) are those in which no visible damage occurs during the staining process or shortly thereafter. However, over an extended period, the stain causes gradual disturbance in the cells and death eventually occurs.

3. *Perturbant stains* (*perturbare:* to disturb thoroughly) are those in which pathological changes occur during the staining procedure or immediately thereafter. However, cell death takes place at some later time.

4. *Disturbant stains* (*disturbare:* to destroy) are those that cause immediate and lethal damage during the staining process. This group of stains cannot be considered vital stains as defined above.

The degree of toxicity of stain solutions depends on the concentrations applied, the object studied, and other conditions. The concentrations used in vital staining are very low. Concentrations of about 1:10,000, or even lower, are sufficient.

Vital stains must *accumulate* inside a cell to effect staining. When the stain penetrates into the cell only to a concentration equilibrium with the external solution, the cell still appears unstained under the microscope. This is because the layer that absorbs the light is too thin, and there is no contrast in absorption intensity between the external stain solution (or water) and the cell interior.

C.   Scope and Application of Vital Staining

Considering the broad definition of vital staining, Drawert (1968, p. 8f.) discussed a variety of problems for which this method may be used.

1.   VITAL STAINING IN MORPHOLOGY AND ANATOMY

The most common purpose of vital staining is to increase the visibility of certain organs, tissues, cells, or cell components when they are in the living state, and when they are small and translucent. Also, organelles often have no color and only slight differences in refractive index. The success of phase contrast in cytology during the recent decades (cf. Haselmann, 1957) is based mainly upon the increase in visibility of the cell organelles by this method. However, vital stains have the advantage that they can stain specific cell elements ("elective staining," Gicklhorn, 1931), so these elements can be identified in the cells or tissue. A combination of vital staining (with fluorochromes) and phase contrast is now also in use (Haselmann and Wittekind, 1957; Gabler and Herzog, 1965).

Vital staining may also be applied in studies of pathological alterations in cell morphology. However, since all of the stains are harmful to a certain degree, extended exposure to them will lead to cell damage.

2. VITAL STAINING AS A PHYSIOLOGICAL TOOL

A closer analysis of vital staining and its selectivity may shed light on physiological mechanisms of uptake, accumulation, and intracellular transport. Under certain conditions ("empty" cell sap, with low salt content) some vital stains can be used to measure the vacuolar pH by determining the threshold pH for stain accumulation (Kinzel and Imb, 1961).

Vital stains also can be applied as cytochemical test substances for the detection of such cell constituents as RNA, DNA, pectin, tannin, flavonoids, and lipids, especially when precipitates are formed. Fluorochrome-labeled antibodies are important in medicine, being used to demonstrate antigen sites in the organism (Nairn, 1969).

The location of positive or negative charges in the cell may be inferred from the application, at a suitable pH, of anionic and cationic vital stains, respectively.

Vital stains interfere with metabolic, developmental, and stimulatory processes by physical (e.g., photosensitization) or chemical action (e.g., interference with basic physiological processes such as assimilation or respiration). Thus, vital staining leads to alterations in the physiology of the cell and the whole organism. Analysis of these physiological alterations may yield information concerning cell function and the processes involved.

The important distinction between living and dead cells often is made possible by tests with vital stains (cf. Wittekind, 1959, p. 11f.; Drawert, 1968, p. 541).

D.   History of Vital Staining

Investigations of the movement of water in the whole plant represent the earliest applications of stains in botany: Magnol (cf. Anonymous, 1706) tested the uptake of water by adding inks or colored fruit juices and using these solutions to water *Polianthes tuberosa* (Liliaceae) and other plants. After a while the colors were observed in the venation of the leaves of these plants. Unger (1848/50) made microscopic observations of the uptake of stains (colored fruit sap from *Phytolaccca decandra*) by individual cells and described them for the first time.

In spite of these early studies, Pfeffer (1886) must be considered to be the father of vital staining of plants. He was the first to apply aniline (coal tar) dyes to living plants (various species of Chlorophyta and Monocotyledonae) in very low concentrations (0.0001 to 0.001%; Pfeffer 1886, p. 184) and to observe the uptake and accumulation of the stain inside the cells.

Later, Küster, Dangeard, and Guilliermond applied vital staining to plants in a wide variety of physiological and cytological investigations. More recent advances in understanding vital staining are based on Strugger's systematic work on the effect of pH of solutions on the uptake of vital stains by the cell and on Höfler's distinction of "empty" and "full" cell saps. At the present time, Drawert and Tsekos are most active in working on the physicochemical basis of the staining mechanism and on the composition, purification, and use of vital stains for many aspects of plant cell analysis.[2]

E.   Classification of Vital Stains

The classification of vital stains corresponding to their chemical structures (for a chemical grouping of biological stains see Lillie, 1969) does not contribute much to an understanding of the staining mechanism. A classification that is physiologically meaningful uses a physicochemical parameter, the ionic characteristics of stains, and the vital stains may thus be classified as (1) basic, (2) acidic, (3) amphoteric, or (4) electroneutral.

Since an auxochromic group of the stain molecule often determines its ionic characteristics, the same chromophoric groups may be found in stains belonging to different categories of the above classification.

The electrical charge of a stain ion can be determined without great difficulty by electrophoresis or capillary analysis. The latter technique can be demonstrated with a piece of round filter paper (p. 360; cf. Ruhland, 1912; Drawert, 1968) when a droplet of the stain solution is placed in its center. A "spreading out" of the water and the stain is observed; when both move at the same speed, and when the water and the stain keep the same front on the filter, the stain in solution is acidic. However, when the stain moves more slowly than the water, and a circular ring of water without stain appears on the filter paper, the stain

---

[2] For citations see the exhaustive list of publications on vital staining in Drawert, 1968.

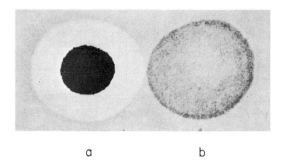

a   b

FIG. 1. Capillary analysis of vital stains: (a) Neutral red (basic); the stain cations are adsorbed to the filter paper and their spreading is slower than that of water. (b) Eosin (acidic); the stain anions are not retained by the negative charges in the filter paper and therefore spread on the filter paper at the same speed as water. (From Strugger, 1949, p. 131, Fig. 90.)

in solution is basic. The negative charges in the filter paper adsorb cations and thus retard the basic stain by preventing it from being carried along with the advancing water front (see Fig. 1).

Electrophoresis is a very accurate method of determining the sign of an ionic charge and the migration speed of ions. This method also is helpful in determining whether a given stain sample is chemically pure or is a mixture of different components such as by-products of processing or added stabilizing substances.

Migration of vital stains in an electric field naturally requires that their solute particles have an electrical charge, that is, the molecule must dissociate. Undissociated molecules will not migrate in an electric field. However, electroneutral stain particles in colloidal solutions may acquire an electrical charge, or, when other ions or colloids are present in the stain solution, the undissociated stain molecules may become adsorbed and thus may migrate in the electric field.

A *cationic or basic stain* has its chromophoric group in the cation, the positively charged group. Since the stain (base) has a low solubility in water, salts (usually chlorides, occasionally also sulfates) are most frequently used. Neutral red is an example of a cationic stain.

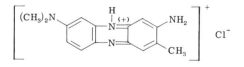

The aqueous solutions of basic stains have an acid reaction, due to the dissociation of the salt.

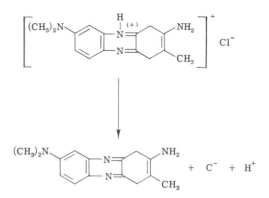

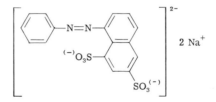

Anionic or acidic stains have the chromophoric group in the anion. Here too, the stain (acid) has a low solubility in water, and inorganic salts (potassium or sodium) are generally used. Orange G is an example of an anionic vital stain.

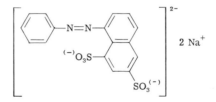

In an aqueous solution the stain salts will undergo hydrolytic dissociation and therefore, the solution will react basically

$$R—COONa + H_2O \rightarrow RCOOH + Na + OH^-$$

Amphoteric vital stains contain both acidic groups (carbonyl, acidic hydroxyl groups) and basic groups in their chromogene. Thus, inner salts may result, as with rhodamine B, a widely used amphoteric vital stain that exists as an undissociated (inner) salt between pH 2 and 11.6 (Drawert, 1937). Harms (1965, p. II/334) considers rhodamine B to be the most harmless vital stain. Also methyl red may be considered to

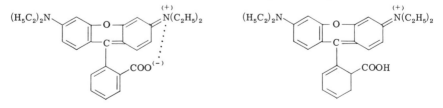

Inner salt   Free acid group present

Rhodamine B

be an amphoteric vital stain because it contains a carboxylic group and an amino group (example: for staining with methyl red see Fig. 8, middle row, left).

Sometimes amphoteric vital stains are called neutral vital stains in the literature. However, this designation is not accurate, since neutral stains are defined as those stains that contain an acidic and a basic chromogene (Drawert, 1968, p. 27).

F. Dissociation

The degree of dissociation is an important factor in vital staining, since the permeation of the stain into the cell depends, to a considerable extent, upon when the stain is ionized or present as an electroneutral molecule.

Vital stains are weak electrolytes. Acidic vital stains undergo electrolytic dissociation ($R—COOH ⇌ R—COO^- + H^+$). For basic vital stains dissociation is best described as protonization, the degree of which depends upon the H^+ concentration in the solution. The vital stain involved may form mono- and divalent ions so that sequences of protonization equilibria are established, as, e.g., for a stain molecule with two amino groups (see reaction below).

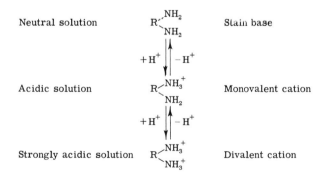

The resulting composition of a vital stain solution at a given concentration and pH, therefore, may be quite complex and several species of particles (ions and undissociated molecule) may be simultaneously present.

For neutral red the molecule of the stain base

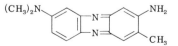

Stain base (yellow)

transforms into three different ion species depending upon the pH of the solution as shown below.

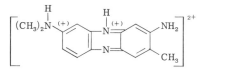 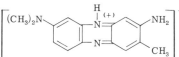

Divalent cation (blue)   Monovalent cation (red)

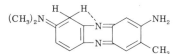

Carbenium form
(between pH 5.5 and 7.5)

Given the pK values (Bartels, 1956a) for each ion species, the composition of a neutral red solution can be calculated for a given pH (see Fig. 2).

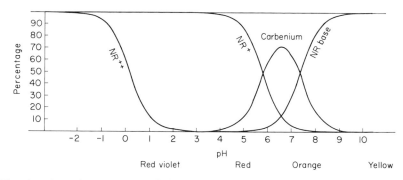

FIG. 2. Contribution of the different forms of neutral red to the composition of a neutral red solution at different pH's (pK values from Bartels, 1956a). *Bottom:* color of the solution. (Original.)

G. Dispersity

This quality refers to the particle size of the vital stain in aqueous solution. The molecules or ions of the vital stain often group together and form dimer, trimer, or polymer associates. Polymer ion associates (polyions) often develop in highly concentrated solutions, under conditions in which the vital stain is almost completely dissociated. The presence of ion associates can be inferred from changes in the absorption or fluorescence spectrum of the vital stain and from anomalies in the electrical conductivity with increasing concentration. The molecular structure of the stain considerable influences the likelihood of ion-associate formation. Acridine orange has a great tendency to form ion associates, while pyronine is not thought to form ion associates. Some vital stains can have molecular aggregates or polyions so large that they behave as colloids. The colloidal nature of vital stain solutions has been often discussed, especially in the earlier literature (cf. Drawert, 1968, p. 56f.).

A solution is *highly disperse* when the solute is present in molecular or monoionic form. When the solution contains colloidal particles only, it has a *low dispersity*, and the particle size is, of course, greater than in molecular dispersion. A solution that contains only one species of particles (e.g., only molecules, ions, or colloids) is *monodisperse*. When particles of different size are present simultaneously in the solution (e.g., molecules and polyions) the solution is *polydisperse*.

Vital stains generally form polydisperse solutions, with particle sizes ranging from molecules to polyassociates (colloids). Little definite information is available concerning the dispersity of a vital stain, but it can be assumed that a stain solution may quite easily change its dispersity. The transition from monomolecular or monoionic species to molecule or ion associates is of special importance for vital stains, since these changes often lead to a considerable shift in light absorption.

The dispersity in a given vital stain solution depends on many factors, e.g., solvent, temperature, concentration, pH and age of the solution, and the presence of other solutes. None of the vital stains available are chemically pure; some need stabilizing additives (mostly salts such as sodium chloride), and others are not purified because their impurities do not hamper the technical and industrial applications for which the dye is generally used. The important aspects of particle size and dispersity in vital stain solutions are discussed thoroughly by Drawert (1968).

H. Color Tone and Metachromasy of Vital Stains

The color tone of a vital stain solution depends on many factors (cf. Drawert, 1968, p. 138f.). A knowledge of these factors may allow a correlation to be made between the color tone found in a given cell component (e.g., vacuole, cytoplasm) and such features as the degree of association of the stain (e.g., monoions versus dimers or polyions) or the type of stain particles (molecules versus ions) involved in the staining process. This information may be indicative of the physicochemical properties of the stained cell component.

The pH influences the color tone by its effect on the degree of dissociation or protonization. In a lipid solution of a vital stain, the presence of polar groups in the solvent may change the color tone. Adsorption of the stain at a surface; additional electrolytes in the stain solution; temperature, oxidation, and reduction processes; and light also influence the color tone considerably. Light, especially of short wavelength, may cause photochemical reactions (mainly oxidations) that lead to a decrease in color intensity that can approach zero. This light effect is of importance in fluorescence microscopy, where a strong UV light source is used.

Analyses of absorption spectra of the stain solution under different conditions and comparison of these data with the absorption spectrum found in the stained living material can yield more significant interpretations of the staining process and of the cell components involved.

When histological material is stained, it is occasionally observed that the color of the stained specimen is significantly different than that of the diluted staining solution. This color change of the stained specimen is called *metachromasy,* which is defined as the reversible shift of the spectrum of an organic stain when it is adsorbed to a structure or when certain substances (the so-called chromotropic substances) are added to the staining solution. Color changes resulting from differences in pH or from different solvents are excluded from this definition. When the color tone in the stained material is the same as that of the staining solution, *orthochromatic staining* has occurred (cf. Drawert, 1968, p. 169f.; Harms, 1965, p. I/86; Richter, 1966, p. 52).

The direction of the shift of the spectrum is used to distinguish between positive metachromasy (when the main absorption band moves toward the shorter wavelength) and negative metachromasy (when the band moves toward the longer wavelength; Lison and Mutsaars, 1950). However, Kinzel (1968b) calls positive metachromasy the shift in color tone found in concentrated solutions containing ion associates and negative metachromasy the change in color tone found in solutions containing phenols and similar compounds. In the case of the most commonly used vital stains the two definitions coincide. The first-mentioned shift in color tone is similar to the one observed in "empty" cell saps, the latter shift to the one found in "full" cell saps.[3]

Positive metachromasy most probably is caused by association of the

[3] The terms full and empty denote the presence and absence, respectively, of substances in the cell sap that react chemically with the vital stain as, for example, tannins or flavon-containing compounds. Since cell saps, obviously, never are empty, these terms are generally used with quotation marks as "full" and "empty."

stain molecules (or ions) at (outer or inner) surfaces of the stained specimen (cf. Richter, 1966). Often the stain molecules or ions are flat and form "coin-like" stacks. Negative metachromasy mostly results from a chemical interaction between the stain and a specific chemical compound of the stained specimen.

1.   Fluorescence and Fluorochromes in Vital Staining

Fluorescence refers to the emission of light by molecules immediately upon absorption of light energy. With very few exceptions (cf. Pringsheim, 1949) the emitted light will always be of a longer wavelength than the exciting light (*Stoke's law*). The wavelength of the emitted light depends on the molecular structure of the fluorescent material and does not change with the exciting wavelength. However, when the wavelength of the exciting light becomes too long, no fluorescence occurs. Thus, to ensure that emission of fluorescent light throughout the entire range of the visible spectrum is possible, the exciting wavelength must be shorter than blue light (i.e., the exciting wavelength must be in the ultraviolet).

Fluorescence microscopy deals with the fluorescence of microscopic specimens. It was initiated in Vienna where C. Reichert developed the first fluorescence microscope in the year 1911 (Höfler 1951, p. 92) and M. Haitinger laid the groundwork for this technique (cf. Haitinger 1938).

At first the carbon arc was used as a source of UV light. More recently the high pressure mercury vapor lamp, which emits ultraviolet light, of a much higher intensity, is used. Both sources radiate considerable amounts of visible light and heat, which must be absorbed by appropriate filters. Aside from light filters, darkfield illumination can be used to prevent direct light from the light source from reaching the objective. When the microscopic specimen is illuminated by darkfield no filter for visible light is needed between the objective and the eyepiece since only the visible fluorescence light emitted from the object will reach the objective (cf. Young, 1961; Nairn, 1969, p. 73). In incident light fluorescence microscopy (cf. Haitinger 1938, p. 16f.) neither the primary nor the secondary (fluorescence) light travels through any thickness of the specimen. Recent developments in optics resulted in considerable improvement in the quality of the microscopic picture (Herzog, 1971). Since the near UV (e.g., the 366 nm Hg line) can be used as exciting light, a condensor of a good grade glass is sufficient, and no quartz lenses are required. To protect the eyes a UV absorbent filter must be placed between the objective and the eyepiece. The near UV is also considerably less damaging for the living cells.

The specimen to be studied may itself contain substances that fluoresce (e.g., berberin in *Berberis*); this is called *primary fluorescence*. Alternatively, it may be necessary to bring the specimen into contact with a fluorescent substance, which then accumulates in specific structures of the specimen and makes them fluorescent upon irradiation with UV (*secondary fluorescence*). The fluorescent substance needed in the latter case is called *fluorescent stain* or *fluorochrome* (as distinguished from *diachrome*, the normal vital stain, that produces staining by its selective light absorption).

Fluorochromes are used in the same way as diachromic vital stains to stain the elements of living cells, but much smaller amounts of fluorochromes are necessary in a given region of the cell to produce visible fluorescence. Thus, only very low concentrations of fluorochromes need be applied in vital staining. This makes possible the use of those fluorescent substances that have an appreciable degree of toxicity.

A number of both basic and acidic fluorochromes are known. One of the most frequently used fluorochromes, acridine orange, was introduced into fluorescence microscopy by Strugger (1940) and Bukatsch and Haitinger (1940; cf. Harms, 1965, p. II/76).

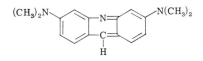

Acridine orange

Acridine orange forms three different ions: trivalent, divalent, and monovalent cations. Between pH 1.5 and 6 almost all acridine orange is in the form of the monovalent cation; as shown below.

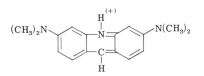

However, dimers of this ion may be formed at higher concentrations of the solution. The monomer ion has a green fluorescence, whereas the dimer fluoresces red. The di- and trivalent ions exist in appreciable amounts only at extremely acidic pH and, therefore, are of no significance in vital staining. Between pH 6 and 12 (pK about 10.5), both the cation and the basic molecule are present in solution in appreciable amounts. Above pH 12 only the basic molecule is present (Zanker, 1952).

Uranin is an example of an acidic fluorochrome (see pp. 353, 367).

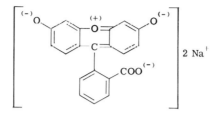

It is a very harmless fluorochrome with almost no diachromic quality. Cell structures appear stained only slightly, at most, with visible light.

II.   Vital Staining of the Different Components of the Plant Cell

Whether vital stain accumulates in a certain cell component (e.g., cell wall, cytoplasm, nucleus, vacuole) depends on the following factors: (1) the cell type, (2) the quality of the vital stain, and (3) the conditions for the staining process (especially the pH value of the staining solution).

The solubility of a vital stain and its changes with external factors, such as pH, determine to a large extent where it will accumulate in the living cell. Stains that are not only soluble in water, but, to a certain extent, soluble in lipids, easily enter the protoplasm from the external solution. Since living protoplasm can be considered, at least to a certain degree, to be a continuous lipid phase (cf. Höfler, 1959), the vital stain may migrate through the protoplasm to the tonoplast and finally pass into the vacuole.

A.   Basic Vital Stains

Undissociated molecules of basic stains differ greatly from their cation counterparts in their solubilities. The ions are hydrophilic (i.e., soluble in water, very slightly soluble in lipids), whereas the undissociated molecule is hydrophobic (i.e., very slightly soluble in water, often highly soluble in lipids). This relationship can be easily demonstrated by shaking aqueous solutions of the vital stain at different pH values with benzene (see p. 360).

A basic stain can enter the protoplasm only when the pH is such that electroneutral molecules of the stain are present in the solution.

1. VITAL STAINING OF THE CELL WALL

Cell walls of both living and dead cells may absorb a vital stain, depending on the type of stain used and the staining conditions. Different staining mechanisms may be involved.

a. Electroadsorptive Cell Wall Staining (By Exchange Adsorption). Plant cell walls consist of cellulose together with some partially acidic carbohydrates (pectins). In addition, the cell wall might be incrusted with lignin, suberin, or other substances.

Cell walls are negatively charged (COO^- groups) due to the acid carbohydrates present there (pectin, polygalacturonic acid residues). Only under very acidic conditions (pH <3) are the carboxyl groups not dissociated. Therefore, at pH values above 3, the negatively charged cell wall is able to adsorb the cations of basic dyes. Two conditions must always be fulfilled:

1. The stain must be ionized at the pH of the solution.

2. The stain cations must compete successfully with the H^+ ions for the negatively charged $-COO^-$ groups in the cell wall (cf. Richter, 1966, p. 11).

The following experiment convincingly demonstrates the importance of the pH of the solution of the vital stain. The upper (interior) epidermis of an onion scale of *Allium cepa* is placed in a solution of neutral red (0.01%) in distilled water (pH 5.5 to 6). The stain accumulates in the cell walls. If, however, the stain is dissolved in calcareous tap water (pH 7 to 7.5) it penetrates into the interior of the cells and accumulates in vacuoles. The cell walls of the living tissue remain unstained.

Neutral red (pK 7.38) will stain the cell walls of living cells in a pH range from 3 to about 6.5, where the cation of the stain predominates. Above pH 6.5 there is an increase in the concentration of undissociated stain molecules. Also at a pH higher than 6.5 neutral red penetrates readily into the interior of the cell and accumulates there, so that only small amounts of neutral red are available for the cell wall ("competition for accumulation," Drawert, 1937). In dead tissue, however, it is possible to stain the cell wall by neutral red at a pH higher than 6.5.

Other stains suitable for staining the cell wall are Nile blue (pK about 8 to 9), brilliant cresyl blue (pK about 9 to 10), toluidine blue, Thionine (pK about 11), and methylene blue (pK about 12 to 13). Those stains that are incompletely ionized or not ionized in the pH range above 3, including chrysoidine (pK about 4) or rhodamine B (pK about 2), are not suitable stains for nonligneous cell walls.

Nonlignified cell walls absorb acridine orange (see p. 340) strongly from pH 3 to about 7 to 8 and fluoresce red. With visible light the cell walls appear yellow, indicating a high degree of adsorption on the acidic groups of the cell wall. When sections of plant tissues are treated with an acridine orange solution of pH less than 3, the nonlignified cell walls fluoresce light green indicating that even at such low pH values traces of acridine orange are adsorbed but the cell walls appear colorless in visible light. The mechanism of uptake of acridine orange at pH less than 3 is not by electrostatic adsorption of cations as for the red fluorescence above pH 3, but is, instead, by *substantive staining* (see below). This is based on van der Waal's forces between the stain and the cell wall and is independent of the pH (Kinzel, 1955; Richter, 1966, p. 52).

Competition for the charged sites in the cell wall may lead to a washing out of the stain after the material is transferred to a buffered salt solution. A greater affinity of the salt cations for the sites in the cell wall leads to an exchange with the stain cation and, therefore, the loss of stain from the cell wall (cf. Höfler and Stiegler, 1947; Kinzel, 1955; Drawert, 1968, p. 373).

b. Staining by Chemical Precipitation (Drawert, 1937, 1968). In some cell walls staining can be obtained even at pH's where there is no dissociation of H^+ ions from the carboxyl groups in the cell wall, and, therefore, all the COO^- groups are saturated. In *Funaria* leaves, for instance, cell walls stain blue with toluidine blue at pH below 3. (Above pH 3, when the COOH groups are dissociated, the staining is violet.) The staining at such a low pH can be explained by chemical reaction of the basic stain with phenolic groups of the cell wall.

Since lignins contain phenolic groups, lignified cell walls stain by chemical precipitation, and basic stains are accumulated strongly and almost independently of the pH value and the state of ionization.

Staining by chemical precipitation is stable and irreversible. Thus application of salt solutions will not wash out the stain as is seen with electroadsorbed stains. After treatment with acridine orange, lignified cell walls fluoresce a bright yellow green. Hydrogen bonds between NH_2 groups of the stain and the phenol OH groups of the lignin probably are responsible for this strong adsorption. This type of staining may be considered to already be some form of substantive staining, since difluorescence (polarization of the emitted light) suggests that the adsorbed stain ions are arranged in a preferred direction (cf. Kinzel, 1955; Richter, 1966, p. 62).

c. Substantive Staining. Substantive staining (also called direct staining) of the cell wall involves deposition of stain molecules or aggregates of stain molecules of a considerable size inside the submicroscopic pore system of the cell wall.

The best example of a direct stain is Congo red. The less dense cell walls of rhizoids of fern prothalli can be stained with Congo red. The more dense walls of the prothallial cells do not stain, even after treatment with osmic acid or other poisons. However, when lipid- or wax-dissolving substances (e.g., boiling alcohol, KOH) are applied to these cell walls, they stain with Congo red. Most probably, the pores are opened or the cuticle is dissolved, so that the Congo red has access to the internal submicroscopic intercellular systems of the cell wall (see Drawert, 1968, p. 228f.).

Direct staining is independent of the pH of the staining solution and the degree of dissociation of acidic groups in the cell wall. Directly stained cell walls often exhibit dichroism, indicating a certain degree of orientation of the stain molecules inside the submicroscopic capillary system. *Dichroism* refers to the change in color intensity of a specimen in polarized light depending upon the direction of the polarization plane.

2. VITAL STAINING OF THE VACUOLE

Accumulation of substances found in the vacuole is largely due to the metabolic activity of the cell. Some *acid stains* may also enter vacuoles by active uptake (Kinzel, 1965; Bolhar-Nordenkampf, 1966). However, the accumulation of *alkaline stains* in vacuoles is a physicochemical process not directly related to the metabolic activity of the cell (Drawert and Endlich, 1956; Bancher and Hölzl, 1960a) and depends on two conditions.

1. The pH value of the external solution should be such that at least the greater fraction of the stain is present as a fat-soluble base.

2. A physicochemical mechanism in the cell sap must operate to retain the stain in the vacuole.

a. Staining Mechanisms. Höfler (1947, 1951, p. 101) noticed that when the basic fluorochrome acridine orange was stored by some cell saps it fluoresced red, whereas in other cell saps it fluoresced green. Soon it became evident that two different mechanisms were involved in the accumulation of the stain. Cell saps that fluoresce green (full cell saps, according to Höfler) contain materials that bind acridine orange chemically. Cell saps that fluoresce red (empty cell saps) accumulate the basic stain acridine orange by the so-called "ion trap" mechanisms.

i. The ion trap mechanism. Although lipophilic substances can diffuse easily through the living protoplasm and into the vacuoles, the living protoplasm presents an almost insurmountable barrier to those particles (ions and molecules) which are water soluble only, i.e., hydrophilic (Bancher and Höfler, 1959). Thus, in a basic stain, uncharged molecules that are present in alkaline solution penetrate easily into the vacuoles of living cells. If a cell sap (which is generally slightly acidic) does not contain materials that bind the basic stain by chemical reaction (empty cell sap), then the penetrating stain molecules will become ionized in the cell sap. Because ions are not fat soluble, they cannot diffuse back through the lipophilic protoplasm. Thus, ions are trapped in the cell sap. Meanwhile new molecules of the stain continue to enter and ionize until a concentration equilibrium of the stain molecules is reached between the external solution and the cell sap (Kinzel, 1954b, 1959). The driving force for the accumulation is the difference in pH between the cell sap and the external solution. The ion trap mechanism was recently verified on a nonliving model system (Bock, 1964).

ii. *Chemical reactions.* The staining mechanism is based on the presence of substances in the vacuole that react chemically with the vital stain. Härtel (1951), Kinzel (1958), and, especially, Bolay (1960), have compared histochemical reactions with vital staining and found that all full cell saps contain either flavone glycosides or tannin-like substances. Phenolic OH groups that can form hydrogen bonds with NH_2 groups of the vital stain are supposedly responsible for the binding of the stain (cf. Bancher and Hölzl, 1963). Thus, if a cell sap proves to be full (for criteria see p. 347) it can be concluded that those substances are present. The accumulation of the vital stain in a full cell sap may give even more detailed information depending on the type of association the vital stain undergoes with the cell sap material. Sometimes these compounds may be in a molecular dispersion (*true diffuse staining*) or in a colloidal dispersion (*false diffuse staining*) in the cell sap. Frequently, however, *segregation* takes place when the reaction product of the vital stain with vacuolar material precipitates. Precipitations frequently occur when the cell material remains for a relatively long period in the staining solution.

α. *True diffuse staining* of full cell sap often leads to metachromasy. Thus, acridine orange fluoresces green in full cell saps. The diachrome neutral red, which is carmine red in acidic aqueous solution and orange to yellow in basic stain solution, changes in full cell saps to violet red. Some blue stains such as brilliant cresyl blue or toluidine blue (which is pure blue in aqueous solutions) appear more or less blue green in full cell saps.

β. *The segregation forms of full cell sap.* Three main types of segregation forms can be distinguished: (1) spherical, or droplet, segregates;

(2) aggregates; and (3) crystal formations such as crystalline precipitations, spherocrystals, or crystal clusters.

1. *Spherical segregates* are the simplest case of a segregation. A liquid segregation occurs resulting in the formation of droplets that later coalesce into bigger spheres (Fig. 3). Occasionally these droplets undergo a slow transformation into spherocrystals.

2. *Aggregates* where the droplets or spheres mentioned above may in time solidify. These may coalesce or stick together to different degrees, depending mainly on their size and frequency. (*a*) *Coalescence products* are the result of the coalescence of relatively large droplets, which are very viscous or which coalesce during solidification. Since coalescence is incomplete, the original shape of the droplets can still be recognized in the coalescence product (Fig. 4a). (*b*) *Dendrites.* When numerous small droplets are present, they may join together in groups of about 20 spheres, but without actually coalescing. This formation sometimes has the appearance of a string of pearls (Fig. 5d). In dendrites the original shape of the little spheres can be recognized from constrictions at more or less regular intervals (Fig. 4b; see also Fig. 6). (*c*) *Granular precipitates* (*Krümel*). When aggregates such as dendrites are composed of spheres so small that they can hardly be seen under the microscope, they are called granular precipitates (Fig. 4c).

3. *Crystal formations* that result directly from the staining process itself are infrequent. However, crystals sometimes do develop from spherical segregations (cf. Bolay, 1960); sometimes these crystals precipitate

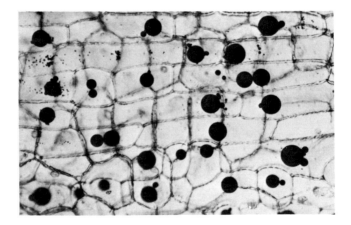

FIG. 3.   Spherical segregates. *Allium cepa,* abaxial (outer) epidermis. After staining with rhodamine B, spheres containing the vital stain appear in the cell sap (Kinzel and Bolay, 1961).

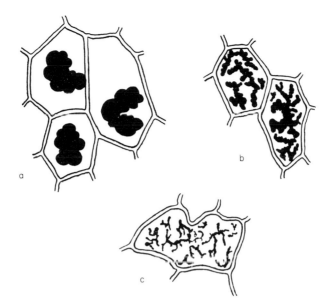

FIG. 4. Different types of aggregates: (a) coalescence products; (b) dendrites; (c) granular precipitates (Bolay, 1960, p. 190).

spontaneously after application of the stain solution so that crystal clusters are formed (Fig. 7).

γ. *False diffuse staining.* Occasionally diffuse staining will take place when stains are applied to cell saps with a high content of tannins. It can be hypothesized that the particle size of the reaction product of the stain and the vacuolar material is of such dimensions that a colloidal solution is possible (Kinzel and Bolay, 1961, p. 197).

b. Characteristics of Empty and Full Cell Sap. It is not too difficult to distinguish between these two types of vacuolar contents. Precipitates that occur within the vacuole after staining always indicate full cell saps. The true diffuse and the false diffuse staining of full cell sap also can be distinguished easily from the staining of the empty cell sap.

Empty cell saps generally show a color tone similar to that found in concentrated stain solutions. Neutral red stains dull brick red, while in full cell saps it stains violet red (see Fig. 8, top row, left), and brilliant cresyl blue and toluidine blue turn violet. Acridine orange fluoresces red (indicating its presence in the dimer form) in empty cell saps but fluoresces bright yellow green in full cell saps.

Similar color changes of a stain can be produced by varying the pH of aqueous solutions. For example, neutral red will change with increasing alkaline reaction from orange red to yellow and with increasing acidic reaction from red to violet and even to blue. Due to the similarity of these color tones, the color appearing in vitally stained cells was mistakenly considered, in earlier work, to indicate the pH of the cell sap. Recent investigations, however, showed that color change in cell saps is a concentration effect. When vital stains are present in rather high concentrations, metachromasy by molecular associations apparently occurs (cf. Drawert, 1968; Kinzel, 1958; Tsekos, 1971, p. 174). A highly concentrated solution of neutral red in a thin layer and an alkaline neutral red solution appear similar to the naked eye, yet their absorption spectra are different (cf. Bartels, 1956a, p. 78, 1956b).

An empty cell sap usually has a slightly acidic reaction, presumably due to the presence of organic acids. A cell sap containing anionic colloids such as pectins theoretically should accumulate basic stains with

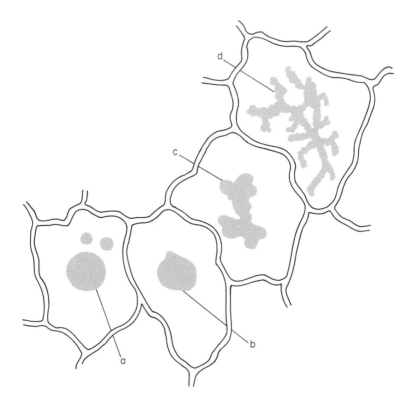

FIG. 5. Segregation forms of full cell sap in cells of the upper epidermis of *Cornus sanguinea:* (a) spherical segregates; (b) spherical segregates not completely rounded; (c) coalescence product; (d) dendrites. (From Kinzel and Bolay, 1961.)

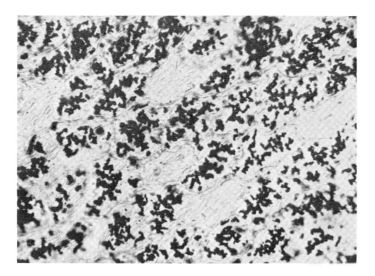

FIG. 6. Formation of dendrites. *Rumex* sp., upper epidermis of leaf. After staining with neutral red, 1:10,000 in tap water (Bolay, 1960, plate 5, Abb. 11).

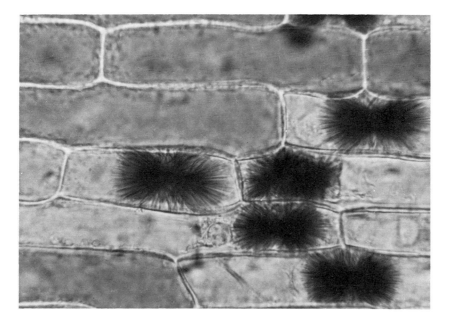

FIG. 7. Crystal clusters. *Polygonatum multiflorum* leaf, upper epidermis. Formation of crystal clusters in the vacuole after staining with rhodamine B, 1:50,000; after about 20 minutes (Bolay, 1960, plate 2, Abb. 4).

positive metachromic color change because of their local association on the surface of those colloids. However, only few cells have a pectin-rich cell sap (e.g., colloidal cell saps of subepidermal petal cells of Boraginaceae; Hofmeister, 1940; Härtel, 1953).

In full cell saps the different forms of segregation are important for purposes of identification (Kinzel and Bolay, 1961). Cell saps with *flavone glycosides* generally give a diffuse staining with basic dyes or give rise to droplets or crystals after segregation. Cell saps with *tannins* develop granular precipitates or a false diffuse staining. The appearance of dendrites usually indicates the presence of tannins or, probably, a mixture of tannins and flavones. The entire range of possible types of segregation, from true diffuse staining to spherical segregates, dendrites, and granular precipitates, all the way to false diffuse staining, might reflect the proportions of flavones and tannins present in the vacuole.

How other plant substances, for example, alkaloids, influence the colors resulting from vital staining has not yet been investigated. Alkaloids with phenol, hydroxyl, or keto groups might possibly react with a basic stain.

The *distinction between the three kinds of diffuse staining* (i.e., diffuse staining of empty cell saps, true and false diffuse staining of full cell saps) is sometimes difficult. False diffuse staining of full cell saps rarely occurs. Vacuoles that show false diffuse staining with basic vital stains are those from the tannin idioblasts in epidermal layers or parenchyma tissue of some Crassulacean (e.g., *Sedum maximum,* see Fig. 8, top row, right) and Papilionacean plants or in *Chrysanthemum, Anagallis,* and *Primula* (see Table 6, Bolay, 1960). Idioblasts are immediately distinguishable by their high color intensity. Probably the color tone varies, depending on the dispersity of the reaction product. With acridine orange, green and sometimes brownish fluorescence results.

To distinguish false diffuse staining from the two other types of diffuse staining, the tannin reaction with an unstained control section is useful. This reaction (1% aqueous solution of caffeine, van Wisselingh, 1915) occurs within a living cell. Caffeine is soluble in water and in lipids and therefore penetrates easily and quickly through the protoplasm into the cell sap, where it produces a precipitate in the presence of tannin. This precipitate consists of small particles exhibiting Brownian movement. True diffuse staining of full cell sap and diffuse staining of empty cell sap can be distinguished by the color tone (see p. 347).

An additional test to distinguish the three types of diffuse staining is performed by applying ammonia (concentration $0.01\,N$) to the stained material. This substance passes easily through living protoplasm into the vacuole (de Vries, 1871), the cell sap becoming alkaline. When the stain (e.g., neutral red) has a low pK and the cell sap is empty, the stain changes from its ionic form to uncharged molecules, and these gradually diffuse through the protoplasm into the external solution. Accordingly, the vacuole loses color. The basic stain molecule of neutral red has a low solubility in water and therefore may precipitate as crystals in the cells when its concentration is high. If ammonia acts on a full cell sap that exhibits true diffuse staining, droplets, dendrites, or granules can form. Cells exhibiting false diffuse staining of full cell sap show no change in their color and no precipitate after treatment with ammonia (Kinzel and Bolay, 1961). When ammonia is applied to unstained cells containing flavones, a yellow color will develop (Geissman, 1955, p. 466).

Intermediates between full and empty cell saps can be said to exist when only a very small amount of the substances that react with a vital stain is present. Such a cell sap first absorbs a stain by chemical reaction. Acridine orange gives green fluorescence that gradually changes with further penetration of the stain to yellow and finally to red. After saturation of the absorption sites of the phenolic material with the vital stain, the cell sap acts toward further penetrating stain as if it were an empty cell sap. Likewise, a few small granules can precipitate at the beginning of the staining process, but as time progresses, an additional diffuse staining may develop gradually with the metachromic colors typical of empty cell saps (Höfler, 1949).

Recently Pop and Soran (1962) observed in cells of the upper epidermis of the bulb scales of *Galanthus nivalis* segregation forms that resembled liquid crystals when stained with neutral red. Histochemical reactions indicate that lipoid substances that somehow are dispersed in vacuoles cause these segregation forms.

Vacuoles of a few cell species do not store basic vital stains, for example, cells in the growth zone of the leaves of *Elodea canadensis* (Kinzel and Pischinger, 1962) and *Chlorobotrys regularis* (von Cholnoky and Höfler, 1950; Höfler and Schindler, 1955). So as not to store basic vital stains, such cell saps must be free of stain-binding phenol derivatives, and, furthermore, the cell sap must have an almost neutral or even a slightly alkaline reaction. Also the buffering capacity of the cell sap must be low enough so that even the first penetrating traces of the basic stain molecules effect a pH shift of such magnitude that the ion trap mechanism cannot function. Such cell saps, therefore, cannot contain any significant amounts of free organic acids; among dissolved substances only sugars or neutral reacting salts can be assumed to be present. However, the composition of these cell saps has yet to be tested by chemical methods.

c. Selection of Vacuolar Vital Stains. Neutral red can be used for staining cell saps at pH 7 and higher. (Full cell saps stain well below pH 7 with neutral red.) Acridine orange and brilliant cresyl blue require somewhat higher pH values; toluidine blue and Thionine require still higher pH values. Nile blue can be used as a weak basic stain solution. However, the metachromasy of this dye is not too prominent. Rhodamine B, a red stain, ionizes only at extremely acid reaction (pH 1 to 2) and cannot be accumulated by empty cell saps unless the cell sap is extremely acid as in some *Begonia* species (Kinzel and Imb, 1961). Otherwise rhodamine B is taken up only by full cell saps and thus may help distinguish full from empty cell saps (Höfler and Schindler, 1955). Rhodamine B also may cause a different kind of segregation than that seen in other stains (Kinzel and Bolay, 1961, p. 186).

3. STAINING OF THE PROTOPLASM

All vital stains that penetrate through the protoplasm into the vacuoles must be present, of course, in the protoplasm at a certain minimal concentration. This concentration, however, is usually so low that the protoplasm itself does not appear to be stained. Only with fluorescence microscopy can traces of certain stains sometimes be seen in the protoplasm. Neutral red (not considered a fluorochrome) may cause a yellow fluorescence of the protoplasm in cells with empty cell saps. However, exposure to UV for too long a time may result in cellular alterations (Drawert and Metzner, 1956).

Acridine orange accumulates in the living protoplasm at pH 8 to 9, probably as a monomeric cation at polar lipid sites (Hölzl and Bancher, 1967), with a green fluorescence color easily visible in the swollen parts of the protoplast at the cell ends that appear during vacuolar contraction. The fluorescence of the nucleus is similar in color tone, but more intense, than that in the cytoplasm, so it can be identified easily. Dead protoplasm fluoresces red with acridine orange of the same pH. Strugger intended to use this drastic change in fluorescence color as a simple and general test for cell viability. He postulated that those cells in which protoplasm fluoresces green with acridine orange are alive, whereas those in which protoplasm fluoresces red are dead. This proved to be true for most bacteria and yeasts, but results with higher plant cells were more complex (cf. Strugger, 1947).

Stains that do not accumulate in the cell sap are more suitable for staining the protoplasm. The most harmless among the vital stains is rhodamine B, however, after prolonged application it may also cause toxic effects (cf. Drawert, 1968). Rhodamine B is highly soluble in lipids and ionizes only at a strongly acidic pH (below pH 3; see Fig. 8, bottom row, left). Rhodamine B will accumulate in the protoplasm of cells with empty cell sap. The red color of the dye weakens or disappears in the protoplasm because some of the stain changes into a leuco compound. However, the yellow fluorescence remains. The protoplasm can be recognized clearly with the fluorescence microscope, especially after plasmolysis.

Vital staining of the protoplasm with diachromes is also possible. Chrysoidine, for instance, will often stain only the protoplasm. However, in cells with full cell saps or very acidic cell saps, the vacuole will take up the vital stain. Vital stains of the triphenylmethane group (e.g., methyl violet) can also be used for staining protoplasm. However, these stains are very toxic and will damage the cells after a short time. All such basic vital stains that are lipid soluble under slightly acidic pH generally are suitable for plasma staining. The plasma of cells with full cell saps cannot be stained with basic vital stains.

The optimum concentration of the stains used for staining protoplasm is 0.01%, the same as for staining vacuoles. The concentration may be lower for those stains that are more toxic. Rhodamine B, chrysoidine, and methyl violet can be dissolved in distilled water, spring water, or nonchlorinated tap water. Neutral red should be buffered to pH 7 to 8 and acridine orange to pH 8 to 9.

B.   Acidic Vital Stains

Acidic dyes stain protoplasm rather than cell walls and vacuoles. In general, stains of the fluorescein group are used. Erythrosin (di- or tetraiodo fluorescein) is applied in a weakly acid solution and stains the protoplasm strongly enough so that it can be easily recognized. Erythrosin, however, is rather toxic (Strugger, 1949) Fluorescein itself is nontoxic but has a low solubility in water. It is most useful in the form of its water-soluble sodium salt (uranin) and yields optimal results in concentrations of 0.05 to 0.01% at pH values of 3 to 5.

When sections of living tissue are treated with uranin solution for 10 to 15 minutes, the protoplasm and cell nuclei fluoresce yellow green. The fact that the cell walls remain colorless can be demonstrated by plasmolyzing the cells (see Fig. 8, middle row, right). When vacuolar contraction or cap plasmolysis occurs, the swollen living protoplasm in the cell ends shows strong fluorescence. Dead protoplasm remains unstained. It will fluoresce only in extremely acidic uranin solution. The uranin-absorbing sites in the protoplasm are not yet known.

Uranin is not definitely bound in the protoplasm. When epidermis pieces of the *Allium* bulb scale are floated for 24 hours or longer on the surface of a uranin solution or after a staining period in tap water, the stain migrates from the protoplasm into the vacuole. This is thought to be related to the metabolic activity of the cell (Höfler *et al.*, 1956; Enöckl, 1960; Bolhar-Nordenkampf, 1966; Bancher *et al.*, 1968).

The dissociation of acidic stains, like basic stains, depends on the pH of the solution. For example, fluorescein is present at basic and neutral pH as a hydrophilic anion; with increased acidity it changes gradually into the more fat-soluble, undissociated acid molecule. Only the latter is able to penetrate living protoplasm. The pH range between 3 and 5 has proved optimal for stains of the fluorescein group. To prepare the solutions, the stain is dissolved in a 0.0067 M solution of KH_2PO_4 to give a final stain concentration of 0.05 to 0.01% and a pH of about 5. The pH may be adjusted by adding small quantities of HCl. The pH values of these solutions are unstable and change with time. Acetate and citrate buffers cannot be used (see Section III,B).

III.   General Rules for Vital Staining

Familiarity with living plant cells and with the procedures used are essential for correct interpretation of observations. A great number of erroneous conclusions, especially in biochemical papers, result from unfamiliarity with the structure and physiology of living cells. An excellent introduction in vital staining is found in "Praktikum der Zell- und Gewebephysiologie der Pflanze" ("Laboratory Manual of Cell and Tissue Physiology of the Plant") by S. Strugger (1949). However, some sections of the chapters on vital staining are obsolete. More up-to-date and indepth information is presented in the monograph on vital staining by Drawert (1968) and the survey articles of Kinzel (1962, 1968a). A few general suggestions regarding methods are given in the following paragraphs, and some examples of vitally stained plant cells are shown in Fig. 8.

A.   Selection and Preparation of the Material

Experimentation with freshwater algae is simplest because such algae do not require any preparation. For instance, *Spirogyra* (tannin-containing cell sap) and *Oedogonium* (empty cell sap), as well as various Desmidiaceae, are suitable. Among the Anthophyta the adaxial (inner, morphologically upper) epidermis of the bulb scale of *Allium cepa* is easily prepared and therefore often used. The onion is freed of dry outer scales according to the directions given by Strugger (1949) and is divided into four sections by two medium cuts with a sharp knife. The tip and the base are cut off so that the scales can be separated. On the inner (adaxial) surface of a fully vital scale (generally the third inward counting from the outermost), several longitudinal and transversal incisions are made with a razor blade, so that the epidermis is cut into rectangular segments of about 5 by 10 mm lengths. Then the scale is placed in spring water[4] and infiltrated by using an aspirator for 5 to 10 minutes. By this procedure the intercellular air is replaced by spring water. In addition, this treatment facilitates the separation of the epidermal layer from the parenchyma underneath by the gentle action of the intercellular air so that the epidermal section can be easily removed with a forceps. The forceps must be applied only at a corner of the section; damage to the cells in the other portion of the epidermal section is definitely avoided. The epidermis should not be bent when it is removed from the tissue underneath. The action of the forceps should be only in the direction away from the scale. The free epidermal sections are transferred immediately into a petri dish containing spring water. Epidermal sections that curl after separation should be discarded (In some onions all epidermal sections curl; such bulbs are not suitable as a source of experimental material.) In the resting onion the cells of the inner epidermis contain empty cell sap. The growing onion in which the scales have begun to separate from each other may have cells with full cell sap in the adaxial epidermis. This change is related to the access of oxygen to these cells. The same change may be introduced intentionally in epidermal sections of resting onions when these sections, as prepared above, are permitted to remain in contact with air in a moist atmosphere for 3 to 5 days (Bancher and Hölzl, 1960b).

The abaxial (outer, morphologically lower) epidermal cells of the onion scale contain full cell sap even in resting onions. This is due to the presence of different flavone glycosides. The abaxial epidermis cannot be removed with a tweezer; cuttings must be prepared in the usual way with a razor blade. The sections must not be too thick (because microscopic observation becomes difficult) or too thin (because too many cells are damaged that react physiologically in an atypical manner).

Epidermal sections can be prepared in the same way from many other plants. Normally, epidermal cells have full cell saps. Empty cell saps are found occasionally in parenchyma and vascular tissue; these sections should be prepared with a microtome, preferably a vibrating one. Vascular tissue must be cut in longitudinal sections, but parenchyma tissue with isodiametric cells can be cut in any direction.

To distinguish living cells from dead cells in higher plants is not difficult, after some experience. Living cells normally contain a crystal clear, ofttimes barely visible groundplasm and a transparent, rather in-

[4] Spring water must be used for its ion content (especially Ca^{2+}); distilled water is harmful to the living protoplasm.

distinct nucleus. Healthy chloroplasts of flowering plants are round and smooth, provided crowding has not altered their shape. If the protoplasm of a cell appears to be coagulated and if the nucleus is quite distinct, one can assume with a high degree of probability that the cell is dead. A sure indication of life is protoplasmic streaming. In cells in which protoplasmic streaming, cannot be seen, plasmolysis in a glucose or KNO_3 solution of suitable concentration (0.5 to 1.0 M, depending on the cell type) may serve as a test for viability (see also Strugger, 1949; Bancher and Höfler, 1959). An osmotic contraction of the tonoplast should never be confused with plasmolysis. The former may occur occasionally in dead or dying cells during plasmolysis. Such contraction of the intact tonoplast often can be identified by the presence of the remains of coagulated protoplasm that may be attached outside the perfectly round tonoplast. Also, it should be noted that substances may

---

FIG. 8. Vitally stained plant cells. *Top row:* Left, different color tone for full (left half of section) and empty (right half of section) cell saps after staining with a basic vital stain. *Allium cepa,* abaxial (outer) epidermis of the bulb scale with intact parenchyma cells underneath the epidermis (right half of section). Stained with neutral red at pH 7.8. The full cell sap of the epidermal cells is stained violet red. The empty cell sap of the parenchyma cells stains a dull brick red (130✕, from Kinzel, 1958, p. 28, Abb. 17). The violet component of the color of the left half of the section is generally more pronounced than shown here. Right, elective vital staining of tannin idioblasts. *Sedum maximum,* leaf upper epidermis. Stained with methyl red (1:10,000; pH 6.3). False diffuse staining (72✕; original).

Middle row: Left, vital staining with an amphoteric vital stain. *Plagiochila asplenioides* (leaf). Stained for several hours with methyl red (amphoteric stain) at pH 5 to 6 and plasmolyzed in 1.0 M glucose. The lipid bodies stain red. The cell wall also adsorbs some stain and assumes a violet color tone. The amphoteric character of the stain and the presence of tannin in the cell walls are the cause for the cell wall staining (240✕; from Zöttl, 1960, p. 472, Fig. 5). Right, vital staining of the protoplasm with an acid fluorochrome. *Allium cepa,* adaxial (inner) epidermis. Stained with uranin (1:10,000) at pH 4.8. Plasmolyzed in 0.8 M glucose. The living protoplasm and especially the nuclei fluoresces with intense green color. The cell wall does not accumulate the stain. The longitudinal walls appear as thin dark lines between the plasmolyzed protoplasts (430✕; original).

Bottom row: Left, vital staining with an amphoteric vital stain. *Allium cepa,* abaxial (outer) epidermis of the bulb scale. Stained with rhodamine B (1:10,000). Plasmolyzed in 1.0 M glucose. The full cell sap accumulates the stain to such a degree that aggregates develop inside the cell sap (170✕; original). Right, vital staining showing positive and negative metachromasy. *Allium cepa,* abaxial (outer) epidermis of the bulb scale. Stained with toluidine blue (1:10,000). The full cell sap of the epidermal cells shows the green blue negative metachromic color tone (left half of section; the green component of the color is generally stronger than shown here), while the empty cell sap of the parenchyma cells underneath (right half of section) exhibits the violet color tone of the positive metachromasy. The spherical segregates in some of the epidermal cells are also indicators for "full" cell saps (200✕; from Kinzel, 1958, p. 28, Fig. 18).

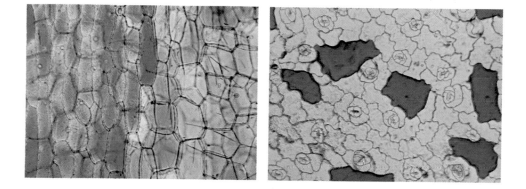

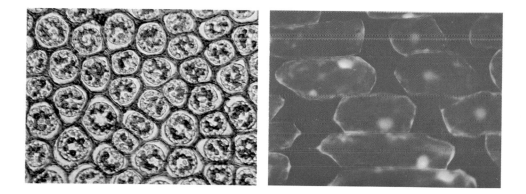

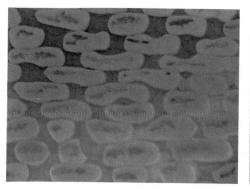
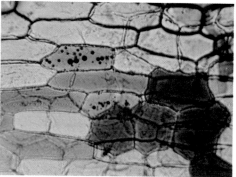

accumulate within the tonoplast during the process of cell death and give the impression of a full cell sap (Burian, 1962a,b).

B.   Preparation of Vital Stain Solutions

As a rule, all stains are dissolved in distilled water to a concentration of 0.01%. A 0.1% stock solution can be maintained for some stains for several months when stored in a cool dark room. When needed, such solutions are diluted tenfold with either distilled water, spring water (or non-chlorinated tap water), or buffer solution (see Appendix C), depending on the particular type of experiment. Alkaline solutions of basic stains do not keep very well. For instance, the stain base of neutral red is only slightly soluble and will eventually precipitate out of solution as microcrystals. To delay the development of this precipitate (which ordinarily does not occur sooner than 20 minutes after mixing the buffered solution) it is recommended that the buffer is diluted before stock solution is added, otherwise precipitation occurs almost immediately. The formation of crystals is delayed in a 0.01% solution. Some stain solutions must be prepared fresh daily or even several times during a day.

To provide a sufficiently large fraction of lipid-soluble stain molecules, solutions of basic vital stains must have a more or less strong alkaline reaction. Neutral red, which has a pK of 7.38 (Bartels, 1956a,b), may be diluted with spring water in a ratio of 1:10, provided the spring water is calcareous and, therefore, to a certain extent, alkaline. Strong basic solutions are required for acridine orange (pH 8 to 9), brilliant cresyl blue (pH 8 to 9), and toluidine blue (pH 10 to 11). These pH values can be best produced by phosphate buffers. Strong alkaline solutions are harmful to the living cell; even exposure to pH 10 to 11 will damage a cell after a period of time. Rhodamine B, which is present in lipid-soluble form in the entire physiological pH range, does not need to be buffered. A concentration of 0.02 to 0.5% is recommended.

Commonly used buffers, especially those that contain organic acids (e.g., acetate, citrate, and phthalate buffer), are very harmful to living cells, especially in the acidic range, because the toxic undissociated molecules of these acids penetrate easily into the protoplasm. Phosphate buffers are harmless and have their greatest buffer capacity between pH 6 and 8; but in physiological work it is often necessary to use phosphate buffers at other pH values. Buffering capacity is practically nil below pH 4.8 (pure primary phosphate) and over pH 9 (pure secondary phosphate).

Four stock solutions are needed for the preparation of buffered vital stain solutions, and the amounts to be mixed for a desired pH are given in Appendix C (see also Strugger, 1949; Kinzel, 1954a). For some of the most frequently used vital stains only two stock solutions are needed.

Vital stain	Parts of 1/15 M solution of		Resulting pH
	KH$_2$PO$_4$	Na$_3$PO$_4$	
Neutral red	6	4	7 to 8
Acridine orange	5	5	8 to 9
Brilliant cresyl blue	5	5	8 to 9
Toluidine blue	4	6	10 to 11
	or 3	7	

To 80 ml distilled H$_2$O and 10 ml of the buffer, 10 ml of the 0.1% vital stain stock solution are added. The final solution contains the vital stain in a concentration of 0.01% (1:10,000) and the buffer in the concentration of $M/150$ (0.0067 M).

C.   Practical Aspects of the Vital Staining Procedures

Sections are always transferred to a large dish (so that ample stain solution is provided) and remain in the stain solution for 10 to 30 minutes, with gentle occasional agitation. The optimal time of staining must be determined by testing each material and may be as long as 1 hour or more. Overstaining must be avoided. After staining is terminated, the sections are washed in a buffer solution of the same pH and buffer concentration as the staining solution (but, of course, without stain). Finally, the sections are mounted on a slide in a sufficiently large droplet of the buffer solution, covered with a cover slip, and observed under the microscope.

When basic stains are applied, it is possible to confuse staining of the cell wall with staining of the interior of a cell. The following points may help indicate where the stain is absorbed. (1) Staining of the cell wall with a basic stain usually occurs when the stain solution used is slightly acidic (pH 3 to 6), or when the cell is overstained. (2) The intensity of the cell wall staining decreases with increasing alkalinity of the vital stain solution. (3) Stained cell walls can be recognized by the intensive coloration of the vertical cell walls. Light has a longer path through the vertical walls—the color is intensified. (4) When plasmolyzed, the protoplast separates from a stained cell wall and thus can be recognized to be unstained. (5) In non-lignified cell walls the stain can be easily extracted from the wall by application of a salt solution (about 0.2 M CaCl$_2$; cf. Höfler and Stiegler, 1947; Kinzel, 1955).

There is very little risk that stained *protoplasm* will be confused with stained *cell sap*. These two types of coloration can be distinguished easily after plasmolysis. In the case of plasma staining, the intensity of coloration is greatest at the periphery of the plasmolyzed protoplast and decreases toward the center; the opposite relationship is observed when the vacuoles are stained.

IV.   Examples for Vital Staining Experiments

A.   Accumulation of Vital Stains in the Plant
(Material: *Elodea* sp.)

About 1 liter each of a solution 1:50,000 (0.002%) methylene blue and a solution 1:20,000 (0.005%) neutral red in spring water are prepared. Two square battery jars or beakers (1 liter in volume) are filled with the methylene blue solution and two others with the neutral red solution. Several shoots of *Elodea* are placed into one jar of neutral red and one jar of methylene blue. White paper is fixed on the back side of the jars to make changes in color intensity more easily visible. The color intensity in the jars containing *Elodea* is compared with the control jars after 4, 6, and 24 hours and after 2, 3, 4, 5, and 6 days. Accumulation of stain in the leaf cells is observed microscopically in leaves detached from the *Elodea* plant at these time intervals. The leaf is transferred to a plasmolyzing solution (e.g., 0.5 M glucose) to test viability of the cells.

After 4 days one shoot of *Elodea* from each stain solution is transferred into a large dish with spring water to determine whether any vital stain is released.

RESULTS

Neutral red will be taken up by healthy plant material within a few hours, while methylene blue uptake may require a longer time. After 3 to 4 days the solutions will be almost completely discolored. Vital stains accumulate in the leaves only in cells with terminated growth (cf. Kinzel and Pischinger, 1962). The type of accumulation inside the cell vacuole changes during uptake. The vacuoles of these leaf cells contain full cell sap. Neutral red is first accumulated as droplets, later perhaps also as a coalescence product. Methylene blue accumulates in vacuoles in diffuse staining or in the form of small crystals. Healthy leaf cells do not discolor after transfer of the stem from the stain solutions into water.

B.   Capillary Analysis of Vital Stain Solutions

The method of capillary analysis developed by Goppelsroeder (1888) is a forerunner of chromatography. A strip of filter paper is placed vertically in a solution of a stain or pigment or other material to be tested. As the liquid rises in the paper strip by capillary action, the substances in the solution gradually separate on the paper, forming a series of bands analogous to the bands observed in column or paper chromatography (Strain, 1942, p. 7).

In the simplified test applied here, a round piece of filter paper is positioned horizontally. A droplet of 1:1000 neutral red solution (made up with distilled water) is placed in the center of the filter paper and the difference (if any) in speed of movement of water and the stain to the periphery is recorded. Readings are taken 2 or 3 seconds after the droplet comes into contact with the filter paper.

The same experiment is performed with a 1:1000 solution of the acidic stain eosin (in distilled water). The relative rates of movement of the fronts of the water and the stain should be observed and recorded.

RESULTS

Since neutral red is dissociated to some degree at the pH of distilled water, its stain cations will be adsorbed at the negatively charged sites of the fibers in the filter paper. Therefore, the spreading of the stain will be retarded with regard to the spreading of the water and the edge of the spreading droplet on the filter paper will be pure water, with neutral red following a few millimeters behind during spreading.

With eosin no electroadsorption will take place. Thus, eosin will advance on the filter paper as fast as the water. However, after a few seconds, a small difference in the rate of spreading between water and eosin will be observed, the water being slightly in advance of the eosin. This is probably caused by the higher diffusion resistance of the large stain molecules as compared to the water molecules.

C.   The Relationship between Lipid and Water Solubility of a Basic Vital Stain at Different pH's

Two milliliters of neutral red solution (1:1000 in distilled water), 18 ml of distilled water, and 2 ml of an approximately 1 N HCl solution are placed into a 25-ml graduated cylinder and mixed well.

About 5 ml of the above mixture is transferred into a test tube and an equal amount of benzene is added. The test tube is shaken vigorously for 15 seconds and then left standing to separate for 5 to 10 minutes.

The same procedure is repeated but instead of 2 ml HCl (a) 2 ml 1 M KOH and (b) 2 ml of distilled water are used. The distribution of neutral red between the water phase and lipid phase (benzene) is recorded.

RESULTS

In the acidic solution almost all the neutral red remains in the water phase (colored red) and the benzene phase shows slight or no color. However, in the test tube with the basic aqueous solution (yellow in color) almost all of the neutral red will be found in the lipid phase. The test tube containing distilled water will have an intermediate degree of stain distribution; some neutral red will be in the aqueous phase, while some will be in the lipid phase.

D.   Differential Vital Staining of Cell Components with Variation of the pH of the Stain Solution (Material: *Allium cepa*—Resting or Dormant)

Ten milliliters of a buffer mixture pH 4.6 to 4.9 (see Appendix C) are diluted with 80 ml distilled water and added to 10 ml of a 1:1000 neutral red solution. The same procedure is followed with a buffer mixture for pH 8.0 to 8.3. The pH values of these stain solutions must be checked with a pH meter. Two rinsing solutions of the same pH [distilled water (10 ml) is used instead of the stain] are also made.

Sections of the inner onion epidermis are prepared as described on p. 354. Two staining dishes are filled, each with one of the two neutral red solutions. Two more dishes are readied with the corresponding rinsing solutions. A few of the epidermal sections are transferred into each stain solution (they should float on the surface of the solution). After 10 to 20 minutes, the epidermal sections are transferred into the dish with the rinsing solution of the same pH, left there for 30 seconds, agitated gently and finally transferred to a droplet of rinsing solution on a slide, covered with a cover slip, and observed under the microscope. Because prolonged enclosure of the epidermis by the cover slip should be avoided, observation should proceed without delay. To distinguish clearly what element of the cell is stained, the epidermal sections may be plasmolyzed with 1.0 M saccharose (or another solution of similar osmotic strength) so that the plasma envelope is separated from the cell wall.

RESULTS

The neutral red will stain only the cell wall at acidic pH. In the basic solution only the vacuole is stained, and vacuolar contraction may develop in some cells. Often cells near the edge of the epidermal section are more heavily stained than the more central cells. This may indicate that some alterations occurred during the preparation of the epidermal sections or the more intense staining may be caused by the better access of the stain to the edges of the epidermal sections. The color of the stained cell wall and the vacuole is a dull brick red.

Transfer of the epidermal section from acidic pH to a buffer solution of basic pH causes a change of the staining site from the cell wall to the cell sap vacuole and vice versa.

When an epidermal section with a stained vacuole is placed in a solution of 0.01 N NH$_4$OH, the vacuole soon loses its color, indicating that, in this instance, only an ion trap mechanism is operating (see p. 344).

E.   Vital Staining of "Full" and "Empty" Cell Saps. Destaining of the Cell Wall (Material: *Allium cepa*)[5]

Several sections of the abaxial (outer) epidermis with at least two intact layers of parenchyma cells in the center part of these sections are made from the third fleshy bulb scale. (The best conditions for observation would be achieved with sections containing only parenchyma tissue in the first half and parenchyma and epidermis in the second half.)

Two buffered neutral red solutions (1:10,000, at pH 4.8 and at pH 8.0 to 8.3) and their corresponding rinsing solutions are prepared (see p. 361). Three to four of the epidermis sections are placed into each of the two neutral red solutions for 10 to 20 minutes, rinsed, and examined in a drop of buffer under the microscope.

Two of the epidermis pieces that were observed at pH 4.8 are placed in a CaCl$_2$ solution (0.2 M).

RESULTS

The epidermal cells of the abaxial side of the scale have full cell sap and stain violet in the basic stain solution; staining may be especially strong in the cells near the edge of the section, since the vital stain has better access there than in the thick part of the center of the section. Since the epidermis is coated with a cuticle, little or no access of stain through the upper surface of the epidermis is possible. The "empty"

[5] See footnote 3, p. 338.

parenchyma cells exhibit a dull brick red color (see Fig. 8, top row, left).

However, when acidic pH is used, the walls of the parenchyma cells and of the epidermal cells will stain. The staining of the full cell sap in acidic stain solution depends upon the concentration of the undissociated molecules which are able to penetrate the protoplasm only. At a pH near 7 there is still a large enough fraction of the stain undissociated so that the full cell sap will accumulate stain.

Sections stained in acidic pH, and placed in the $CaCl_2$ solution exhibit destained cell walls. The stain cations that were bound by electroadsorption to the negatively charged sites of the cell wall are replaced by Ca^{2+} ions, which have a greater affinity for these sites.

F. Vital Staining with Toluidine Blue
(Material: *Allium cepa*)

A toluidine blue solution (1:10,000, pH 11) and a rinsing solution with the same pH are prepared.

Sections of the abaxial epidermis of the onion scale are placed into the stain solution for 20 minutes. Thereafter the epidermal sections are transferred to a droplet of rinsing solution on a slide for microscopic observation.

RESULTS

The full cell sap of the epidermis stains green blue, while the empty cell sap of the parenchyma is violet. In some of the epidermal cells there will be a precipitation of the stain in the form of droplets of an intense blue color (see Fig. 8, bottom row, right).

G. Elective Vital Staining of Guard Cells
(Material: *Zebrina pendula*)

Several thin tangential sections of the lower side of the leaf are prepared and kept floating on spring water.

The neutral red solutions (1:10,000, buffered at pH 4.6 to 4.9 and pH 8.0 to 8.3) with the corresponding rinsing solutions are prepared.

The epidermal sections are placed in each of the two solutions in a staining dish, rinsed after about 20 minutes of staining, and transferred to a droplet of the corresponding rinsing solution on a slide for microscopic observation.

RESULTS

At acidic pH the stain accumulates, most commonly in the form of droplets, in the full cell sap of the guard cells.

Frequently subsidiary cells have a full cell sap. Here often granular precipitates result and show intensive Brownian movement. The other epidermal cells are empty, and their cell sap does not take up the stain at acidic pH values. The cell walls are stained brownish-red to orange, as usual. They are destained after placing the sections for 20 minutes in a 0.2 M CaCl$_2$ solution (see Section E).

In the basic neutral red solution the vacuoles of epidermal cells stain dull brick red. The staining of the vacuoles of guard cells is the same in basic as in acidic pH.

H. Destaining of Cells with "Empty" Cell Sap after Experimental Change of Its pH (Material: *Allium cepa*)[6]

Several sections of the abaxial (outer) epidermis are made as described above (p. 362). A neutral red solution 1:10,000 at pH 8.0 to 8.5, the corresponding rinsing solution, and a 0.01 M NH$_4$OH solution are prepared. Three epidermal sections are placed for 20 minutes in the neutral red solution, rinsed, and observed in a droplet of buffer solution under the microscope.

Next the cuttings are placed into a dish with 0.01 N ammonia solution and observed after 5 to 10 minutes.

Results

The abaxial epidermal sections show the usual type of staining in basic neutral red. Full cell saps of the epidermal cells are stained violet, but often only close to the edge where the stain has easy access. Parenchyma cells have an empty cell sap and a dull brick red color.

After treatment in 0.01 N ammonia solution, the parenchyma cells become completely destained. The color of the vacuole of the epidermal cells changes slightly because of a shift in pH. Often the epidermal cells of the central part of the section (which were unstained before the transfer of the section into the ammonia solution due to the inability of the vital stain to gain access) now become stained the color of the full cell sap at that pH. Obviously the stain that was released from the empty cell sap of the parenchyma cells was accumulated and absorbed by the full cell sap of the epidermal cells.

[6] See footnote 3, p. 338.

I. Vital Staining with Rhodamine B
(Material: *Allium cepa*)

A solution of rhodamine B (1:10,000) in distilled water is prepared. No buffer is needed.

Sections of the adaxial (inner) and the abaxial (outer) epidermis of the third bulb scale are prepared (see p. 354) and floated in a staining dish for 10 minutes on the staining solution. Next, the epidermal sections are transferred into distilled water for rinsing and then are placed in a droplet of distilled water on a slide. Visible light and UV (fluorescence) are used for observation.

The abaxial epidermal sections are treated similarly.

Plasmolysis (transfer into a droplet of 1 M mannitol on the slide) may be produced in order to better distinguish cell wall, protoplasm, and vacuole.

RESULTS

The protoplasm of the adaxial epidermis is stained little or not at all in visible light (see p. 352). When observed with UV the protoplasm shows a yellowish fluorescence.

The full cell sap of the abaxial epidermis accumulates rhodamine B with a strong red fluorescence, and any fluorescence in the protoplasm can hardly be recognized (see Fig. 8, bottom row, left). This experiment illustrates the competition for the stain between full cell sap and the protoplasm. The phenolic compounds in the full cell sap have greater affinity for the stain than has the protoplasm. Only in a cell with empty cell sap (which contains no such substances of great affinity) will the stain accumulate in the protoplasm.

J. Vital Staining of the Protoplasm with a Diachrome
(Material: *Allium cepa*)

A 1:10,000 chrysoidine solution with a pH 7.5 to 7.6 and a rinsing solution of the same pH and adaxial (inner) epidermal cuttings (see p. 354) are prepared. The sections are placed in the stain solution for about 20 minutes and transferred to a dish with the rinsing solution for 30 seconds. Next the material is placed in a droplet of the rinsing solution on a slide for microscopic observation. To recognize the stained cell parts clearly, plasmolysis may be elicited.

RESULTS

Chrysoidine stains in cells with empty cell saps only the protoplasm with yellow color. Most probably the monovalent ions and the base molecule are accumulated in microvolumes of the cytoplasm containing polar lipids (Tsekos and Kristen, 1969, p. 586; Kristen, 1970, p. 446). In cells with full cell sap the stain is accumulated in the cell sap [competition for the stain (see Section I) results].

K.   Vital Staining with a Basic Fluorochrome (Material: *Allium cepa*)

Two acridine orange solutions (1:5000 or 1:10,000, pH 3.8 and pH 8.5) with corresponding rinsing solutions of the same pH values are prepared.

Adaxial (inner) epidermal sections from the bulb scale (see p. 354) and sections from the abaxial epidermis are prepared and floated on each of the two acridine orange solutions for 10 to 20 minutes.

The sections are then washed thoroughly in three consecutive dishes of the rinsing solution (well agitated) to eliminate any traces of adhering stain solution and mounted on a slide in a droplet of the respective rinsing solution for observation by fluorescence microscopy.

Three to four sections of the adaxial epidermis stained for 10 minutes at pH 3.8 are placed into a solution of 0.2 or 0.5 M CaCl$_2$ in distilled water and observed after about 20 minutes.

RESULTS

At pH 3.8 the cell walls of the adaxial epidermis are stained red. The protoplast does not absorb the vital stain at this pH, and the vacuole is not stained because the cell wall competes successfully for the stain. In the abaxial epidermis the cell sap fluoresces green (the characteristic color for full cell saps). The cell walls of the abaxial cells and the intact parenchyma cells underneath are stained as in the adaxial epidermis. The protoplasm is unstained.

At pH 8.5 in the adaxial epidermis the cell wall is unstained. Vacuolar contractions often occur, and the protoplasm fluoresces green. The vacuole accumulates the stain with a red fluorescence. The nucleus can always be seen clearly, since it shows a bright green fluorescence. In plasmolyzed cells (1.0 M mannitol) the green fluorescence of the protoplasm is much more visible, especially at places where the swollen protoplasm is detached from the cell wall. This can be observed better at a higher magnification (objective 40 or 63 X dry).

At pH 8.5 in abaxial epidermis the cell sap of the epidermal cells shows a green fluorescence, and the empty sap of the parenchyma cells a red fluorescence. There is little or no staining of the cell wall.

Following treatment with $CaCl_2$ the cell walls are destained because acridine orange is replaced by Ca^{2+} on the active sites of the cell wall. Occasionally, however, the cells are plasmolyzed and the protoplasm shows a green fluorescence. Probably the addition of the Ca^{2+} causes a transfer of the acridine orange from the cell wall to the protoplasm. It is not certain whether this is merely an effect of a change in pH by transfer of the epidermis into the more basic $CaCl_2$ solution or whether this is an effect of the calcium ion itself (see also Kinzel, 1953, p. 216).

L. Application of an Acidic Fluorochrome (Material: *Allium cepa*)

A solution of uranin (1:10,000) buffered at pH 4.8 and the corresponding rinsing solution are prepared. Several sections of the adaxial epidermis are floated on the stain solution for 10 minutes.

After careful rinsing of an epidermal section, it is mounted in a droplet of the rinsing solution on a slide. At this time protoplasmic streaming can still be observed.

To better recognize the protoplasmic envelope, the cells are plasmolyzed (0.8 M mannitol).

A droplet of a 1% $HgCl_2$ solution (which is a strong poison) is placed on the right edge of the cover slip on the slide. It is perfused under the cover slip by a strip of filter paper applied at the left edge of the cover slip. In this way the mannitol is gradually replaced by $HgCl_2$ solution.

Other stained epidermal sections are floated for about 20 hours in a dish on spring water (or nonchlorinated tap water). The dish is covered to avoid evaporation.

RESULTS

After the 10-minute staining period, the protoplasm shows an intensive green fluorescence. Often the protoplasm is slightly swollen. Protoplasmic streaming can be observed occasionally. Plasmolysis clearly proves that the protoplasm is the site of the fluorochrome accumulation, the cell walls remaining completely dark (see Fig. 8, middle row, right). Where the cells are dead, there is no fluorescence of the protoplasm.

After the 1% $HgCl_2$ solution is applied, cell death can be followed easily, beginning at the side where the solution of $HgCl_2$ has first access. Gradually the protoplasm destains, first at the longitudinal walls of the cells and later in the swollen protoplasm of the cell ends, until finally the whole cell appears dark. The whole process may take 3 to 5 minutes for one cell.

When an epidermal section, which is stained with uranin, is floated for about 20 hours on spring water, the stain moves from the protoplasm into the vacuole, and vacuolar contractions disappear. This may need metabolic energy but the mechanism for this transport process is not known yet (see p. 353).

ACKNOWLEDGMENTS

The authors wish to thank Professor Dr. George J. Fritz, University of Florida, Gainesville, Florida and Mrs. Virginia Pedeliski, University of Vienna, Austria for suggested revisions in English and style.

Appendix A.   Characteristics of Some Vital Stains

Stain	Acridine orange	Neutral red	Brilliant cresyl blue	Toluidine blue	Chrysiodine	Rhodamine B
Orthochromatic color tone (in diluted aqueous solution)	Green fluorescence	Carmine red	Blue	Blue	Yellow	Carmine red
Positive metachromic color tone (in concentrated aqueous solution or in empty cell saps)	Red fluorescence	Dull brick red	Violet	Violet	—	—
Negative metachromic color tone (in full cell saps)	Green fluorescence	Violet red	Blue-green	Blue-green	—	—
Staining of non-lignified cell walls at pH[a]	3–7	3–6.5	3–7	3–8	—	—
Staining of empty cell saps occurs at pH[a]	>8	>7	>8	>9	Only if the cell sap is extremely acidic	
Staining of full cell saps at pH[a]	>6	>5	>6	>7	>4	>3
Staining of the protoplasm (only in cells with empty cell saps) at pH[a]	(>8)	(>7)	—	—	>4	>3

[a] The pH values involved refer to the stain solution and not to the acidity of the cell sap.

Appendix B. Possibilities for Diagnosis of the Cell Sap Content by Vital Staining

Observed phenomena	Possible conclusions
A. Basic vital stains do not accumulate in the cell sap at any pH of the stain solution	Test if cell is alive. If so, most likely the cell sap contains only sugars and related compounds and neutral salts
B. Basic vital stains accumulate in the cell sap diffusely with positive metachromic color tone (Appendix A) when the pH of the external solution is above a threshold value characteristic for a given cell type	In addition to the above, the cell sap contains free organic acids or anionic colloids (empty cell saps)
C. Basic vital stains accumulate in the cell sap with negative metachromic color tone (a) diffusely or (b) with precipitate formation in different forms of segregation or submicroscopically (false diffusion staining)	The cell sap contains phenolic compounds (full cell saps)
(1) True diffuse staining or aggregates or crystal formations occur	Probably flavone glycosides and related compounds predominate in the cell sap
(2) Granular or submicroscopic precipitation of the stain occurs	Probably tannin-like compounds predominate in the cell sap

Appendix C.   Preparation of Buffer Solutions[a]

pH range	Quantity of solution[b,c]			
	I	II	III	IV
2.0–2.2	9.5	0.5	—	—
3.4–3.9	0.5	9.5	—	—
4.6–4.9	—	10.0	—	—
5.6–5.7	—	9.5	0.5	—
5.8–6.1	—	9.0	1.0	—
6.3–6.5	—	8.0	2.0	—
7.0–7.1	—	5.0	5.0	—
7.5–7.6	—	2.0	8.0	—
8.0–8.3	—	4.5	—	5.5
9.8–10.1	—	5.0	—	5.0
10.7–10.8	—	3.0	—	7.0
11.2–11.3	—	—	3.0	7.0

[a] After Strugger, 1949, p. 134; Kinzel, 1954a.

[b] Stock solutions: I $N/10$ $(0.1\ N)$ HCl; II $1/15\ M$ $(0.067\ M)$ primary potassium phosphate KH_2PO_4; III $1/15\ M$ $(0.067\ M)$ secondary sodium phosphate Na_2HPO_4; IV $1/15\ M$ $(0.067\ M)$ tertiary potassium or sodium phosphate K_3PO_4 or Na_3PO_4. This gives a total of 10 ml buffer mixture (example: line 1: 9.5 ml + 0.5 ml).

[c] For the preparation of the *staining solution* 80 ml of distilled water are mixed well with the buffer mixture to obtain the desired pH. Next 10 ml of the stain stock solution (1:1000) are added. To prepare the *rinsing solution* 90 ml of distilled water are added to the buffer mixture and the stain solution is omitted. The alkaline pH ranges may be to some extent inaccurate, since Kinzel (1954a) using carefully prepared stock solutions, obtained different pH values for the buffer mixtures.

REFERENCES

Anonymous (1709). *Acad. Roy. Sci. Paris, Histoire,* pp. 44–49.

Bancher, E., and Höfler, K. (1959). *In* "Grundlagen der allgemeinen Vitalchemie in Einzeldarstellungen" (H. Linser, ed.), Vol. VI, pp. 1–184. Urban und Schwarzenberg, Wien-Innsbruck.

Bancher, E., and Hölzl, J. (1960a). *Oesterr. Bot. Z.* **107,** 18–38.

Bancher, E., and Hölzl, J. (1960b). *Flora (Jena)* **149,** 396–425.

Bancher, E., and Hölzl, J. (1963). *Protoplasma* **57,** 33–50.

Bancher, E., Hölzl, J., and Schiffbauer, R. (1968). *Protoplasma* **66,** 327–337.

Bartels, P. (1956a). *Z. Phys. Chem.* [N. F.] **9,** 74–94.

Bartels, P. (1956b). *Z. Phys. Chem.* [N. F.] **9,** 95–105.

Bock, U. (1964). *Flora (Jena)* **154,** 99–135.

Bolay, E. (1960). *Oesterr. Akad. Wiss. Math.-Naturw. Kl. Sitzber. Abt. I* **169,** 269–317.

Bolhar-Nordenkampf, H. (1966). *Protoplasma* **62,** 133–156.

Bukatsch, Fr., and Haitinger, M. (1940). *Protoplasma* **34,** 515–523.

Burian, K. (1962a). *Protoplasma* **55**, 156–176.

Burian, K. (1962b). *Protoplasma* **55**, 607–631.

Butterfass, T. (1960). *Ber. wiss. Biol.* **147**(1), 3.

Cappell, D. F. (1929). *J. Pathol. Bacteriol.* **32**, 595–707.

Cholnoky, B. von, and Höfler, K. (1950). *Oesterr. Akad. Wiss. Math.-Naturw. Kl. Sitzber. Abt. I* **159**, 143–182.

deVries, H. (1871). *Arch. Néerl. Sci. Exactes Nat.* **6**, 117–126.

Doan, C. A., and Ralph, P. H. (1950). *In* "McClung's Handbook of Microscopical Technique" (R. M. Jones, ed.), pp. 571–585. Hoeber, New York.

Drawert, H. (1937). *Flora* (*Jena*) **132**, 91–124.

Drawert, H. (1968). *In* "Protoplasmatologia" (L. V. Heilbrunn and Fr. Weber, eds.), Vol. II/D/3, pp. 1–749. Springer, Wien.

Drawert, H., and Endlich, B. (1956). *Protoplasma* **46**, 170–183.

Drawert, H., and Metzner, I. (1956). *Protoplasma* **47**, 359–383.

Enöckl, F. (1960). *Protoplasma* **52**, 344–375.

Foot, N. C. (1950). *In* "McClung's Handbook of Microscopical Technique" (R. M. Jones, ed.), pp. 564–570. Hoeber, New York.

Gabler, F., and Herzog, F. (1965). *Appl. Opt.* **4**, 469–472.

Geissman, T. A. (1955). *In* "Moderne Methoden der Pflanzenanalyse" (K. Paech und M. V. Tracey, eds.), Vol. III, pp. 450–498. Springer, Berlin.

Gicklhorn, J. (1931). *Ergebn. Physiol.* **7**, 549–685.

Goppelsroeder, F. (1888). *Mitt. K. K. Technol. Gew. Mus. Wien.* **2**, 86–114; **3**, 14–19, Beilage 1–78.

Haitinger, M. (1938). "Fluorescenz-Mikroskopie." Akadem. Verlagsges. Leipzig. [The second edition (1959) has many shortcomings and is not recommended (Drawert, 1968, p. 14; Butterfass, 1960).]

Harms, H. (1965). "Handbuch der Farbstoffe für die Mikroskopie." Staufen Verlag, Kamp-Lintfort (Germany).

Härtel, O. (1951). *Protoplasma* **40**, 338–347.

Härtel, O. (1953). *Protoplasma* **42**, 83–89.

Haselmann, H. (1957). *Z. Wiss. Mikrosk.* **63**, 140–155.

Haselmann, H., and Wittekind, D. (1957). *Z. Wiss. Mikrosk.* **63**, 216–226.

Herzog, F. (1971). Personal communication.

Höfler, K. (1947). *Mikroskopie* **2**, 13–29.

Höfler, K. (1949). *Mikroskopie,* **1.** Sonderband "Beiträge zur Fluoreszenz-mikroskopie," pp. 46–70.

Höfler, K. (1951). *Biol. Gen.* **19**, 90–113.

Höfler, K. (1959). *Ber. Deut. Bot. Ges.* **72**, 236–245.

Höfler, K., and Schindler, H. (1955). *Protoplasma* **45**, 173–193.

Höfler, K., and Stiegler, A. (1947). *Mikroskopie* **2**, 250–258.

Höfler, K., Ziegler, A., and Luhan, M. (1956). *Protoplasma* **46**, 322–366.

Hölzl, J., and Bancher, E. (1967). *Protoplasma* **64**, 157–184.

Hofmeister, L. (1940). *Protoplasma* **35**, 161–186.

Jones, R. M. (1966). "Basic Microscopic Techniques." University of Chicago Press, Chicago, Illinois.

Karrer, P. (1950). "Organic Chemistry." Elsevier, New York.

Kinzel, H. (1953). *Protoplasma* **42**, 209–226.

Kinzel, H. (1954a). *Protoplasma* **43**, 441–449.

Kinzel, H. (1954b). *Protoplasma* **44**, 52–72.

Kinzel, H. (1955). *Protoplasma* **45**, 73–96.

Kinzel, H. (1958). *Protoplasma* **50**, 1–50.

Kinzel, H. (1959). *Ber. Deut. Bot. Ges.* **72**, 253–261.

Kinzel, H. (1962). *Oesterr. Apoth. Ztg.* **16**(38), 573–578.

Kinzel, H. (1965). *Ber. Deut. Bot. Ges.* **78**, 23–27.

Kinzel, H. (1968a). *Abh. Deut. Akad. Wiss. Berlin Kl. Med.* **4a**, 205–215.

Kinzel, H. (1968b). Private communication.

Kinzel, H., and Bolay, E. (1961). *Protoplasma* **54**, 177–199.

Kinzel, H., and Imb, R. (1961). *Protoplasma* **53**, 422–437.

Kinzel, H., and Pischinger, I. (1962). *Protoplasma* **55**, 555–571.

Kristen, U. (1970). *Protoplasma* **71**, 443–449.

Lillie, R. D. (1969). "H. J. Conn's Biological Stains." Williams & Wilkins, Baltimore, Maryland.

Lison, L., and Mutsaars, W. (1950). *Quart. J. Microsc. Sci.* **91**, 309–313.

Lowenhaupt, B. (1970). *Stain Technol.* **45**, 29–34.

Lowenhaupt, R. W., and Lowenhaupt, B. (1966). *J. Cell Biol.* **31**, 70A.

Nairn, R. C. (1969). "Fluorescent Protein Tracing." Williams & Wilkins, Baltimore, Maryland.

Pfeffer, W. (1886). *Untersuch. Bot. Inst. Tübingen* **2**, 179–329.

Pop, E., and Soran, V. (1962). *Flora (Jena)* **152**, 91–112.

Pringsheim, P. (1949). "Fluorescence and Phosphorescence." Wiley (Interscience), New York.

Richter, H. (1966). *Bot. Stud.* **17**, 71 p.

Ruch, F. (1968). Quantitative Fluorescence Microscopy. *Summary and Reports 3rd Int. Congr. Histochem. Cytochem. New York 1968,* p. 228.

Ruhland, W. (1912). *Jahrb. Wiss. Bot.* **51**, 376–431.

Runge, W. (1966). *Science* **151**, 1499–1506.

Schwantes, H. O. (1965). *Mikroskopie* **20**, 291–327.

Strain, H. (1942). "Chromatographic Adsorption Analysis." Wiley (Interscience), New York.

Strugger, S. (1937). *Flora (Jena)* **131**, 113–128.

Strugger, S. (1940). *Jena Z. Med. Naturwiss.* **73**, 97–134.

Strugger, S. (1947). *Naturwissenschaften* **34**, 267–273.

Strugger, S. (1949). "Praktikum der Zell- u. Gewebephysiologie der Pflanze," 2nd ed. Springer, Berlin.

Tsekos, I. (1970). *Protoplasma* **71**, 173–190.

Tsekos, I., and Kristen, U. (1969). *Ber. Deut. Bot. Ges.* **82**, 577–588.

Unger, F. (1850). *Denkschr. Akad. Wiss. Wien. Math.-Naturw. Kl.* **1**, 75–82.

van Wisselingh, C. (1915). *Beih. Bot. Zbl.* **32**, 155–217.

Wittekind, D. (1959). *Mikroskopie* **14**, 9–25.

Young, M. R. (1961). *Quart. J. Microsc. Sci.* **102**, 419–449.

Zanker, V. (1952). *Z. Phys. Chem.* **199**, 225–258.

Zöttl, P. (1960). *Protoplasma* **51**, 465–506.

Chapter 11

Synchrony in Blue-Green Algae[1]

HARALD LORENZEN AND G. S. VENKATARAMAN

*Pflanzenphysiologisches Institut der Universität Göttingen, Germany and
Division of Microbiology, Indian Agricultural Research Institute, Delhi, India*

I. Introduction

The importance of synchronized populations of cells for obtaining information on the different steps of cell development has recently increased. Some results and problems have already been discussed in some review papers. A few of them dealing with single, photosynthesizing organisms are mentioned here (Kuhl and Lorenzen, 1964; Lorenzen, 1970; Pirson and Lorenzen, 1966; Schmidt, 1966; Senger and Bishop, 1969; Tamiya, 1966).

Only a few papers have appeared in which blue-green algae are synchronized. This is surprising in view of the theoretical and practical importance of these prokaryotic organisms. They are self-sufficient cells able to carry on photosynthesis (chlorophyll *a* and phycocyanin) and do not possess nuclei, mitochondria, or plastids. Many strains have the capacity to fix N_2, and evidence has been presented that *Anacystis* can undergo

[1] This paper is dedicated to Prof. Drs. h.c. Kurt Mothes, Halle, on occasion of his 70th birthday in 1970.

genetic transformation (see Shestakov and Khyen, 1970). The blue-green algae play an important role in the paddy fields and in the processing of sewage. They probably originated about a billion years ago. The entire cell resembles a chloroplast in contemporary eukaryotic cells. The cells are relatively small and tend to produce chains of a few or a large number of cells. The above characteristics may account for the fact that few synchronizing procedures have been developed to date. A more recent summary of the general conditions for culturing blue-green algae in the laboratory is given by Carr (1969), and isolation procedures are described by Koch (1965).

There are two possible approaches to the problem of establishing mass synchrony in these populations. With the selection technique, a synchronous culture is started from cells in the same stage of the cell life cycle which have been separated in quantity from a large culture in balanced, exponential, but asynchronous growth. When isolated, these cells go through one or more divisions in some degree of unforced synchrony. In the induced systems, the balanced state is purposely disrupted by exposing the population to environmental changes, which establish, under favorable conditions of serial dilution, a semicontinuous type of synchronous culture. The present chapter deals mainly with the experimental methods used to induce synchrony of cell division in *Anacystis nidulans* and in heterocyst production in two strains of *Nostoc*.

II.   Synchrony in *Anacystis*

A.   Method of Lorenzen and Venkataraman

Organism

Anacystis nidulans Nr. 1402-1 from the culture collection of the Institute of Plant Physiology, University of Göttingen was used for this study.

Culture Medium and Conditions

The synthetic inorganic medium used for culturing the alga is a modification of that of Kratz and Myers and contains (in molarity)

$NaNO_3$, 1×10^{-3}	$Ca(NO_3)_2$, 5×10^{-5}
$MgSO_4 \cdot 7\ H_2O$, 1×10^{-3}	K_2HPO_4, 5×10^{-3}
H_3BO_3, 1×10^{-5}	$MnCl_2 \cdot 4\ H_2O$, 1×10^{-5}
$ZnSO_4 \cdot 7\ H_2O$, 1×10^{-6}	$CuSO_4 \cdot 5\ H_2O$, 1×10^{-7}
$(NH_4)Mo_7O_{24}$, 1×10^{-7}	Fe-EDTA 0.5, ml/liter

The pH of the medium is about 8 (Bothe 1968). The alga is grown in a unialgal condition in a light thermostat (Lorenzen, 1959) at 32°C under continuous illumination (15,000 lux), provided by a bank of two Philips TL 40 W 1/32 and one TL 40 W 1/55 tubes at a distance of 11 cm from the culture vessels. The cultures (250 ml) are continuously aerated with 1.5% (by volume) CO_2 in air mixture.

SYNCHRONIZATION TECHNIQUE

Under continuous illumination at 32°C, the algal population doubles in 5 hours, with an initial lag of about 1 hour. Periodic dilution of the algal suspension at the end of every 5 hours with fresh medium to the original cell number results in a repeated doubling (Fig. 1). The picture is essentially the same when the alga is grown at 26°C, with the exception that the doubling time is increased to 12 hours (Fig. 2). These successive rhythmicities after periodic dilutions of the algal suspension do not, however, produce synchrony of cell division, with a consequence

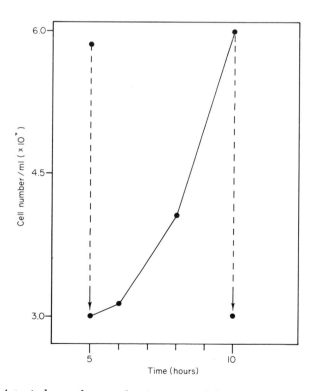

Fig. 1. A typical growth curve for *Anacystis nidulans* in continuous light (15,000 lux) at 32°C with serial dilutions after doubling every 5 hours.

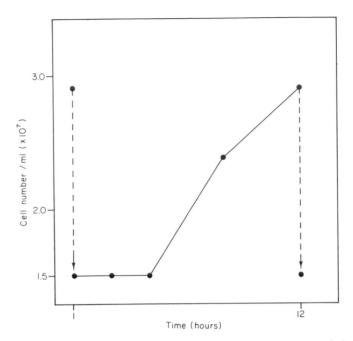

Fig. 2. A typical growth curve for *Anacystis nidulans* in continuous light (15,000 lux) at 26°C with serial dilutions after doubling every 12 hours.

that the algal population is found to consist of a mixture of different proportions of single and dividing cells.

When the alga is grown at 32°C under 15,000 lux for 48 hours and allowed to stand overnight (ca. 12 hours) in the dark, the supernatant portion of the suspension is found to contain a high proportion (~98%) of single cells. Starting with these cells, synchronous cell division in the cells can be induced by subjecting the cells to differential temperatures, followed by periodic dilution at the end of each cycle, as described below.

The single cells are grown for 8 hours at 26°C under continuous illumination (15,000 lux). The cells grow in length, but do not divide during this phase. The cells are then exposed to 32°C, with all other conditions kept constant. Two hours after the shift to 32°C, most of the cells reach division (division index 98%). The complete separation of the divided cells into individual single daughter cells occurs in the next 4 hours at 32°C. The algal suspension is then diluted with fresh medium to the original cell number (density 2×10^7 cells/ml) and again subjected to a cycle of 8 hours at 26°C and 6 hours at 32°C. Three successive synchronization cycles, with periodic dilutions at the end of each cycle, are

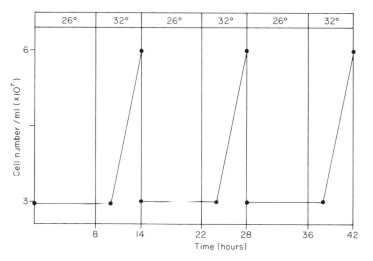

FIG. 3. Three successive 14-hour cycles of completely synchronized *Anacystis nidulans* (8 hours 26°C, 6 hours 32°C) with serial dilutions at the end of each cycle.

shown in Fig. 3. Synchronized cell division can also be induced in a mixed population of cells, if the population is subjected to this cyclic temperature–dilution rhythm for at least four cycles (Lorenzen and Venkataraman, 1969). It should be mentioned again that at both temperatures used for the cycle, 26° and 32°C, *Anacystis* can complete its

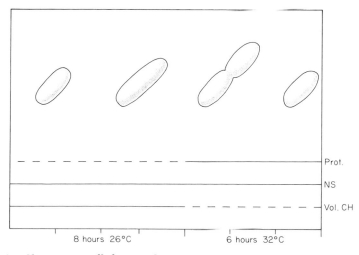

FIG. 4. Changes in cell form, volume, and in some cellular components during one representative cycle of temperature-induced synchrony in *Anacystis nidulans*. Prot, proteins; NS, nucleic acids; Vol, volume; CH, carbohydrates.

cell cycle in normal fashion (Figs. 1 and 2). This is in contrast to some of the other temperature-induced synchronizations in which a temperature that blocks normal development is used, with a subsequent release of cells following their return to a physiological temperature.

Comments on the subcellular morphology of the alga and some biochemical changes during the synchronized cycles of growth of *Anacystis* were reported by Venkataraman *et al.* (1969) and by Venkataraman and Lorenzen (1969). Figure 4 summarizes the most salient results of changes in cell form and shows the main phases of increases in volume, carbohydrate, nucleic acids, and proteins.

B. Method of Herdman *et al.*

Organism

Anacystis nidulans is used for this study.

Conditions

Medium according to M. M. Allen (1968) (in gm/liter)

$NaNO_3$, 1.5	$Na_2SiO_3 \cdot 9\ H_2O$, 0.058
K_2HPO_4, 0.039	EDTA, 0.001
$MgSO_4 \cdot 7\ H_2O$, 0.075	Citric acid, 0.006
Na_2CO_3, 0.02	Fe citrate, 0.006
$CaCl_2$, 0.027	Microelements[2] 1 ml

The pH of the medium is 7.8. Temperature 38°C, illumination with two 20 W warm white fluorescent lights, and aeration with 5% CO_2 in air.

Synchronizing Procedure

A culture at midlog phase (mean generation time, 120 minutes) is deprived of light, and the aeration stopped. Within the first 30 minutes the cell number increases by 15%, then no further change in cell number is observed. After 24 hours normal growth conditions are reestablished, and during the succeeding 6 hours two bursts in cell number increase occur (Fig. 5). The first starts after about 2 hours and lasts approximately 1 hour, and all cells of the culture (density 7×10^7 cells/ml) participate. After continuing growth for an additional hour, a second synchronized division begins and lasts for 75 minutes, but only two-thirds of the daughter cells of the first division are viable. Theoretical considerations attribute this to an effect on DNA replication.

[2] Composition (gm/liter): H_3BO_4, 2.86; $MnCl_2 \cdot 4\ H_2O$, 1.81; $ZnSO_4 \cdot 7\ H_2O$, 0.222; $Na_2MoO_4 \cdot 2\ H_2O$, 0.391; $CuSO_4 \cdot 5\ H_2O$, 0.079; $Co(NO_2)_2 \cdot 6\ H_2O$, 0.0494.

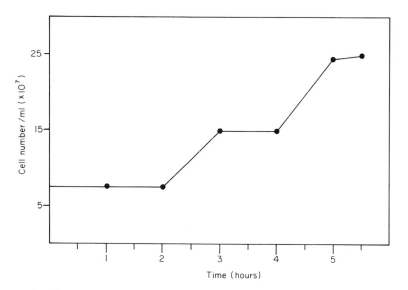

FIG. 5. The course of increase in cell number of *Anacystis nidulans* under normal growth conditions (38°C) according to the synchronizing method of Herdman *et al.* (1970).

C.   General Considerations

When we consider both methods of synchronization of *Anacystis*, the results published so far seem to indicate minimal success. There are still discrepancies between increase in cell number and syntheses of macromolecules.

In the method introduced by Herdman *et al.* (1970), during the first cycle there is evidently a lag of 30 minutes that is the result of the pre-treatment. RNA synthesis then begins, with the synthesis of DNA beginning 40 minutes later. DNA synthesis extends over 60 minutes, but the increase in DNA content is only 50%. Using the other method, we achieved a doubling in nucleic acids during each cycle.

Prerequisites for good synchrony are conditions that allow the cells to double their mass, but this doubling does not induce division synchrony automatically. A mass doubling with *Anacystis* has been achieved also at a higher light intensity (about 20,000 lux) and a starting cell density of up to 8×10^7 cells/ml with a temperature schedule of 32°C for 2 hours and 40°C for 1 hour. After a 1:1 dilution with fresh medium, another cycle was started immediately. A 3-hour cycling using 41° instead of 40°C under otherwise unchanged conditions and a cell number of 1.5×10^8 cells/milliliter also produced a doubling within each cycle. It is clear that division synchrony is obtainable with these treatments. However, it seems impossible to achieve synchrony of the cell cycle if the macro-molecular content is not doubled or if it increases in an uneven manner. In the first method mentioned, the conditions, regular changes of temperature during continuous illumination and periodic dilutions, bring about synchronization of a nonsynchronous culture.

Very recently, a paper was published by Lindsey *et al.* (1971) in which the authors reported a method of synchronizing *Anacystis nidulans* (TX-20) for two successive doublings without dilution. The technique used was a combination of three light–dark cycles of 9:15 hours (39°C, 1500 fc, culture well shaken) followed by a phase of continuous light (1500 or 250 fc). Four hours after the last light phase, and within one hour, the cell number (4×10^7 cells/ml) doubled; this happened again at eight hours with completion of the doubling before ten hours. Subsequently there was a rapid loss of synchrony. It is remarkable that illumination was by mixed fluorescent and incandescent light sources.

III.   Synchrony of Heterocyst Production in *Nostoc*

An important morphological structure in nitrogen-fixing blue-green algae is the heterocyst, which has recently been shown to be a site of nitrogen fixation (Fay *et al.*, 1968; Fogg, 1969). The production of these heterocysts is suppressed by an exogenous source of ammonial nitrogen (Fogg, 1949; Talpasayi and Kale, 1967), with a consequent suppression of nitrogen fixation in these algae. In the laboratory of G. S. Venkataraman it has been possible to synchronize the production of heterocysts in these algae by using ammonial nitrogen as a synchronizing agent. Two strains of *Nostoc* from the Culture Collection of Microalgae, Indian Agricultural Research Institute, were grown in the same medium employed for *Anacystis nidulans* (Lorenzen and Venkataraman, 1969), omitting the nitrogen source. Nitrogen, when used, was added in the form of ammonium chloride (20 mg/100 ml). Shifts in media with and without nitrogen were made by transferring the algal mass after repeated washing by centrifugation and suspending it in the respective media with and without nitrogen. Heterocysts were counted microscopically and expressed as the percentage of vegetative cells. Figure 6 shows three cycles of synchronized heterocyst production induced by alternate addition and omission of nitrogen in the medium. The total percentage of heterocysts in the presence of exogenous nitrogen remained fairly constant for 24 hours. When the algae were transferred to nitrogen-free medium, there

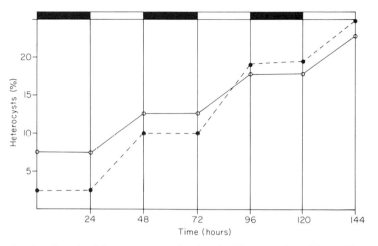

Fig. 6. Synchronized heterocyst production in *Nostoc* induced by addition and omission of ammonium chloride (20 mg/100 ml) as the nitrogen source in the medium. *Nostoc muscorum*, ◯———◯, *Nostoc sp.*, ●-----●. Dark area is with ammonium chloride; light area, without ammonium chloride. Temperature 28°C, 5000 lux.

was a sharp rise in the percentage of heterocysts, which again leveled off in the presence of nitrogen. The significance of such synchronized production of heterocysts in these species in terms of their biological efficiency seems obvious. We must, however, refer to the paper of Smith and Evans (1970), who were able to show that a nitrogenase could also be isolated from vegetative cells of *Anabaena cylindrica*. With a sonication procedure it is possible to selectively break the vegetative cells leaving the heterocysts intact.

IV. Discussion

From the biological point of view there are some criteria that should be met in ideal synchronization (Lorenzen, 1970). Transferring the experience with other unicellular, photosynthetic organisms to the blue-green algae, we have to improve our techniques considerably to gain more satisfactory results. We feel that it is necessary to establish completely synchronized cultures, retained in synchrony along with periodic dilution. The organism should grow with the shortest possible life cycle and the interval of increase of cell number during the cycle should be as short as possible. Only in such cases are we sure to have a synchronous population that enables us to collect information regarding the behavior of a normal cell during its development.

A type of synchrony in *Chlorogloea fritschii* was established by Fay *et al.* (1964) using light and temperature. After 20 days at 35°C and illumination with 5000 lux, some 80% of the cells obtained a diameter of about 7 μ; 80% of the cells obtained a diameter of 3 μ after another 20 days of culture in fresh medium, at 25°C and illumination with 300 lux. These changes could not be correlated with the growth cycle.

The failure to synchronize blue-green algae with the light–dark changes applied successfully to other photosynthetic unicellular algae may depend on the great differences in physiological mechanisms operating during the dark phase. In particular, the blue-green algae lack a normal (high) respiration (Biggins, 1969), which is very important with respect to intracellular changes of carbohydrates and proteins (which may take part in phasing of the cells). Hayashi *et al.* (1969), using *Anacystis*, performed a light–dark experiment combined with shifts of temperature (25° and 38°C) and achieved a doubling of cells (number and macromolecules) within 7 hours; also during log phase growth, in continuous light, the doubling required about 7 hours. Contrary to what other authors have reported, there was absolutely no increase in cell number in the dark. The cell number increased without any lag during the whole length of light phase (cf. Fig. 1). There was in effect no division synchrony whatsoever.

We are hopeful that this outline of preliminary results may encourage others to develop better methods, which are needed to obtain more knowledge about the life cycle processes of this very important group of organisms.

References

Allen, M. M. (1968). *J. Phycol.* **4**, 1–4.
Biggins, J. (1969). *J. Bacteriol.* **99**, 570–575.
Bothe, H. (1968). Dissertation, 125 pp. University Göttingen, Germany.
Carr, N. G. (1969). In "Methods in Microbiology" (J. R. Norris and D. W. Ribbons, eds.), Vol. 3b, pp. 53–77. Academic Press, New York.
Fay, P., Kumar, H. D., and Fogg, G. E. (1964). *J. Gen. Microbiol.* **35**, 351–360.
Fay, P., Stewart, W. D. P., Walsby, A. E., and Fogg, G. E. (1968). *Nature (London)* **220**, 810–812.
Fogg, G. E. (1949). *Ann. Bot. (London)* [N.S.] **13**, 241–259.
Fogg, G. E. (1969). *Verh. Int. Verein. Limnol.* **17**, 761–762.
Hayashi, F., Ishida, M. R., and Kikuchi, T. (1969). *Res. Reactor Inst. Kyoto Univ.* **2**, 56–66.
Herdman, M., Faulkner, B. M., and Carr, N. G. (1970). *Arch. Mikrobiol.* **73**, 238–249.

Koch, W. (1965). *Zentralbl. Bakteriol. Parasitenk. Infektionskr. Hyg. Suppl.* **1**, 415–431.

Kuhl, A., and Lorenzen, H. (1964). *In* "Methods in Cell Physiology" (D. M. Prescott, ed.), Vol. 1, pp. 159–187. Academic Press, New York.

Lindsey, J. K., Vance, B. D., Keeter, J. S., and Scholes, V. E. (1971). *J. Phycol.* **7**, 65–71.

Lorenzen, H. (1959). *Flora (Jena)* **147**, 382–404.

Lorenzen, H. (1970). *In* "Photobiology of Microorganisms" (P. Halldal, ed.), pp. 187–212, Wiley (Interscience), London.

Lorenzen, H., and Venkataraman, G. S. (1969). *Arch. Mikrobiol.* **67**, 251–255.

Pirson, A., and Lorenzen, H. (1966). *Annu. Rev. Plant Physiol.* **17**, 439–458.

Schmidt, R. R. (1966). *In* "Cell Synchrony: Studies in Biosynthetic Regulation" (I. L. Cameron and G. M. Padilla, eds.), pp. 189–235. Academic Press, New York.

Senger, H., and Bishop, N. I. (1969). *Prog. Photosynthesis Res.* **1**, 425–434.

Shestakov, S. V., and Khyen, N. T. (1970). *Mol. Gen. Genet.* **107**, 372–375.

Smith, R. V., and Evans, M. C. W. (1970). *Nature (London)* **225**, 1253.

Talpasayi, R. R. S., and Kale, S. (1967). *Curr. Sci.* **36**, 218–219.

Tamiya, H. (1966). *Annu. Rev. Plant Physiol.* **17**, 1–26.

Venkataraman, G. S., and Lorenzen, H. (1969). *Arch. Mikrobiol.* **69**, 34–39.

Venkataraman, G. S., Amelunxen, F., and Lorenzen, H. (1969). *Arch. Mikrobiol.* **69**, 370–372.

AUTHOR INDEX

Numbers in *italics* refer to the pages on which the complete references are listed.

SUBJECT INDEX

A

Acetoorcein, 93
Acridine orange, 337, 340, 343–345, 347, 351–353, 357, 366–368
Acrosome, 89
Actinomycin D, 270–272
Agarose, 42
Agglutination, 85, 87, 89
Alanine, 14, 63
Alcian green, 51
Alconox, 69
Algae, 306, *see also* specific types
culturing of, 374
Allantoic fluid, 81
Allium cepa, 342, 361–367
Allograft, 129
American Type Culture Collection, 9, 80
Amethopterin, 256
Amino acids, 40
solution of, 14
stock solution, 17
Aminobenzoic acid, 15
Aminopterin, 91
Amoeba proteus, 282
Amphibian oocytes, 76
nuclei of, 168
Anabaena cylindrica, 381
Anacystis, 373
Anacystis nidulans, 374–376, 382
medium for, 378
synchrony in, 377–380
Anaphase, 112
Animals
preservation of specimens, 6–7
procurement of, 2–4
Antechinus flavipes, 131
Antechinus swainsonii, 130–131, 150, 158–159, 161, 163
Antibiotics, 16
AntiFoam B, 232
AR-10 stripping film, 29–31
Arginine, 14, 60
Asbestos filter pads, 9
Ascites hepatoma, 181
Ascites tumor cells, 79, 92

Ascorbic acid, 15
Asparagine, 14, 62
Aspartic acid, 14, 63
Aspergillin, 43
ATP, 35
ATPase, 192, 194–195
Autoradiographic emulsions, gelatins in, 296
Autoradiographs, 36
degraining of, 33
Autoradiography, 29, 31, 35, 278
in electron microscopy, 289
Auxotrophic mutants, 91, 113
8-Azaguanine, 91

B

Bacitracin, 81
Bacto-tryptone, 224, 226
Bandicoots, 130–131
Basic fuchsin, 23
Bats, 24
BBL Micro test II, 20
Beer-Lambert law, 314
Berberin, 340
Bettongia cuniculus, 135
Bettongia lesueur, 135
Bettongia penicillata, 135
Binucleate, 87, 100–101, 120
Biotin, 15, 60, 226
Blastomeres, 88
Blood, 4, 149
Blue-green algae, 373–374, 380, 382
synchrony in, 373
Bone marrow, 12–13, 23–24, 59, 133
preparations of, 23, 27
Bromodeoxyuridine, 45, 91
Buffer solutions, *see also* specific buffers
preparation of, 370
Bull sperm, 86

C

Cadaverine, 120
Caenolestidae, 128, 141
Caluromys, 129, 141
Caluromys derbianus, 141–143
Carbofuchsin, 23, 25

396